Recent Trends on Electromagnetic Environmental Effects for Aeronautics and Space Applications

Christos D. Nikolopoulos
Department of Electronic Engineering, School of Engineering, Hellenic Mediterranean University, Greece

IGI Global
PUBLISHER of TIMELY KNOWLEDGE

A volume in the Advances in Mechatronics and Mechanical Engineering (AMME) Book Series

Published in the United States of America by
IGI Global
Engineering Science Reference (an imprint of IGI Global)
701 E. Chocolate Avenue
Hershey PA, USA 17033
Tel: 717-533-8845
Fax: 717-533-8661
E-mail: cust@igi-global.com
Web site: http://www.igi-global.com

Library of Congress Cataloging-in-Publication Data

Names: Nikolopoulos, Christos D., 1981- editor.
Title: Recent trends on electromagnetic environmental effects for aeronautics and space applications / Christos D. Nikolopoulos, editor.
Description: Hershey, PA : Engineering Science Reference, [2021] | Includes bibliographical references and index. | Summary: "This book provides relevant theoretical frameworks and the latest empirical research findings in the area of Electromagnetic Compatibility and Electromagnetic Interference (EMC/EMI) for aerospace industry and will help professionals who want to improve their understanding of the emissions and immunity regarding aerospace design engineering"-- Provided by publisher.
Identifiers: LCCN 2020045761 (print) | LCCN 2020045762 (ebook) | ISBN 9781799848790 (hardcover) | ISBN 9781799874911 (paperback) | ISBN 9781799848806 (ebook)
Subjects: LCSH: Electromagnetic compatibility. | Electromagnetic interference in aeronautics. | Space vehicles--Electronic equipment--Protection.
Classification: LCC TK7867.2 .R43 2021 (print) | LCC TK7867.2 (ebook) | DDC 629.135--dc23
LC record available at https://lccn.loc.gov/2020045761
LC ebook record available at https://lccn.loc.gov/2020045762

This book is published in the IGI Global book series Advances in Mechatronics and Mechanical Engineering (AMME) (ISSN: 2328-8205; eISSN: 2328-823X)

British Cataloguing in Publication Data
A Cataloguing in Publication record for this book is available from the British Library.

Advances in Mechatronics and Mechanical Engineering (AMME) Book Series

J. Paulo Davim
University of Aveiro, Portugal

ISSN:2328-8205
EISSN:2328-823X

Mission

With its aid in the creation of smartphones, cars, medical imaging devices, and manufacturing tools, the mechatronics engineering field is in high demand. Mechatronics aims to combine the principles of mechanical, computer, and electrical engineering together to bridge the gap of communication between the different disciplines.

The **Advances in Mechatronics and Mechanical Engineering (AMME) Book Series** provides innovative research and practical developments in the field of mechatronics and mechanical engineering. This series covers a wide variety of application areas in electrical engineering, mechanical engineering, computer and software engineering; essential for academics, practitioners, researchers, and industry leaders.

Coverage

- Nanomaterials and nanomanufacturing
- Micro and nanomechanics
- Sustainable and green manufacturing
- Biologically Inspired Robotics
- Intelligent Sensing
- Bioengineering Materials
- Manufacturing Methodologies
- Mechanisms and machines
- Vibration and acoustics
- Computational Mechanics

IGI Global is currently accepting manuscripts for publication within this series. To submit a proposal for a volume in this series, please contact our Acquisition Editors at Acquisitions@igi-global.com or visit: http://www.igi-global.com/publish/.

The Advances in Mechatronics and Mechanical Engineering (AMME) Book Series (ISSN 2328-8205) is published by IGI Global, 701 E. Chocolate Avenue, Hershey, PA 17033-1240, USA, www.igi-global.com. This series is composed of titles available for purchase individually; each title is edited to be contextually exclusive from any other title within the series. For pricing and ordering information please visit http://www.igi-global.com/book-series/advances-mechatronics-mechanical-engineering/73808. Postmaster: Send all address changes to above address.

Titles in this Series

For a list of additional titles in this series, please visit:
http://www.igi-global.com/book-series/advances-mechatronics-mechanical-engineering/73808

Evaluation and Applications for Ultrasonic Methods of Non-Destructive Testing
Ashish Khaira (Samrat Ashok Technological Institute, India)
Engineering Science Reference • © 2021 • 215pp • H/C (ISBN: 9781799845645) • US $195.00

Handbook of Research on Advancements in Manufacturing, Materials, and Mechanical Egineering
Leonid Burstein (Independent Researcher, Israel)
Engineering Science Reference • © 2021 • 462pp • H/C (ISBN: 9781799849391) • US $295.00

Genetic Algorithms and Remote Sensing Technology for Tracking Flight Debris
Maged Marghany (Syiah Kuala University, Indonesia)
Engineering Science Reference • © 2020 • 334pp • H/C (ISBN: 9781799819202) • US $195.00

Humanizing Cities Through Car-Free City Development and Transformation
Rahma M. Doheim (Assiut University, Egypt & University of Business and Technology, Saudi Arabia) Alshimaa Aboelmakarem Farag (Zagazig University, Egypt) and Ehab Kamel (University of Central Lancashire, UK)
Engineering Science Reference • © 2020 • 305pp • H/C (ISBN: 9781799835073) • US $215.00

Airline Green Operations Strategies Emerging Research and Opportunities
Yazan Khalid Abed-Allah Migdadi (Qatar University, Qatar)
Engineering Science Reference • © 2020 • 231pp • H/C (ISBN: 9781799842552) • US $185.00

Recent Technologies for Enhancing Performance and Reducing Emissions in Diesel Engines
J. Sadhik Basha (International Maritime College Oman, Sohar, Oman) and R.B. Anand (National Institute of Technology, Tiruchirappalli, India)
Engineering Science Reference • © 2020 • 298pp • H/C (ISBN: 9781799825395) • US $215.00

Practical Approach to Substrate Integrated Waveguide (SIW) Diplexer Emerging Research and Opportunities
Augustine Onyenwe Nwajana (University of East London, UK) and Kenneth Siok Kiam Yeo (Universiti Teknologi Brunei, Brunei)
Engineering Science Reference • © 2020 • 171pp • H/C (ISBN: 9781799820840) • US $195.00

Diverse Applications of Organic-Inorganic Nanocomposites Emerging Research and Opportunities
Gabriele Clarizia (Institute on Membrane Technology, National Research Council, Italy) and Paola Bernardo (Institute on Membrane Technology, National Research Council, Italy)
Engineering Science Reference • © 2020 • 237pp • H/C (ISBN: 9781799815303) • US $195.00

IGI Global
PUBLISHER of TIMELY KNOWLEDGE

701 East Chocolate Avenue, Hershey, PA 17033, USA
Tel: 717-533-8845 x100 • Fax: 717-533-8661
E-Mail: cust@igi-global.com • www.igi-global.com

Editorial Advisory Board

Table of Contents

Detailed Table of Contents

Nowadays, a wide range of space missions accommodate ever-stricter electromagnetic cleanliness requirements arising either from the need for more precise measurements or from the implementation of highly sensitive equipment. Therefore, the establishment of a methodology that ensures the minimization of the electric and/or magnetic field in specific areas inside or outside the spacecraft structure is crucial. Towards this goal, the current chapter proposes that utilizing the results of a process completed during the early design stages of a mission, that is, the measurement and characterization of each implemented device, the desired elimination of the field can be achieved. In particular, the emerged electromagnetic signatures of the units are proven essential for the proposed methodology, which, using a heuristic approach, defines the optimal ordinance of the equipment that leads to system-level electromagnetic field minimization in the volume of interest. The dimensions of the devices and the effect of the conductive surfaces of the spacecraft's hull are also taken into account.

SpaceWire is a point-to-point bit shipping protocol for high-speed data communication links and networks providing equipment compatibility and seamless component reusability. It has found great application in many space missions reducing the development cost, offering architectural flexibility, and improving reliability. This chapter delves into the standard describing the SpaceWire, focusing on the lower levels that play a key role in the electromagnetic behavior of the system and concern cable assemblies, shielding, bonding, and grounding. Findings regarding emissions affecting spacecraft components are presented as well as other EMC issues that have an impact on the system performance. Recent developments on the modelling of the cable of the system with a focus on radiated emissions of SpW systems are also presented and discussed.

Electromagnetic environmental shielding is one of the main research topics for the development of aeronautical and space applications. Numerous research groups around the world study the problems that space systems and astronauts experience when these are subjected to space radiation. Despite the progress made so far, different proposals of advanced materials have been continuously proposed throughout the history of space career to protect space systems and astronauts against the solar particle events (SPE), cosmic rays galactic (GCRs), and proton-electron radiation (PERs). This chapter presents the recent advances made about space environmental shielding and that have been reported so far to visualize the future perspectives that this type of research must carry out so that future space voyage is completely reliable for space systems and astronauts. This research area is fully current, and its experimental success will depend on the work done by all space researchers and professionals.

Electromagnetic compatibility has emerged in the last decades as one of the most important aspects of product design. Space equipment, in particular, due to their increased sensitivity requirements and a greater risk of failure, are required to comply with a variety of EMC requirements. The scope of this chapter is to provide a detailed overview of these requirements according to the two main standards applicable, MIL-STD461G and ECSS-E-ST-20-07C, describing the proposed methodologies, the associated testing instrumentation, and the tailoring capabilities. The limitations of these methodologies and relative research, aiming to validate or improve them, are also presented highlighting potential deficiencies. This chapter aims to serve as a compact guide to EMC testing of space equipment according to the requirements of two of the most active organizations of space engineering.

Nowadays, a wide range of space missions accommodate ever-stricter electromagnetic cleanliness requirements arising either from the need for more precise measurements or from the implementation of highly sensitive equipment. Therefore, the establishment of a methodology that ensures the minimization of the electric and/or magnetic field in specific areas inside or outside the spacecraft structure is crucial.

Towards this goal, the current chapter proposes that utilizing the results of a process completed during the early design stages of a mission, that is, the measurement and characterization of each implemented device, the desired elimination of the field can be achieved. In particular, the emerged electromagnetic signatures of the units are proven essential for the proposed methodology, which, using a heuristic approach, defines the optimal ordinance of the equipment that leads to system-level electromagnetic field minimization in the volume of interest. The dimensions of the devices and the effect of the conductive surfaces of the spacecraft's hull are also taken into account.

This chapter is composed by three parts. The first is an introductory part, providing general information about magnetism and related phenomena. Magnetic materials are also discussed and presented. Afterwards, the magnetic field and various measurement techniques are discussed. In the second part, different magnetic sensors used in a laboratory or space are presented. Magnetic sensors that are discussed include anisotropic magneto-resistance (AMR), giant magneto-resistance (GMR), giant magneto-impedance (GMI), flux-gate, and superconducting quantum interference device (SQUID). Although some of them may be outdated and well known, they are widespread, and they still pose an excellent choice for certain applications. Advances in magnetometers also presented in order to provide the reader with the recent trends in the field. Magnetic cleanliness is an important factor both in calibration and in normal operation of a system; in the third part, current techniques to isolate a system from the external magnetic field providing cleanliness are discussed.

The forward and the inverse problem of a thin, circular, loop antenna that radiates in free space is modeled and solved by using soft computing techniques such as artificial neural networks and adaptive neuro fuzzy inference systems. On the one hand, the loop radius and the observation angle serve as inputs to the forward model, whereas the radiation intensity is the output. On the other hand, the electric field intensity and the loop radius are the input and output, respectively, to the inverse model. Extensive numerical tests indicate that the results predicted by the proposed models are in excellent agreement with theoretical data obtained from the existing analytical solutions of the forward problem. Thus, the

Preface

Electromagnetic Compatibility, as well as regulatory compliance issues are subjects of great importance in electronics engineering. Avoiding problems regarding an electronic system's operation is very important, in space missions and in satellite structures however is critical. The vast majority of these problems can be traced to electromagnetic field disturbances, as interference from unintended sources and other electromagnetic phenomena. As a result, stringent requirements are to be met in terms of electromagnetic emissions level. The inclusion of this electromagnetic environment in the design of a multi - millions mission can lead to a system able to withstand whatever challenge the harsh environment throws at it. Failure to do so, may lead to important data corruption or loss, destruction of expensive instruments, waste of resources and even a total mission failure. The main scope of this book is to address the possible major emissions contributors and provide theoretical and practical background in order to model and design space systems with regard to immunity of spacecraft components and the elimination of electromagnetic interference.

Nowadays circuits for communication and automation, instruments and electronic components in general operate in close proximity and consequently affect each other adversely. As a main system design objective is to provide to the extent possible acceptable emissions and susceptibility levels. Thereafter electromagnetic compatibility has two aspects, the ability of a device to react to unwanted electromagnetic energy (susceptibility to noise) and emissions that can cause potential interference to other products (radiated & conducted). The opposite of susceptibility is immunity which defines the electromagnetic environment where the device can operate satisfactory without performance degradation. The majority of these acceptable levels of emissions are determined and controlled by regulations established from various governments' agencies and regulatory bodies; in some cases like space missions and although the devices are certified and they meet these EMC specifications, additional stringent requirements are mandatory for a successful system operation. Chapter 1 and 3 examine such cases providing new measurement approaches for modeling purposes.

This book aims to provide the relevant theoretical framework and the latest empirical research findings in the area of Electromagnetic Compatibility and Electromagnetic Interference (EMC/EMI) for space industry. Moreover, main EMC standards and relative handbooks are presented along with testing methodologies and procedures covering various areas of expertise. The target audience of this book is composed of professionals, spacecraft designers, science and data processing managers, electrical & mechanical engineers, EMC testing engineers and researchers working in the field of aerospace industry and they want to improve their understanding of the emissions and immunity regarding aerospace design engineering. Moreover, the book provides insights and support executives concerned with the management of expertise, knowledge, information and organizational development in different types of work communities and environments (Academia and industry).

Chapter 1 provides modeling methodology and measurement techniques for accurate representation of the electromagnetic signature of spacecraft's units in terms of extremely low frequency electric field emissions. Modeling methodology exploits custom measurement setups in order to provide capable dataset for solving the inverse electromagnetic problem, producing accurate model validated in an extrapolated distance. Furthermore, measurement position uncertainty issues are discussed along with other contributing phenomena in a system level architecture. Moreover, it analyzes and discuss modeling methods and techniques of static and time varying magnetic fields generated by spacecraft equipment, regarding magnetic cleanliness requirements in space missions. The analysis includes specific issues concerning the test measurement methodology both for DC and AC magnetic fields (also including induced magnetic field measurements). The developed methods that are illustrated involve the implementation of stochastic techniques on the magnetic dipole modeling inverse problem and are validated via real and simulated test measurement results.

Chapter 2 presents a short description of the SpaceWire protocol with emphasis on the physical layer and reviews the types of the SpaceWire emphasizing to the most commonly used micro-miniature D-type connector. Also in this chapter, bonding and grounding techniques are discussed with regards to EMC/EMI aspects. Concluding, it focuses on parameters that affects the electromagnetic behavior of the cable, as cable assembly, the differential characteristic impedance, the skew, and the insertion & return losses.

Chapter 3 presents the recent advances made with respect to space environmental shielding that have been reported so far and visualizes the future perspectives that this type of research must carry out so that future space voyage is completely reliable for space systems and astronauts. This research area is fully current, and its experimental success will depend on the work done by all space researchers and professionals.

Chapter 4 provides a detailed overview of the EMC requirements for space applications and focuses to the two main Standards applicable, MIL-STD461G and ECSS-E-ST-20-07C, describing the proposed methodologies, the associated testing instrumentation and the tailoring capabilities. The limitations of these methodologies and relative research, aiming to validate or improve them, are also presented highlighting potential deficiencies. This chapter aims to serve as a compact guide to EMC testing of Space equipment according to the requirements of two of the most active organizations of space engineering.

Chapter 5 proposes that utilizing the results of a process completed during the early design stages of a mission, i.e. the measurement and characterization of each implemented device, the desired elimination of the field can be achieved. In particular, the emerged electromagnetic signatures of the units are proven essential for the proposed methodology, which, using a heuristic approach, defines the optimal ordinance of the equipment that leads to system-level electromagnetic field minimization in the volume of interest. The dimensions of the devices and the effect of the conductive surfaces of the spacecraft's hull are also taken into account.

Chapter 6 revises general consideration about magnetism and magnetic field while examines different types of magnetic field sensors with their applications in measuring magnetic field. Moreover, this chapter provides active and passive techniques in order to create a clean magnetic room or to provide electromagnetic shielding at low frequencies.

Chapter 7 deals with the forward and the inverse problem of a thin, circular, loop antenna that radiates in free space is modeled and solved by using soft computing techniques such as artificial neural networks and adaptive neuro fuzzy inference systems. On the one hand, the loop radius and the observation angle serve as inputs to the forward model, whereas the radiation intensity is the output. On the other hand, the electric field intensity and the loop radius are the input and output, respectively, to the inverse model. Extensive numerical tests indicate that the results predicted by the proposed models are in excellent agreement with theoretical data obtained from the existing analytical solutions of the forward problem. Thus, the employment of artificial intelligence techniques for tackling electromagnetic problems turns out to be promising, especially regarding the inverse problems that lack solution with other methods.

Chapter 8 aims at presenting the current International Standards and the recent bibliography regarding the transient phenomena associated with space systems, with emphasis on the EMC requirements. The first section of the chapter focuses on the description of the physical mechanisms causing transient electromagnetic phenomena in space. The second section reviews the procedures proposed for testing the immunity of space equipment against these transients, as described in the current space Standards and in recent bibliography as non-Standardized testing procedures. Finally, the last section investigates possible mitigation approaches and design guidelines against these electromagnetic disturbances, essential for every space system designer.

This book attempts to present current and emerging trends in research and modeling of EMC/EMI aspects in Space Missions. Features include:

- Electromagnetic Cleanliness for Space Missions
- Standards, Specifications and Best Practices
- Instrumentation and Measurement Techniques for Space Applications
- Measurement uncertainty/ Quality control in test laboratories
- Uncoupling Test Facilities for ELF field Measurements: Methodologies
- Electromagnetic Interference between spacecraft's components
- Transient Phenomena (emissions, susceptibility)
- EUT Electromagnetic Modeling
- SpaceWire: Types & Comparison
- SpaceWire Structure Harnesses Identification and Evaluation
- Shielding Techniques in Spacecraft/ Satellite

It is also hoped that this book will serve as a comprehensive reference for graduate students who wish to enhance their knowledge of all aspects of EMC/ EMI for space systems engineering.

Christos D. Nikolopoulos
Department of Electronic Engineering, School of Engineering, Hellenic Mediterranean University, Greece

Chapter 1
Recent Advances on Measuring and Modeling ELF-Radiated Emissions for Space Applications

Christos D. Nikolopoulos
https://orcid.org/0000-0003-1344-4666
Department of Electronic Engineering, School of Engineering, Hellenic Mediterranean University, Greece

ABSTRACT

Nowadays, a wide range of space missions accommodate ever-stricter electromagnetic cleanliness requirements arising either from the need for more precise measurements or from the implementation of highly sensitive equipment. Therefore, the establishment of a methodology that ensures the minimization of the electric and/or magnetic field in specific areas inside or outside the spacecraft structure is crucial. Towards this goal, the current chapter proposes that utilizing the results of a process completed during the early design stages of a mission, that is, the measurement and characterization of each implemented device, the desired elimination of the field can be achieved. In particular, the emerged electromagnetic signatures of the units are proven essential for the proposed methodology, which, using a heuristic approach, defines the optimal ordinance of the equipment that leads to system-level electromagnetic field minimization in the volume of interest. The dimensions of the devices and the effect of the conductive surfaces of the spacecraft's hull are also taken into account.

INTRODUCTION

Since the very beginning of the first space missions in 1957, the scope of the majority of these missions is to measure extraterrestrial electromagnetic fields and particle population in plasma (Benkhoff et al., 2010; Drinkwater et al., 2003; Scheeres et al., 1998; Escoubet & Schmidt, 2000; Antonucci et al., 2011; Müller et al., 2013; Bayle et al., 2016). GOCE, EXOMARS, Rosetta, Cluster, BepiColombo, LISA and

DOI: 10.4018/978-1-7998-4879-0.ch001

Solar Orbiter, are some examples that aim to facilitate these measurements, carrying particle detectors and sensitive sensors (e.g. fluxgate or search-coil magnetometers) in their satellites or space probes. A lot of effort is carried out in the early design phases of a mission in order to place these sensors in the optimum electromagnetically clean environment. Moreover, and regarding the payload of these missions that is necessarily sensitive to electric and magnetic fields, requires stringent electromagnetic cleanliness with emphasis to random and periodic AC electric and magnetic field variations in frequency and time domain. From the perspective of extremely low frequency (ELF) electric cleanliness and AC & DC Magnetic cleanliness, these instruments are meant to measure slow time- variant fields (both electric and magnetic), equivalent to frequencies not higher than 250 KHz, based on recent study of (Vaivads et al., 2016). The targeted frequency range is not usually studied in standard ElectroMagnetic Compatibility (EMC) space engineering techniques (ECSS-E-ST-20-07C rev-1 Space engineering Electromagnetic compatibility, 2012) and thus the developed test measurement methods of this activity, that are presented in this chapter, are rather unique.

The acceptable mission threshold both for radiated AC electric (E-field) and magnetic (B-field) emissions in terms of spectral density in case of THOR mission are presented in Figures 1 and 2 respectively. Table 1 & 2 tabulated the mission requirements. It should be noted that in order to compare them with the measured radiated emission, these values shall be rescaled to the measured distance according to the equation 1. In the same Figures, radiated emissions of a Command and Data Handling Management unit (CDMU) from the Gravity Field and Steady-State Ocean Circulation Explorer (GOCE) mission as measured in Thales Alenia Space (TAS-I) premises on the Pre-verification of THOR mission Electromagnetic Cleanliness Study (*ESA contract 4000118770/16/NL/BW*) are also present and compared with the mission requirements both for steady state and transient behavior. It is apparent that the unwanted radiated electromagnetic fields from spacecraft's equipment cannot be higher than the accepted limits (predefined mission requirements).

$$F_{req}^{\,rescaled} = F_{req}\left(\frac{R_{req}}{R_{meas}}\right)^{3} \tag{1}$$

where R_{req} the distance between the magnetic (or electric) sensor and the satellite's (or spacecraft's) closest edge, referenced in the spacecraft's coordinate system (in this case, 6.3 m is assumed), F_{req} defines the magnetic (or electric) cleanliness requirements at the location of the sensitive or/and measuring instruments (sensors) and R_{meas} is the distance that the sensitive or/and measuring instruments are located during the measurement of the device under test (DUT). It should be noted, that according to the Equation 1 and under the quasi-static regime, the electric field is falling with the $\frac{1}{r^3}$ distance law. Therefore, the rescaled mission requirements shall be calculated at the measurement distance where, in our case, for the electric field the capturing antenna monopole was at 1 m while in the magnetic test campaign the magnetometers were placed in the distance of 50 cm.

In order to conform to the demanding requirements, both characterization of the units as radiating sources and the accurate prediction of the emissions are vital for the design and position arrangement of the equipment inside the spacecraft structure. The EDM technique (Equivalent dipole modeling) has been proved to be very efficient and robust in characterizing electromagnetic emissions from radiating elements under the quasi – static regime (C. Obiekezie et al., 2013). Basic modeling requires well defined parameters especially focused on the localization and identification of the equivalent sources. Moreover, heuristic approaches for optimizing the best candidate solutions remain the best practice with limited computational resources (C. S. Obiekezie et al., 2014). In the case of the ELF regime it is valid to assume that the near electric field can be described by a quasi-static formulation. Under this assumption the equivalent source reconstruction can be achieved by using equivalent current densities or electric dipoles with specific position and moments. Both methods are equal and the differences between them (current densities or dipole moments) are negligible. Given the fact that the electromagnetic fields from a dipole moment can be expressed in detail and the exact distributions of current and voltages can be described in a completely equivalent manner using dipole moments, a Multi-Frequency Electric and Magnetic Dipole Model methodology (MFEDM) has been proposed by (Koutantos et al., 2017). The novelty in the proposed MFEDM technique consists of the modeling of ELF electric field (and AC Magnetic Field) variations assigning an equivalent (electric o/and magnetic) dipole source for every frequency bin. The proposed methodology is able to solve the equivalent inverse problem assuming that the magnitude of each frequency component is the corresponding electrostatic field value. An iteration of this process for all frequency samples results to the full MFEDM model.

Representatives (test cases) in the said frequency area (Table 2.) (Lapeña et al., 2017) are usually model its steady state emissions with one oscillating dipole per frequency with 6 parameters (three spatial contributors and another three for the moments p_x, p_y, p_z) for the electric contribution and with another six for the magnetic behavior respectively (m_x, m_y, m_z). The corresponding equations and the theoretical background will be presented shortly in the next subsection.

Table 1 Representative Sources for Radiated Electric Emissions (CDMU case)

Common measurement procedure for ELF E-field emissions are performed via probes (specified for near field acquisition), or sensors or monopole antennas capturing only the field's amplitude without vectoral analysis. For each discrete frequency bin of the measured radiated emissions, a magnitude value is assigned to each measurement point. Thus, a number of quasi-static inverse problems are formed – one for every frequency bin. Valuable frequencies (for modelling point of view) are these whose emissions levels are way above the noise profile, rendering this way the necessity to define the mean noise reference threshold. This way, for every frequency bin of the measured spectrum above this threshold, each DUT can be modeled as one electric dipole. A repeated loop of this process for all measured spectrum concludes to a set of equivalent dipole models that compose the full ELF spectrum model of the DUT.

This methodology proposed by Baklezos et al., (2016) is using an heuristic approach (exploiting the Differential Evolution Algorithm - DE), for the prediction of dipole's parameters (position and moments) that propagate an electric field identical as the measured one (in various measurement positions). The source identification (dipole parameters) given the electric field measured values constitutes an inverse electromagnetic problem. After the source identification (the solution to the inverse problem) comes the validation process and in order to enable this, a second set of measurements is necessary. The correct extrapolation of the electric field on a greater distance, as compared against the second set of measurements, will validate the model. This yields the solution for a specific frequency. Repeating the aforementioned process over the whole range of frequencies that are of interest, a matrix of the dipole characteristic values for all frequencies is attained forming the DUT's Multi-Frequency Electric Dipole Model.

Figure 1. Amplitude Spectral Density (ASD) of radiated ELF electric field radiation compared to the mission Test limits (Source: ESA THOR Missions Requirements Document, 2015). upper: Steady state radiated emissions, Lower: Transient behavior.

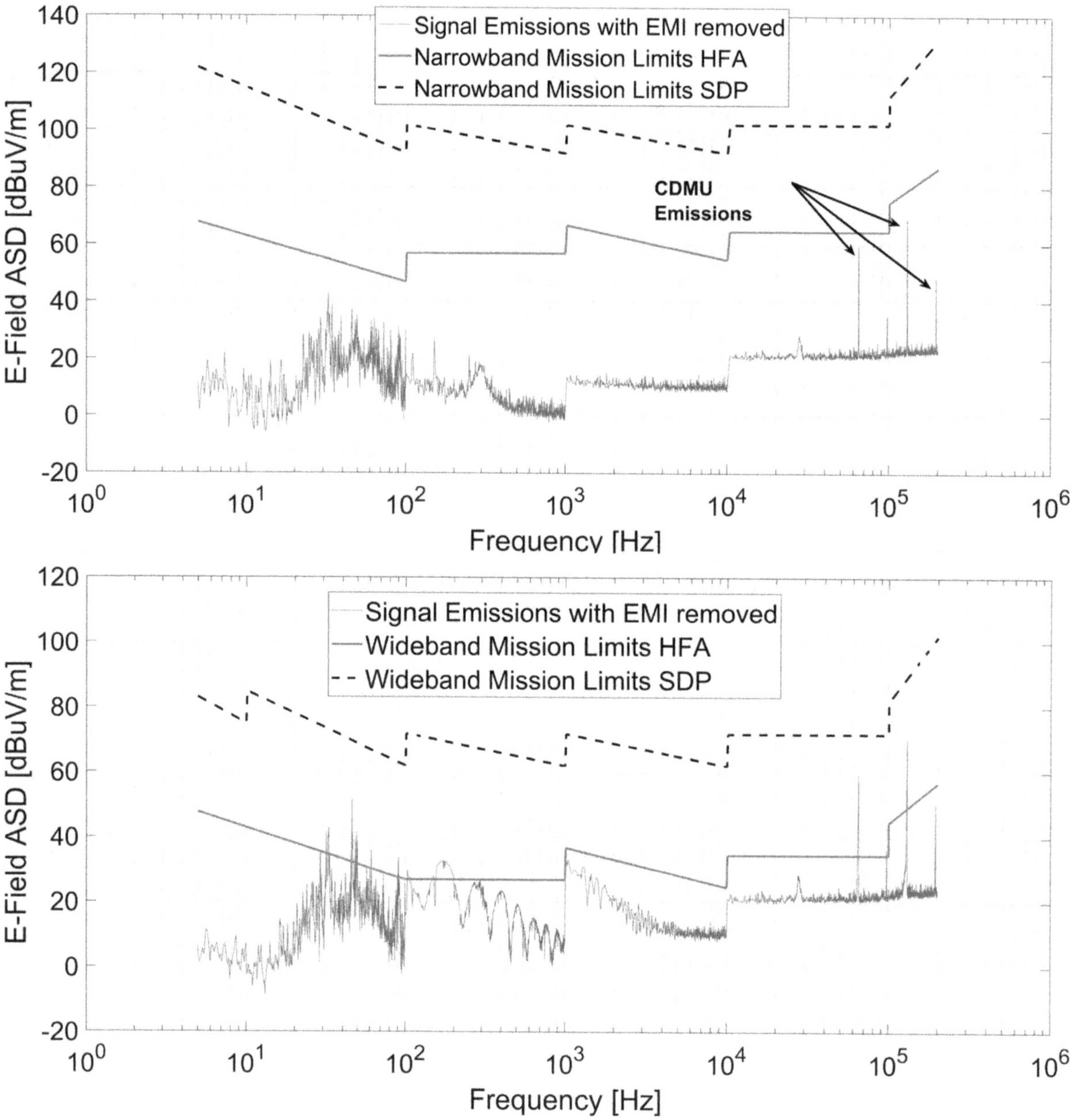

THEORETICAL BACKGROUND

Electric Modeling Considerations

It is valid to assume that the contributors (electric sources) inside a DUT and for this frequency range

Figure 2. Amplitude Spectral Density (ASD) of radiated AC magnetic field radiation compared to the mission Test limits (Source: ESA THOR Missions Requirements Document, 2015). Left: Steady state emission, right: Transient behavior.

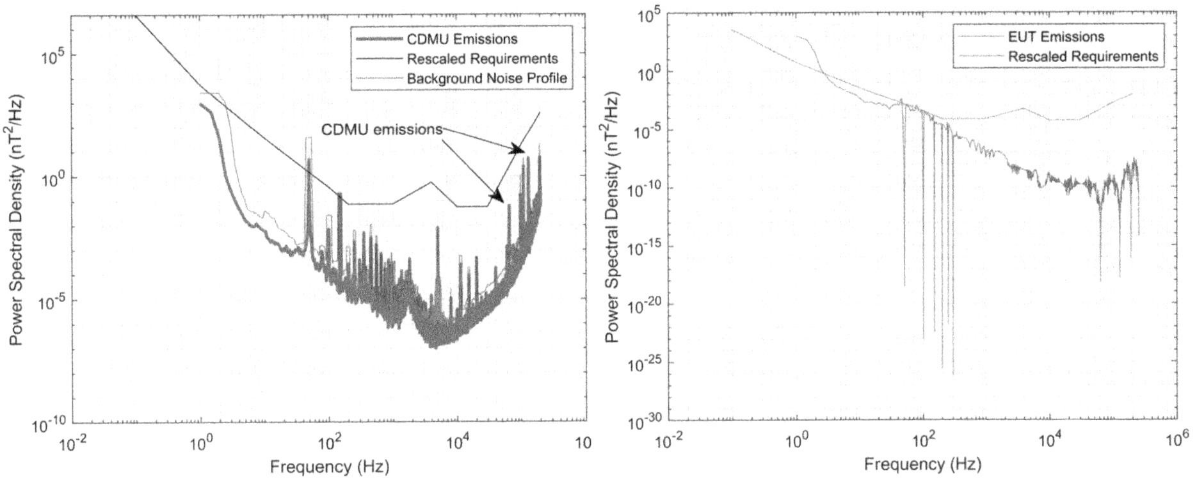

(up to 200 kHz) can be represented by a couple of electric dipoles (Koutantos et al., 2017). However, in some cases the previous statement cannot be valid as the case of electric currents floating on spacecraft's harness where for higher frequencies they can be represented and modelled as travelling wave antennas.

Considering a DUT placed at position $\overrightarrow{X}_{EUT}$ that radiates at extremely low frequency (ELF) frequency range and its emissions are measured at a point $\overrightarrow{X}_m$, under the quasi static regime it is safe to calculate the radiated emissions according to (Jackson, 1962):

$$E(\vec{n},t) = \frac{e^{-i\omega t}}{4\pi\varepsilon_0}\left\{k^2(\vec{n}\times\vec{p})\times\vec{n}\frac{e^{ikr}}{r} + \left[3\vec{n}(\vec{n}\cdot\vec{p}) - \vec{p}\right]\left(\frac{1}{r^3} - \frac{ik}{r^2}\right)e^{ikr}\right\} \quad (2)$$

Where $\vec{r} = \overrightarrow{X}_m - \overrightarrow{X}_{EUT}, r = |\vec{r}|, \vec{n} = \frac{\vec{r}}{r}$ and $\vec{p}$ is the electric moment vector. It should be noted that the electric field fall-off (of the dipole) follows the r^3 distance power law.

F (kHz)	Indicative Sources
0.4 kHz	Tracking Frequency
65 kHz	DC/DC B push pull driver
81 kHz	DC/DC A push pull driver
130 kHz	MPPT switching
162 kHz	DC/DC A clock
250 kHz	TM data

Figure 3. Flow diagram of the Multi Frequency - EDM process (Koutantos et al., 2017)

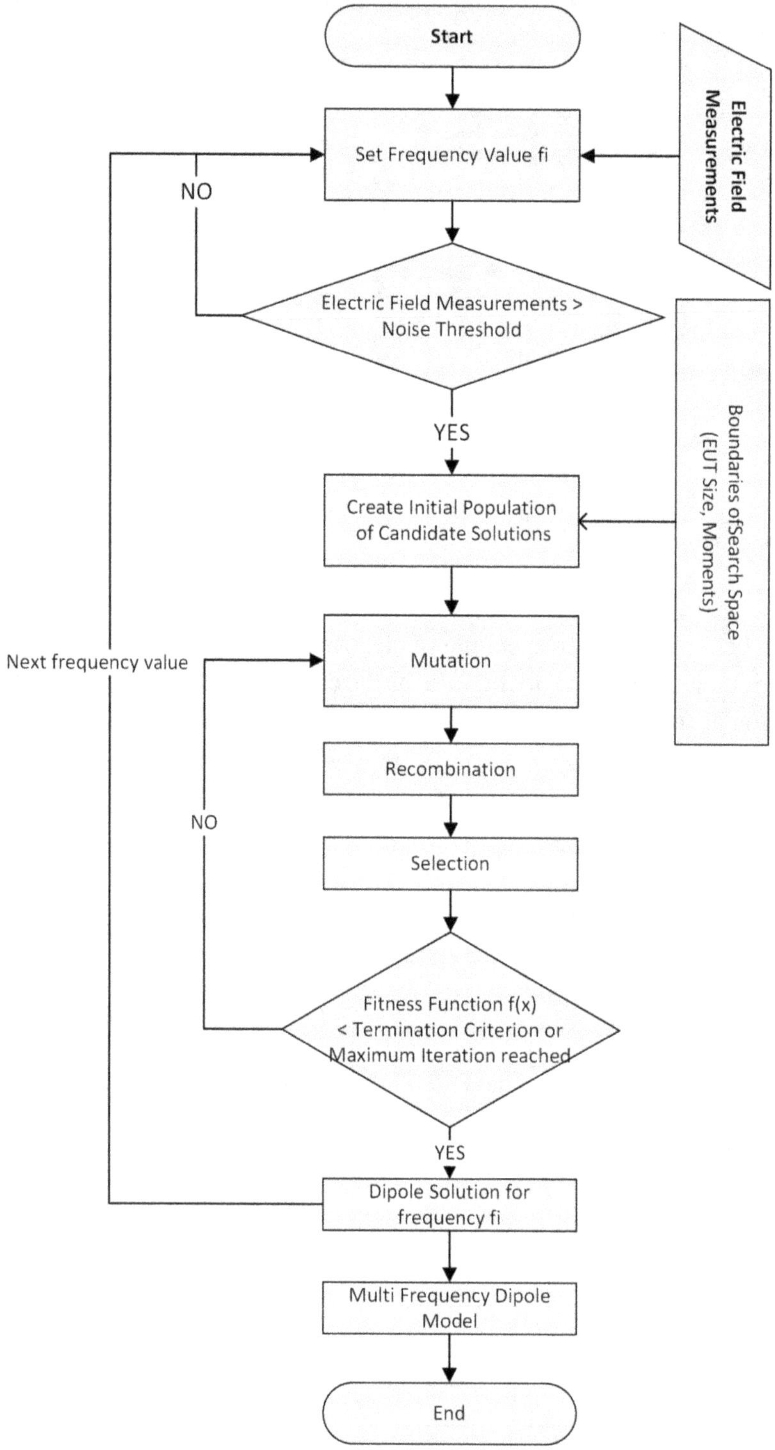

Wavelengths corresponding to ELF fields are in the order of kilometers and since the area of interest is in close proximity to the DUT (several meters), near field approximation is well justified. Near field approximation ($kr \ll 1$) reduces Eq. 2 to:

$$\boldsymbol{E}\left(\vec{\boldsymbol{n}},\boldsymbol{t}\right)=\frac{\boldsymbol{e}^{-i\acute{E}t}}{4\grave{A}\mu_0}\left\{3\vec{\boldsymbol{n}}\left(\vec{\boldsymbol{n}}\cdot\vec{\boldsymbol{p}}\right)-\vec{\boldsymbol{p}}\right\}\frac{1}{\boldsymbol{r}^3} \tag{3}$$

The electric field, with the exception of its oscillations in time, is just the static electric dipole field (Jackson, 1962). Hence the electric field can be written in the form of:

$$\boldsymbol{E(t) = E_{static}\cos(\omega t)} \tag{4}$$

Fourier analysis represent a time domain signal as a sum of n sinusoidal components according to:

$$\boldsymbol{f}\left(\boldsymbol{t}\right)=\frac{1}{2}\boldsymbol{a}_0+\sum\nolimits_{n=1}^{\infty}\boldsymbol{a}_n\cos\left(\boldsymbol{nt}\right)+\sum\nolimits_{n=1}^{\infty}\boldsymbol{b}_n\sin\left(\boldsymbol{nt}\right) \tag{5}$$

Under this regime, the complete electric behavior of a DUT (in case of a frequency domain measurements) could be modeled with one electric dipole per frequency yielding this way a Multi-Frequency Electric Dipole Model (MFEDM).

For each frequency bin an electric dipole (located in the position x, y, z) produces an electric field at measurement point (x_m, y_m, z_m) with the following time invariant components in each axis:

$$\left.\begin{aligned} Ex &= \frac{1}{4\pi\varepsilon_o}\left[\frac{3\left(x_m-x\right)\cdot C}{r^5}-\frac{p_x}{r^3}\right] \\ Ey &= \frac{1}{4\pi\varepsilon_o}\left[\frac{3\left(y_m-y\right)\cdot C}{r^5}-\frac{p_y}{r^3}\right] \\ Ez &= \frac{1}{4\pi\varepsilon_o}\left[\frac{3\left(z_m-z\right)\cdot C}{r^5}-\frac{p_z}{r^3}\right] \end{aligned}\right\} \tag{6}$$

Where:

$$c=p_x\left(x_m-x\right)+p_y\left(y_m-y\right)+p_z\left(z_m-z\right)$$

And

$$r=\sqrt{\left(x_m-x\right)^2+\left(y_m-y\right)^2+\left(z_m-z\right)^2}$$

and the amplitude measured for frequency ωi for every measurement point Nm_j) is:

$$\left|E_{static}\right|_{i,j} = \sqrt{E_x^{\ 2} + E_y^{\ 2} + E_z^{\ 2}} \tag{7}$$

The main idea in this work is to resolve the inverse elecrtomagnetic problem exploiting the N_m measurements of the electric field magnitude for every measured frequency bin ωi as an input dataset. This way the calculation of the 6 parameters (position and electric moments) of the equivalent source – electric dipole (Koutantos et al., 2017) is feasible.

The proposed methodology uses a heuristic approach, the Differential Evolution algorithm, attempting to solve the inverse electromagnetic scattering problem, in other words to identify the six (6) variables of the dipole's source that emits the exact same radiated emissions as the measured ones. The algorithmic flow diagram is tabulated in Figure 3 and the heuristic (Differential Evolution) methodology implementation is discused in detail in the following subsection.

It should be noted that in general case a DUT may also have simultaniusly a magnetic behaviour. However, usually the magnetic moments are adequate small and the electric field produced by the magnetic moment is negligible.

$$E(x,t) = \frac{-Z_0 k^2 e^{-i\omega t}}{4\pi}(n \times m)\frac{e^{ikr}}{r}\left(1 - \frac{1}{ikr}\right) \tag{8}$$

The field dependence in this case is in the order of r^2:

$$E(x,t) = \frac{1}{4\pi\varepsilon_0}\left\{\frac{\dot{m}(t_0) \times n}{r^2} + \frac{\ddot{m}(t_0) \times n}{cr}\right\} \tag{9}$$

A two-step methodology is performed in order to acquire an accurate and robust model. First, the magnetic field is measured, characterizing this way the DUT's magnetic signature and estimating the magnetic moment of the equivalent magnetic dipole. In case this value can produce a substantial electric field distribution, then equation 8 should be taken also into consideration in the Electric Dipole Moment model extraction.

Magnetic Modeling Considerations

In consistency with the electric considerations the magnetic contributions (magnetic sources) inside a device (DUT) can be represented by a couple of magnetic dipoles or quadrupoles (Paul, 2006), although there are cases of magnetic field sources inside a spacecraft that cannot be represented by magnetic dipoles (e.g. closed loop current routes, harness topologies, or the electric currents flowing in the devices and solar arrays shall calculated analytically using Biot - Savart law. For demonstration purposes, the DUT is placed at the origin of the Cartesian coordinate system and its magnetic contribution can be modeled as a number of M magnetic dipoles positioned at $\mathbf{r}'_i = (x_i, y_i, z_i)$, i=1,2,…,M, with magnetic moments:

$$\mathbf{m}_i = m_{i_x}\hat{\mathbf{x}} + m_{i_y}\hat{\mathbf{y}} + m_{i_z}\hat{\mathbf{z}} \tag{10}$$

Moreover, N_{MAG} measurement positions are located at $\mathbf{r}_{0j} = \left(x_{0j}, y_{0j}, z_{0j}\right)$, j=1,2,…,N_{MAG}. The magnetic field produced from the magnetic dipole i at the measurement position j is analyzed in vector components as (Jackson, 2003):

$$Bx_{ij} = \frac{\mu_0}{4\pi}\left[\frac{3\left(x_{0j}-x_i\right)}{\left|\mathbf{r}_{0j}-\mathbf{r'}_i\right|^5}L_{ij} - \frac{m_{i_x}}{\left|\mathbf{r}_{0j}-\mathbf{r'}_i\right|^3}\right]$$

$$By_{ij} = \frac{\mu_0}{4\pi}\left[\frac{3\left(y_{0j}-y_i\right)}{\left|\mathbf{r}_{0j}-\mathbf{r'}_i\right|^5}L_{ij} - \frac{m_{i_y}}{\left|\mathbf{r}_{0j}-\mathbf{r'}_i\right|^3}\right] \tag{11}$$

$$Bz_{ij} = \frac{\mu_0}{4\pi}\left[\frac{3\left(z_{0j}-z_i\right)}{\left|\mathbf{r}_{0j}-\mathbf{r'}_i\right|^5}L_{ij} - \frac{m_{i_z}}{\left|\mathbf{r}_{0j}-\mathbf{r'}_i\right|^3}\right]$$

where the dot product L_{ij} is:

$$L_{ij} = m_{i_x}\left(x_{0j}-x_i\right) + m_{i_y}\left(y_{0j}-y_i\right) + m_{i_z}\left(z_{0j}-z_i\right) \tag{12}$$

and $\left|\mathbf{r}_{0j}-\mathbf{r'}_i\right|$ is the distance between the measurement position and origin of the magnetic source:

$$\left|\mathbf{r}_{0j}-\mathbf{r'}_i\right| = \sqrt{\left(x_{0j}-x_i\right)^2 + \left(y_{0j}-y_i\right)^2 + \left(z_{0j}-z_i\right)^2} \tag{13}$$

The total magnetic contribution from the M magnetic sources (dipoles) at the jth measurement position is calculated by the superposition of each independent contributor:

$$B_j = \sum_{i=1}^{M} Bx_{ij}\hat{x} + \sum_{i=1}^{M} By_{ij}\hat{y} + \sum_{i=1}^{M} Bz_{ij}\hat{z} \tag{14}$$

The magnetic field vector is expressed in components (B_x, B_y, B_z), since the sensors (fluxgate magnetometers) measures static and alternating magnetic fields (until the 10 kHz) in three axes. However, several types of magnetometers, measure the magnitude of the magnetic field:

$$|\mathbf{B}_j| = \sqrt{Bx_j^{\,2} + By_j^{\,2} + Bz_j^{\,2}} \tag{15}$$

Implementation of the equations (11) – (14) results to the calculation of the magnetic signature and thus the characterization of a typical magnetic source.

DE Algorithm Application for Solving the Inverse Radiation Problem

The Differential Evolution algorithm is a population based, derivative-free, Heuristic Direct Search employing a Global Optimization process, first introduced by (Storn & Price, 1997). DE process is starting with a number of initial candidate solutions (initial population) that are randomly generated from a normal or uniform distribution within the space of valid solutions (dimensions of the DUT for the spatial parameters and indicative valid ranges for the three values of the electric moments) and are subjected to iterations of recombination, evaluation, and selection. It usually encodes decision variables as floating point numbers and manipulates them with simple arithmetic operations such as addition, subtraction, and multiplication. The recombination approach (mutation and crossover) involves the creation of new candidate solution vectror (dipole components) based on the weighted difference between two randomly selected population members added to a third population member (Brownlee, 2011).

DE has a specific terminology that caracterizes the adopted configuration. This is encoded as *De/x/y/z*, where *x* represents the solution to be perturbed (with two configurations: random or best). The *y* denotes the number of difference vectors used (Brownlee, 2011) in the perturbation of *x*. The difference vector is the "distance" between two randomly selected although distinct members of the population. Finally, *z* denotes the recombination operator performed such as "*bin*" for binomial and "*exp*" for exponential. The weighting factor *F* controls the amplification of differential variation and the crossover weight CR probabilistically controls the amount of recombination. The DE optimization algorithm is well documented and used for many problems so it will not be discussed any further. In (Dragoi et al., 2012) and (Oliveri et al., 2012) authors have presented a wide range of DE applications in electromagnetics. For more details on the internals of the DE algorithm, the reader is directed to the rich DE literature.

The implementatin of DE where it is heuristically trying to solve the inverse radiation problem, is applying the mutation strategy of *DE/rand/*1 with the inclusion of a technique called dither (Dawar & Ludwig, 2015). Dither is the selection of factor *F* from the interval [0.5, 1.0] randomly for each generation or for each difference vector, a technique that is shown to improve convergence behavior significantly. Crossover weight CR is selected as 0.9 and the population size *NP* is set at 50. As fitness function the RMS error of the field of the solution dipole, as calculated from equation 16, is selected.

The extracted model is composed by the 6 parameters (the solution variables of the algorithm) which are mandatory in order to describe a dipole source at each frequency bin. As depicted in Figure 3, DE is used in order to match the input data from the N_m values of the electric field measurements for each frequency with the produced field from a model dipole. For every frequency step, the DE provides the optimal set of 6 parameter's values, that are produced according to the methodology described before, that minimize the RMS error value depicted in Eq. 16. The iteration of this process for all frequency values yields the complete MFEDM model.

$$(RMS) = \sqrt{\frac{\sum_{i=1}^{N_m}\left(E_{meas}(i) - E_{model}(i)\right)^2}{N_m}} \tag{16}$$

The E_{model} values are calculated according to equation 7.

Ambient Conditions

In order to have a reference, the ambient levels shall be measured. The ambient measurements are defined as the background emission levels, when:

a) All the function and supervisory units of the Electrical Ground support equipment (EGSE) are powered and in operation.
b) The DUT is non-powered.
c) A resistive load is connected to the Test Equipment with a load corresponding to the power consumption of the equipment under test.
d) During the ambient measurements a LISN shall be connected to the power terminal of the resistive load.
e) The ambient electromagnetic emission levels shall be at least 6 dB below the specified limit.

In the event that the ambient measurements exceed the relevant limit-6dB every effort shall be made to reduce or eliminate those emissions, if this cannot be achieved the source of the emissions should be identified and reported in the EMC Test Report. It should be noted that it may be necessary to address the configuration of the test equipment and its harnessing, as far as possible this should also be installed such that ground loops are minimized, harnesses may require to be over-shielded, aluminum foil can be used for this if necessary.

The impedance of power sources providing input power to the DUT shall be controlled by Line Impedance Stabilization Network(s) (LISN(s)) for all measurement. LISN's shall not be used on output power leads.

RADIATED EMISSIONS MEASUREMENTS AND MODELLING RESULTS

Proposed Measurement Technique for ELF Electric Field

The proposed measurement procedure of Nikolopoulos et al., (2019) is in accordance to the MIL-STD-461 regarding DUT and acquisition sensor (monopole antenna) placement. The initial DUT position and orientation as well as the global coordinate system are depicted in Figure 4. Identical methodology is used for both steady state and transient acquisition with minor adjustments. In order to ensure steady state emission, the acquisition process has to be started after the maximum estimated transient period has passed, and the current distribution has been stabilized. However, in order to capture transient phenomena, the acquisition process must be started before the DUT is "powered-on". It should be stated that in

Figure 4. Measurement setup in the global-antenna X-Y plane (60^0 plane), (b) Measurement setup in the global-antenna X-Y plane (45^0 plane). (Nikolopoulos, Baklezos, & Capsalis, 2019)

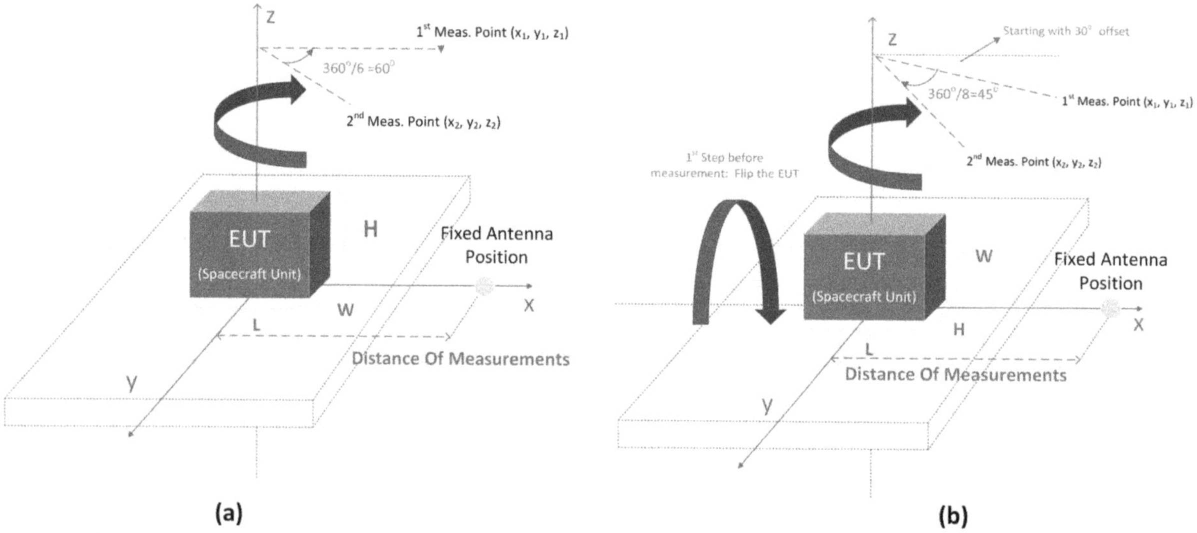

case that transient phenomena are to be captured, the steady state emission measurement is obligatory and must precede the transient one.

This is necessary in order to identify steady state emissions that are also present in the transient acquisition.

During normal measurement procedure, the following measurement steps shall be performed:

- One (1) measurement with a monopole antenna (e.g. R&S AM524) according to MIL-STD-461-F/G with the DUT non-operational (powered off) in order to have an estimate of the background noise.
- Six (6) measurements with the R&S AM524 antenna according to MIL-STD-461-F/G in the XY plane (according to global coordinate system), rotating the DUT clockwise with an angular step of 60^0.
- Next the DUT is flipped (a one-time 90^0 rotation around X-axis), and
- Eight (8) measurements with the monopole antenna according to MIL-STD-461-F/G in the XY plane (according to global – antenna coordinate system) rotating the DUT clockwise with an angular step of 45^0 starting with an angular offset of 30^0. This offset is added to the measurement procedure in order to ensure that all measurement points are different.

The proposed measurement setup provides more than 12 field values (the necessary number of measurements according to Koutantos et al., 2017). This measurement procedure is carefully chosen in order to provide a denser spatial sampling of the field emissions. Each of the 14 measurements can be considered as a "stand-alone" MIL-STD-461 measurement

Indicative Measurement and Modelling Results for ELF Electric Sources

As stated in the introductory part, the necessity for the development of an alternative measurement setup in order to acquire the sufficient dataset for modelling spacecraft unit ELF emissions, arose during the measurement phase of THOR Pre-verification study at the premises of Thales Alenia Space Italy (TAS-I) and led to the proposed measurement methodology as described in the previous section.

Figure 5. Measured noise figure with EMI sources present at TAS-I facilities (anechoic tent).

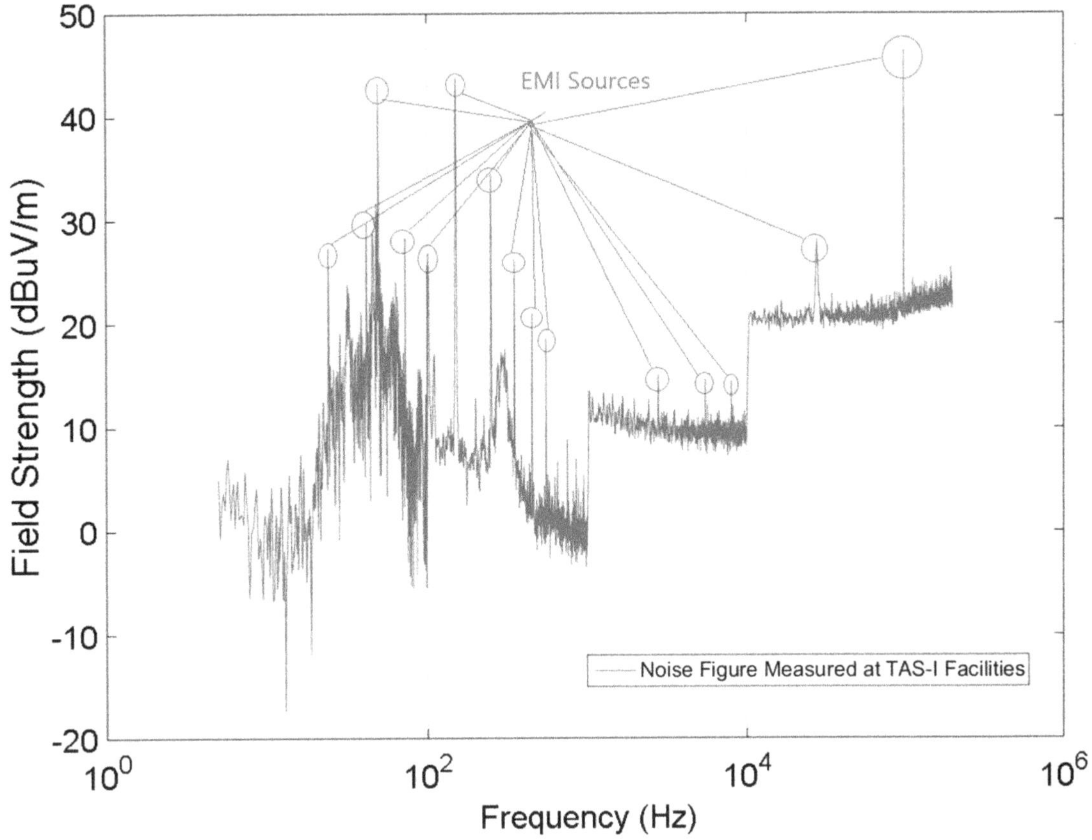

In the framework of that measurement phase multiple spacecraft units were measured in the frequency range of 5Hz-200KHz (per ESA's mission study requirements) employing the proposed methodology, including the Command & Data Handling Management unit (CDMU). The equipment for data acquisition is tabulated in Table 2. For this unit and for one measurement point indicative results are depicted in Figure 6 (steady state) and Figure 8 (transient emissions).

In order to provide an accurate model for every DUT, a reliable characterization of the emissions is necessary. For this reason, a baseline noise profile (without an DUT active) was established for every measurement dataset according to step 1 of the proposed measurement methodology. Background Noise in the Anechoic Tent Thales Alenia Space-Italia facilities, during the measurement campaign, is depicted in Figure 5. As it is marked (red colour) a lot of EMI sources are present. The frequency bins of these EMI sources are excluded from the modelling process.

Table 2. Test Measurement Equipment

Equipment Under Test	CDMU
Receiver	N9038A MXE
Monopole Antenna	R&S AM524

Steady State Emissions

As described in the measurement methodology, steady state emissions are captured and depicted in Figure 6. Steady state DUT emissions (with red colour marks) are identified and the set of 14 field Strength values (of the 14 different measurement positions) for each of the identified frequency bins (65KHz, 130KHz, 195KHz) are fed to the modelling algorithm (Capsalis et al., 2019).

Figure 6. Measured DUT (CDMU) radiated emissions in the case of Steady State at TAS-I facilities (anechoic tent).

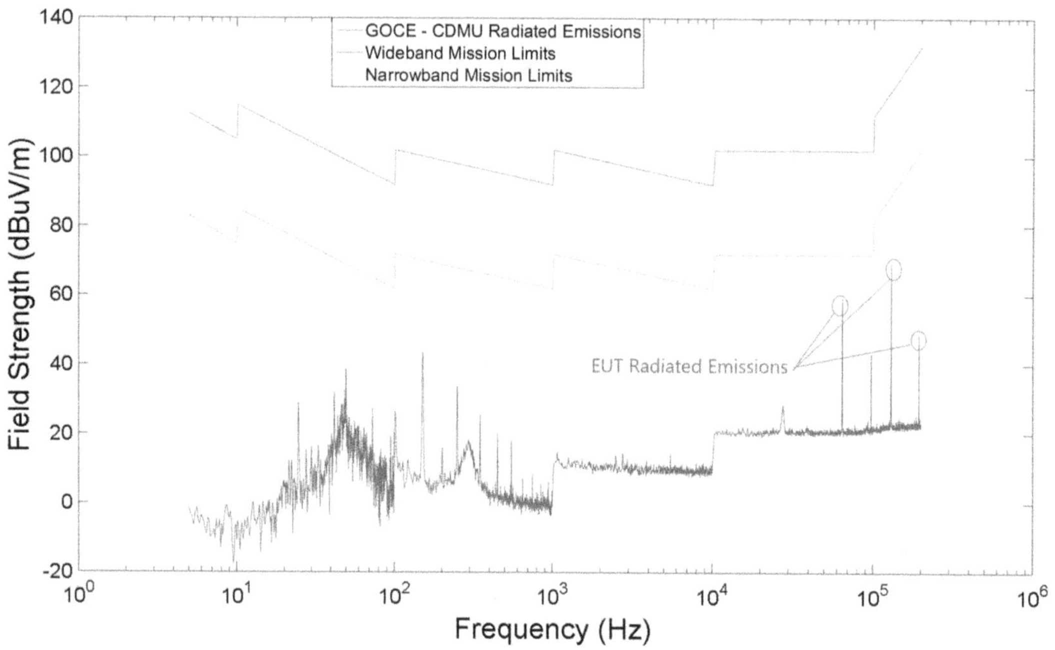

For every one the identified frequency bins, the algorithm attempts to solve the inverse electromagnetic problem and to correspond the 14 spatially distributed field values to a dipole source whose parameters (position and electric moment) can produce field values that minimize the difference from the 14 measured field values used as input. Indicatively modelling results for the frequency bin of 195 KHz are depicted for the 14 distinct measurements of the methodology in Figure 7. The produced dipole model's field emissions fit the measured emissions with a maximum difference of 1.5 dBuV/m and an average difference of 0.57 dBuV/m). These values are more than acceptable for evaluation of the unit's emission purposes since the commonly accepted difference among EMC measurements usually is 3-6dB.

Moreover, during the measurements, it was noted that the DUT was placed on a conductive plane (grounded table). In order to include the effect of the conductive plane on the measured ELF electric field values, the algorithm includes image theory adding the contribution of the image dipole (Chatzineofytou et al., n.d.).

Transient Electric Emissions

Figure 7. Measured field values in comparison with the Modeled-ones.

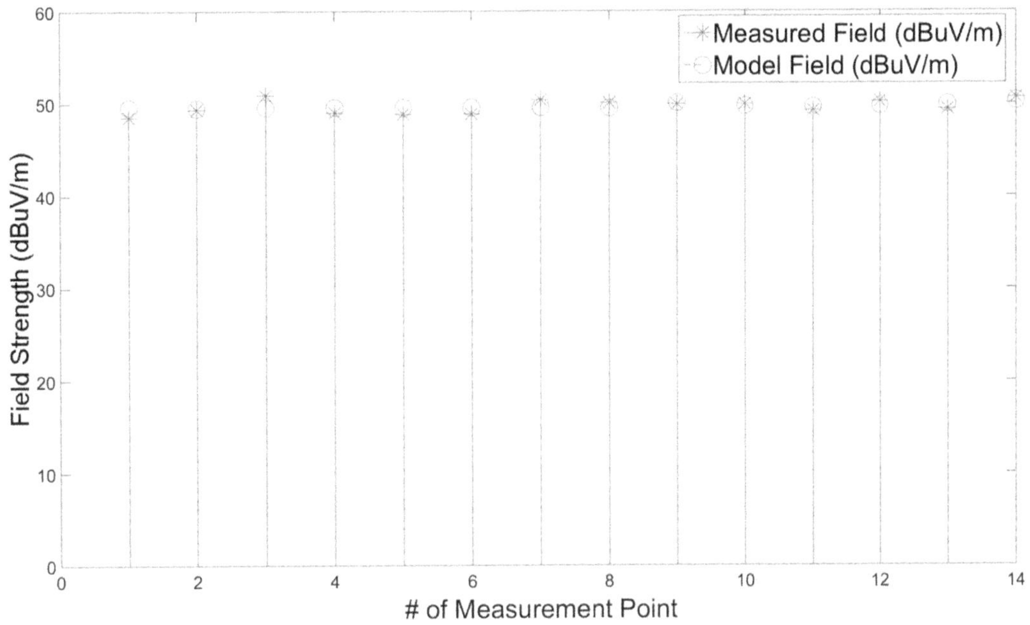

Transient phenomena according to the proposed methodology as described in this Section, are captured carefully starting the acquisition process before the DUT is powered-on. Figure 8 depicts the transient measurement results for 1 measurement out of the 14 acquired in TAS-I facilities for the CDMU unit. It should be noted that since the DUT remains in the powered-on state during acquisition, steady state emissions are also present in the measurement results (Figure 8), but already characterized and identified in the previous step of the proposed methodology. For clarity, it should be stated once more that transient acquisition must follow the steady state emission measurements.

In order to model the transients, the phenomenon is approached in a two-step process. A number of fictitious quasi-static dipoles is assumed in order to determine the spatial dependency of the field emissions. The emitted electric field is then considered to be oscillating according to a pulse signal. This

Figure 8. Measured DUT (CDMU) radiated emissions in the case of transient behavior at TAS-I anechoic tent.

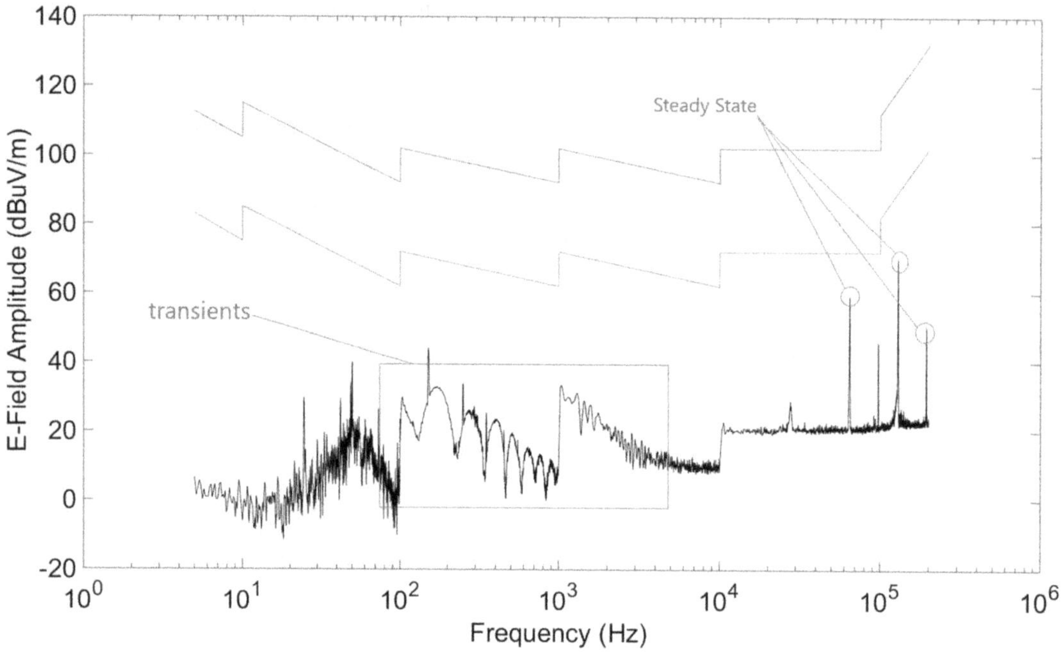

way the frequency dependency of the signal is separated from the spatial dependency. So, it is possible to have multiple transient phenomena each modelled with a static dipole producing DC field and one pulse with amplitude equal to this DC field. The number of dipoles used to model and the number of model pulses are always equal.

When solving the inverse problem, the process takes place in reverse order. First the appropriate pulse parameters in order to best fit the measured spectral behavior are estimated for each of the measured datasets. The key parameter of the pulse representation for this work is the amplitude A since it provides an equivalent "static measurement" of the unit at the measurement point. An aggregation of all the amplitudes A_i across all the measurement points yields a single quasi-static inverse scattering problem which can be handled like in (Baklezos et al., 2017). So, from the amplitudes of these pulses (from the 14 measurement datasets) the DC electric dipole parameters are extracted. The combined parameters are the overall model of the transient phenomenon; the overall model consists of the position coordinates of the dipole source x,y,z, combined with the electric moments p_x, p_y, p_z yield the quasi-static field at an arbitrary measurement/observation point which presents the spectral dependency of a pulse spectrum with width τ, rise time τr, fall time τf.

Two important remarks should be made at this point. First, actual measurements do not present the smoothness of the pulse spectrum. So, in order to provide a more reliable dataset for the algorithm to model (to extract the pulse parameters) the measurement dataset's spectral envelope is initially calculated and passed on for modelling. The algorithm is taking into consideration as input parameter the number of Envelope Samples.

Secondly, in practice, the synchronization between the OFF and the ON state of each DUT may be slightly off among the 14 measurements. For this reason, the model employs the mean value of the three

Figure 9. ELF transient electric behavior modeling process

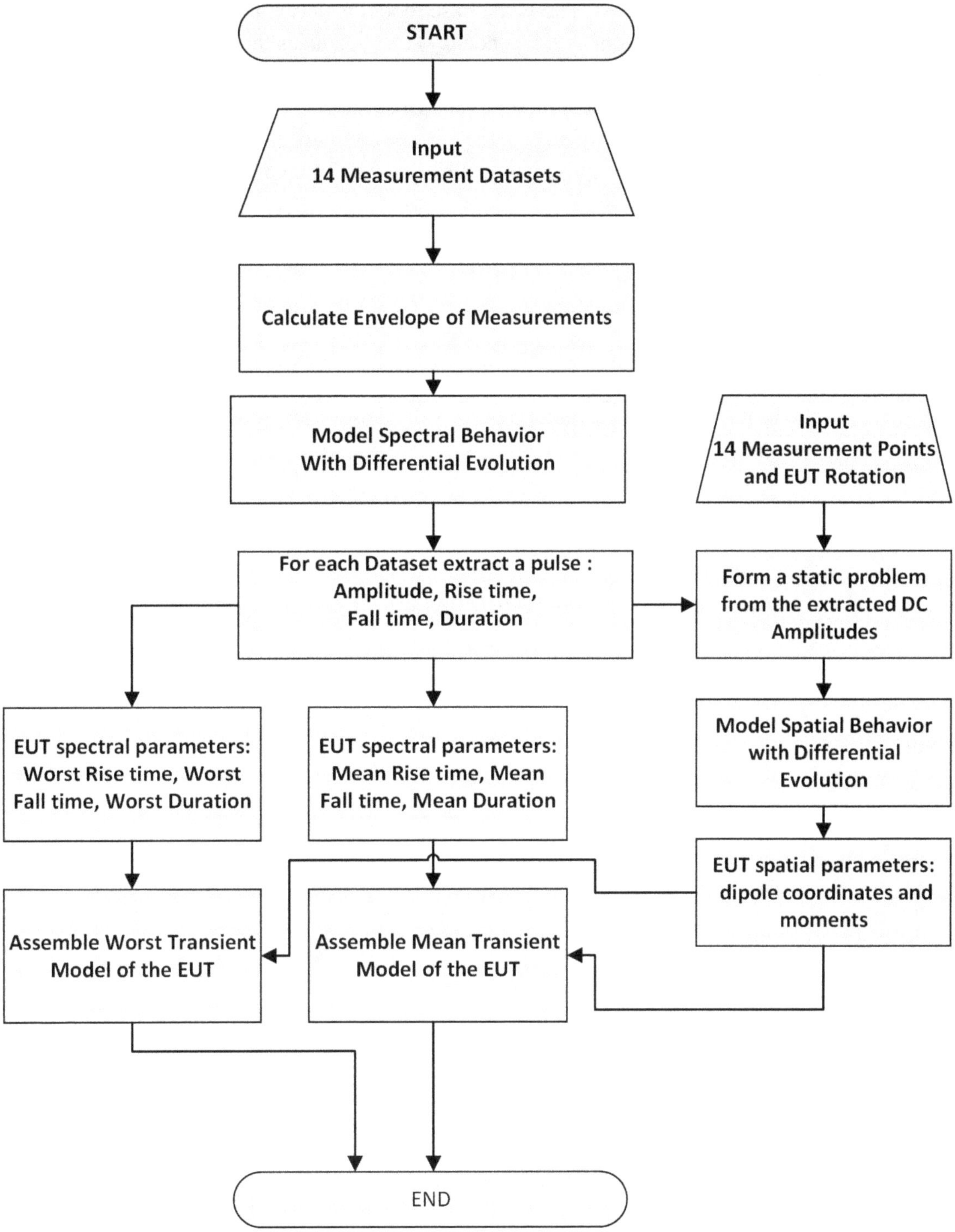

(3) pulse parameters from the pulse solutions (width/ duration τ, rise time τr, fall time τf). Additionally, to mean model, another one is extracted. The worst-case model is defined as the one set of parameters (among the ones produced for the frequency dependency) that produces the higher emission levels.

The overall modeling process is depicted of Figure 9 in order to clarify the flow of steps followed in order to assemble a complete ELF transient electric behavior model.

Initially for the CDMU, the envelope of each of the 14-measurement dataset is modelled regarding its spectral behavior aiming to identify the pulse parameters that best fit the measured "event". The results are tabulated in Table 3.

Table 3. Pulse Model Parameters

Model	**Pulse Parameter**		
	Duration (s)	*Trise (s)*	*Tfall (s)*
Mean	8.523837792	0.008653053	0.00856127
Worst	9.102296734	0.008749457	0.00815637

Figure 10. CDMU model results for the 100-1000Hz frequency range at xy-plane at 60⁰.

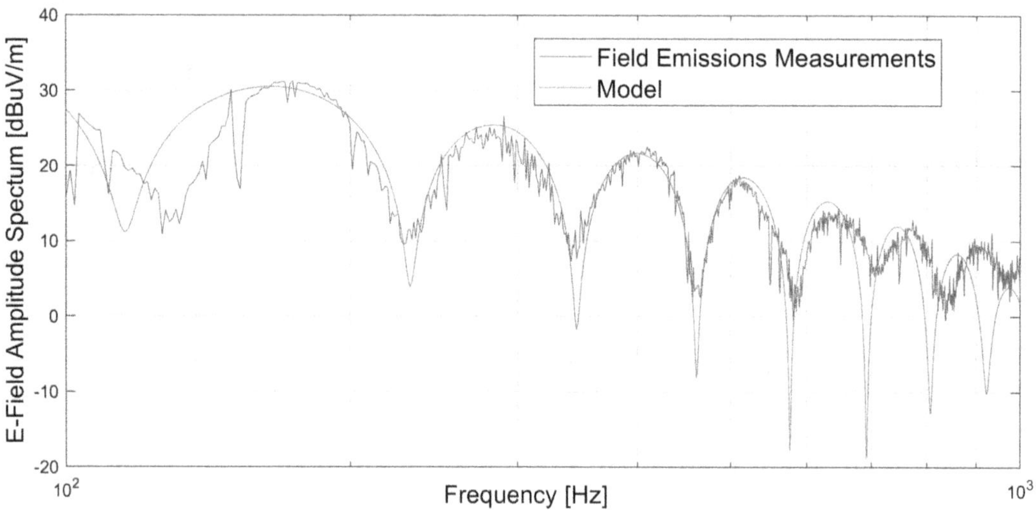

The corresponding mean model results are depicted in Figure 10 against measurements and are found in excellent agreement. The extracted dipole model parameters are tabulated in Table 4.

Table 4. Dipole Model Parameters

Dipole Parameter					
x (m)	*y (m)*	*z (m)*	*px (Cb*m)*	*py (Cb*m)*	*Pz (Cb*m)*
0.1361	-0.23	0.0508	2.93E-11	1.00E-10	4.04E-11

AC Electric Field from Transmission Lines (from Conducted to Radiated Emissions)

Harnesses can act as radiating source due to common and differential mode currents flowing on their shields and depending on the operational frequency under investigation can be treated as dipole antenna or travelling wave antenna. Therefore, the cable contribution to the AC electric field must be also taken into consideration for system level electric cleanliness. Additionally, the main idea is to correlate the conducted emissions spectrum of a unit with the spectrum of its radiated emissions (in both cases emissions may emerge at the same frequencies). Moreover, a study is in order to analyse the conducted emissions of a unit under test and correlate them with the radiated emissions of the same unit, especially to identify the frequencies that exhibit emissions.

The harness, depending on the operational frequency, can be modelled as an infinitesimal dipole or a travelling wave antenna with a flowing current distribution. Conducted common mode emissions lead to the current flowing on the harnesses shield, which is the driving force of radiation.

In the case of extremely low frequencies the harness, due to its extremely low length compared to the wavelength, can be treated as one (or more based on the actual geometry of the harness) infinitesimal electric dipole. Under this approximation the near fields of the harness in the near region is:

$$\vec{E}\left(r,\theta,\varphi\right) \approx -j\omega\mu_O \bullet \frac{I_O l}{2\pi k^2 r^3} \bullet \left| cos\theta\hat{r} + \frac{sin\theta}{2}\hat{\theta} \right| \bullet e^{-jkr} \tag{17}$$

and

$$\vec{B}\left(r,\theta,\varphi\right) = \mu_O \bullet \vec{H}\left(r,\theta,\varphi\right) \approx \mu_O \bullet \frac{I_O l}{4\pi r^2} \bullet sin\theta\hat{\varphi} \bullet e^{-jkr} \tag{18}$$

I_O is the current and l is the length of the harness.

For higher frequencies (well in the MHz range) the harness translates to a straight conductor with a length one to many wavelengths and the region of interest is in this case in the far field. So, the traveling wave approach will be used to describe the behavior of the harness (a long wire) in the far field. According to this approximation, the current distribution of the forward traveling wave along the harness can be represented by

$$\boldsymbol{I}_f = I_z\left(z'\right)e^{-\gamma\left(z'\right)z'}\hat{z} = I_O e^{-\left[a\left(z'\right)+jk_z\left(z'\right)\right]z'}\hat{z} \tag{19}$$

$\gamma\left(z'\right)$ is the propagation coefficient ($\gamma\left(z'\right)=a\left(z'\right)+jk_z\left(z'\right)$ where $a\left(z'\right)$ is the attenuation costant and $k_z\left(z'\right)$ is the phase constant associated with the traveling-wave). There are losses due to leakage as well as wire and ground losses. However, these losses are usually very small, especially when the radiating medium is vacuum/air so for simplicity are neglected. So, the current distribution can be approximated by

$$\boldsymbol{I}_f = I_O e^{-jk_z(z')z'}\hat{z} \tag{20}$$

where $I(z') = I_O$ is assumed to be constant.

Then the far field is described by

$$\vec{E}(r,\theta,\varphi) \approx j\omega\mu_O \bullet \frac{lI_O e^{-jkr}}{4\pi r} \bullet e^{-j\left(\frac{kl}{2}\right)(1-cos\theta)} sin\theta \frac{sin\left[(kl/2)(cos\theta-1)\right]}{(kl/2)(cos\theta-1)}\hat{\theta} \tag{21}$$

and

$$\vec{B}(r,\theta,\varphi) = \mu_O \bullet \vec{H}(r,\theta,\varphi) \approx\approx jk\mu_O \bullet \frac{lI_O e^{-jkr}}{4\pi r} \bullet e^{-j\left(\frac{kl}{2}\right)(1-\cos\theta)} \sin\theta \frac{\sin\left[(kl/2)(\cos\theta-1)\right]}{(kl/2)(\cos\theta-1)}\hat{\varphi} \tag{22}$$

In either source scenarios (depending on the frequency range), the exact geometry and ground plane contribution must be taken into consideration along with image theory.

The radiated electric field can be calculated according to the previous considerations considering also the following parameters: the orientation of the cable, the height of the cable from the ground and the dielectric properties of the ground. Based on this assumption/hypothesis and for known geometric assortments and layouts of the system under test, the radiated emissions can be predicted based on the knowledge of the current flowing on the shield of the harness as employed in Nikolopoulos, Baklezos, Tsatalas, et al., (2019) and Baklezos et al., (2020). The current parameter may be estimated by either a) *measurement of the radiated emissions* of the harness if available/possible and then extracting the current $I_O(f)$ parameter solving the inverse electromagnetic problem based on equation 17 (dipole), equation 21(travelling wave) taking into consideration the exact geometry of the setup used or b) *measurement of the conducted emissions* $I_C(f)$ of the devices under test on the harness and/or power cord. Then the current $I_O(f)$ parameter is assigned as:

$$I_O(f) = a(f) \bullet I_C(f) \tag{23}$$

$a(f)$ parameter accounts for the shielding of the harness.

Based on the form of conducted emissions measurement results $I_C(f)$ as well as information regarding each specific equipment under test involved in the system along the harness, the modeling of the current $I_O(f)$ may be carried out either on a "bin" by "bin" (independent frequencies) mode as in Nikolopoulos, Baklezos, Tsatalas, et al., (2019) or on a structured (pulse) mode as in Baklezos et al., (2020).

The knowledge/estimation of the $I_O(f)$ as well as the exact parameters/characteristics (geometry/materials) of the system layout will enable the prediction of the radiated emissions of the harness.

AC Magnetic Field from Transmission Lines (from Conducted to Radiated Emissions)

Considering the frequency range of interest and the typical relative magnetic permeabilities of many materials, the relation between conducted emission currents (nominal or measured) and resulting magnetic fields is expected to be rather straight forward, i.e. following a (quasi-)static relation like the Biot-Savart law.

In order to calculate the magnetic field of a finite current element of length dl that carries a steady current I, Biot - Savart law can be used:

$$dB = \frac{\mu_0}{4\pi}\frac{Idl \times R}{|R|^3} \tag{24}$$

where $R = r - r'$ is the distance vector between the **r'** (source point) and **r** (observation point), and the integral is evaluated over the path C in which the electric current flows.

The time dependent generalization of the Biot-Savart may be expressed:

$$\mathbf{B}(\mathbf{r},t) = \frac{\mu_0}{4\pi}\int\left[\frac{1}{R^2}\mathbf{J}(\mathbf{r}',t') + \frac{1}{cR}\frac{\partial \mathbf{J}(\mathbf{r}',t')}{\partial t'}\right] \times \hat{\mathbf{R}}d^3r' \tag{25}$$

where $t' = t - R / c = t - |r - r'| / c$, c is the speed of light and **J** is the current density. In the frequency range under investigation, the second term of this equation can be safely neglected, since the current density is slowly varying over time (quasi-static approximation) and the magnetic field can be calculated:

$$\mathbf{B}(\mathbf{r},t) = \frac{\mu_0}{4\pi}\int\left[\frac{1}{R^2}\mathbf{J}(\mathbf{r}',t')\right] \times \hat{\mathbf{R}}d^3r' \tag{26}$$

Proposed Measurement Technique for DC and AC Magnetic Field

The measurement setup is presented as a schematic in Figure 11. The configuration consists of 12 magnetometers placed symmetrically in two circles capable to measure simultaneously the components of static (DC) and low frequency magnetic fields up to 10 kHz with quite good sensitivity. Past projects indicate that a usual and adequate number of magnetic dipoles for modeling spacecraft equipment is limited to 5. As already mentioned in the previous section of electric dipole modeling, six parameters need to be determined by the algorithm for each dipole source. Therefore, at least 30 independent parameters (in the five-dipole regime) are required in order to accurately estimate the radiated model. In the case that the measuring sensor provides 3 values (three-axial Fluxgate magnetometers), the number of the sensors located at different positions (independent equations) should be at least 10. In the proposed facility, however, 12 sensors are used for redundancy purposes. As shown in Figure 11, the magnetometers form two circles above and below the turntable in order to capture the DUT's magnetic signature from several possible angles. In this configuration, both circles have a radius of r=20 cm and are positioned at different heights (z=10 cm and z= -10 cm for the upper and lower circle respectively). The angle difference between two successive magnetometers is fixed at 60°. The choice of observation distance of the magnetometers is appropriate for capturing the magnetic signature of sources with weak magnetic behavior, but may be rescaled according to SNR demands and dimensioning of the unit limitations. Ideally, measurements on a sphere around the DUT would ensure the total coverage of the DUT's magnetic signature, but considerably additional measurement points are required. In the present work, the upper and lower planes of the configuration in Figure 11 indicate that the magnetometers are located provide better coverage of the magnetic signature of the unit from several possible angles. Other configurations may be implemented with the use of multiple magnetometers – for instance three circles, helix, etc. Finally, it is worth mentioning that no calibration procedure for the position of the 12 magnetometers is implemented.

Figure 11. Measurement setup employing 12 magnetic sensors

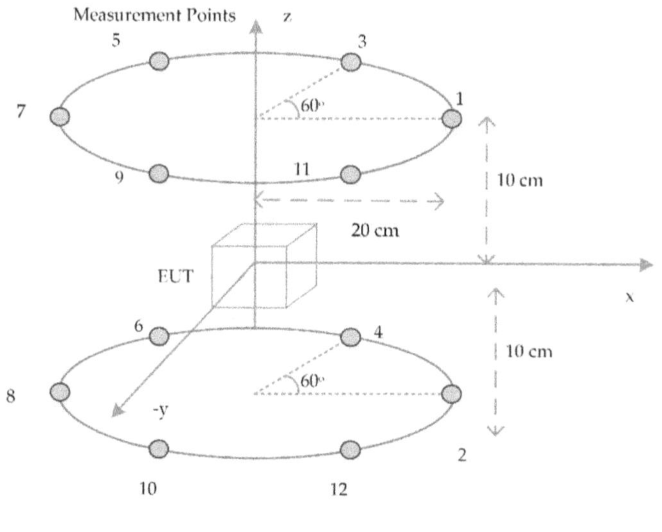

The noticeable aspect of this configuration is the simultaneous capture from all 12 fluxgate sensors of the magnetic field of an DUT in the presence of Earth's magnetic field. Specifically, existing MCF facilities use a pair of Helmholtz coils to compensate the DC component of Earth's magnetic field that has approximate amplitude of around 50 μT. However, disturbances of the ambient field during the rotation of the turntable that the DUT is placed on (Earth's magnetic field variations, noise from power lines, nearly technical equipment, etc.) commonly require several repetitions. On the contrary, the test methodology of this measurement setup includes three successive measurements: (i) ambient field measurement before the placement of the DUT ($\mathbf{B}_{bef}$), (ii) capture of DUT's magnetic field ($\mathbf{B}_{EUT}+\mathbf{B}_{ambient}$), and (iii) removal of the DUT and ambient field measurement ($\mathbf{B}_{aft}$).

In principle, the total measurement time is very short (in most cases less than 1 min) in order to avoid significant environmental field disturbances. The average value of the ambient fields before and after the DUT measurement:

$$\frac{\mathbf{B}_{bef} + \mathbf{B}_{aft}}{2} \tag{27}$$

is then used to effectively compensate the ambient field during the DUT measurement.

In principle, the location of the real magnetic sources inside an DUT, as well as their magnetic moments, are a priori unknown. Consequently, in order to validate the proposed setup and modeling methodology a series of simple test cases are measured and characterized in the following paragraphs, namely a single dipole source and two dipole sources. It should be noted that the position of the dipole sources on the table is known with significant accuracy (in the order of mm), whereas their magnetic moments are a priori unknown. The accuracy of the prediction is mainly determined by the relative deviation between the estimated position of the test object and the real position of the dipole, as well as the goodness of fit criterion G_{rel}. Low values of the position deviation between the real and the modeled magnetic dipoles (typically less than 5 mm) and of the goodness of fit evaluation criterion (typically less than 10%) will ensure the accuracy of the prediction and validate the measurement configuration and the modeling process.

Induced DC Measurements

Apart from a pair of Helmholtz coils to compensate Earth's magnetic field, existing MCFs typically employ another pair of Helmholtz coils that generates a (spatially) uniform external magnetic field at the volume of the DUT exhibiting conducting and/or magnetic properties, in order to measure and characterize the unit's induced magnetic behavior (ECSS-E-ST-20-07C rev-1 Space engineering Electromagnetic compatibility, 2012). Typical levels of the DC applied magnetic field are +/- 50 μT and +/- 100 μT. It should be noted that several tests are conducted with different amplitude levels of the applied field for purposes of testing the linear response of the unit. In fact, various units are composed of magnetic materials (e.g. iron) and their response can be highly non-linear, exhibiting hysteresis effects. In these cases, a more detailed characterization of the unit is required and magnetic measurements are performed with several amplitude levels of the applied field. The tests are also conducted with applied external field in all 3 directions to measure the induced magnetization of the DUT along the 3 orthogonal axes.

Without loss of generality, it is assumed that the induced magnetic response can be modeled with a single equivalent induced magnetic dipole and its magnetic moment aligns to the direction of the external magnetic field. Thus, when an external field ($\mathbf{B}_{app}$+) is applied vertically, the direction of the induced magnetic dipole moment stays constant during the DUT's rotation. If the external field is applied vertically in the opposite direction ($\mathbf{B}_{app}$-), the remanent magnetic dipole's orientation stays constant. The induced dipole moments, however, will align to the new orientation of the field. In this way, the separation of the induced and remnant magnetic field contributions can be practically achieved. Specifically, the measured magnetic field values regarding the two configurations of the test are ($\mathbf{B}_{remn+ind}+$) for positive and ($\mathbf{B}_{remn+ind}-$) for negative orientation of the applied field.

The induced magnetic field measurements may be performed by employing the multiple magnetometer configuration depicted in Figure 11 and its associated test measurement methodology with the exception that a pair of Helmholtz coils must be used to generate the external magnetic field and control its uniformity. Thereafter, the measured induced magnetic field values can be modeled according to the well-established MDM methods and the Heuristic (DE, PSO) technique may be employed to estimate the magnetic moments $\left(\mathbf{m}_{\mathbf{B}_{app}}+\right)$ and $\left(\mathbf{m}_{\mathbf{B}_{app}}-\right)$ depending on the orientation of the magnetic field ($\mathbf{B}_{app}$), as described in a previous section. Then, vector analysis can be used to isolate the induced magnetic dipole moment of the DUT, as depicted in Figure 12.

Figure 12. Induced magnetic moment measurement and characterization

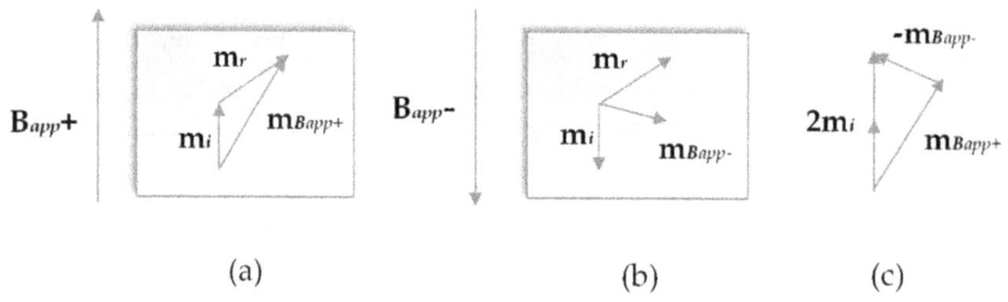

The difference between the moments $(\mathbf{m}_{\mathbf{B}_{app}}+)$ and $\left(\mathbf{m}_{\mathbf{B}_{app}}-\right)$ basically yields twice the induced magnetic moment, while the remnant magnetic moments cancel out:

$$m_i = \frac{(m_{B_{app}}+) - (m_{B_{app}}-)}{2} \tag{28}$$

The characterization of the induced magnetic behavior of each DUT that will be mounted on the spacecraft is essential and should be considered in missions that demanding levels of magnetic cleanliness are required. Moreover, the induced magnetic properties will allow the accurate prediction of the in-flight performance of the equipment, since the total magnetic field generated by each DUT is the superposition of the individual contributions of all its magnetic sources. Finally, the generated magnetic field from surrounding equipment at the location of an DUT that exhibits conducting and/or magnetic properties is basically treated as external field that may additionally cause its induced response. The unit to unit interaction is consequently a critical issue in most space programs and the proposed modeling methodology is analyzed in the following subsection.

MEASUREMENTS AND MODELING OF A UNIT'S AC INDUCED MAGNETIZATION

In the present subsection, a baseline methodology is illustrated that may be used to carry out the AC induced magnetic dipole modeling analysis. Magnetic polarizability, also called AC magnetic susceptibility, is the dipolar response per unit volume from the unit under test to an external applied field $\mathbf{H}_a$. The induced dipole moment can be expressed (ECSS-E-ST-20-07C rev-1 Space engineering Electromagnetic compatibility, 2012):

$$\mathbf{m}_{AC} = V \chi_{AC} \mathbf{H}_a \tag{29}$$

where V is the volume of the DUT and χA_C is the AC magnetic susceptibility. In principle, most of the DUTs consist of conducting materials and exhibit different responses to an applied magnetic field depending on their shape, their conductivity σ and the frequency and amplitude of the external field. Moreover, at high frequencies, the induced (eddy) currents on the conducting surfaces of the unit possibly produce an AC induced magnetic behavior that tends to oppose the external field.

The response of these DUTs may be modeled and represented by an AC induced dipole moment $\mathbf{m}A_C$ and linearly expressed with respect to the external field via the frequency dependent 3×3 magnetic polarizability tensor χAC. For each direction of the applied field Ha, an induced magnetic dipole model may be generated, according to the AC magnetic field verification method described in the previous section. The unit's response to an applied field of arbitrary orientation can be therefore estimated as a linear combination of responses corresponding to each principal axis x,*y*,*z*.

The measurement procedure is quite similar to the DC induced magnetic measurements. The external magnetic field should be as uniform as possible (spatially) in the volume of the DUT. Three sets of measurements of the DUT's magnetic signature are performed with external applied field $H_{a_x}, H_{a_y}, H_{a_z}$, as shown in Figure 13. These measurements shall be performed for several frequencies of the external field from DC to several 10 KHz. Furthermore, the typical levels of the external field are about 50 μT and 100 μT peak-to-peak in order to investigate the linearity of the DUT's response with respect to the amplitude of the applied field.

Figure 13. AC induced magnetic measurement and characterization

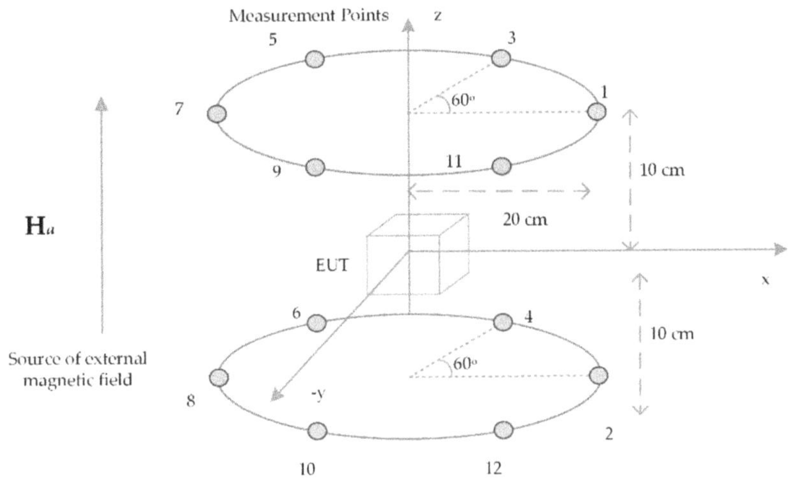

The induced magnetic field $\mathbf{B}_{ind}$ of the DUT is expected to emerge at the same frequency of the respective external magnetic field. The AC induced magnetization of the DUT may then be modeled in the frequency domain with a single magnetic dipole source oscillating at the frequency of the external applied field by employing the developed heuristic-based method (DE, PSO), as described in the previous section.

MEASUREMENT UNCERTENTIES DUE TO INACCURATE SENSOR POSITION

In the work of (Koutantos et al., 2017), authors have proposed the Multi-Frequency Electric Dipole Model (MFEDM) method to describe the spectral dependence and predict extremely low frequency field emissions. Small variations in the relative distance of the source and the measurement points alter the field measurement results. This, in return, leads to different models produced by the proposed method. Consequently, the precision of this method highly relies on the accuracy of the relative measurement distance. The correlation between relative measurement distance uncertainty and the electric field distribution is a key parameter to a successful model prediction.

One way to calculate the measurement uncertainty due to the sensor position inaccuracy is to perform the electric field measurement multiple times and estimate the mean value of the field with a standard deviation. In order to achieve this, the knowledge of field distribution due to distance displacement shall be known.

In this section relative measurement distance varied according to several distributions in order to evaluate this correlation. The results indicate that the electric field distribution follows always normal distribution independently with the way the sensor position is distributed.

Figure 14. Moving sensor position with uniform distribution (b) Corresponding E-Field Distribution

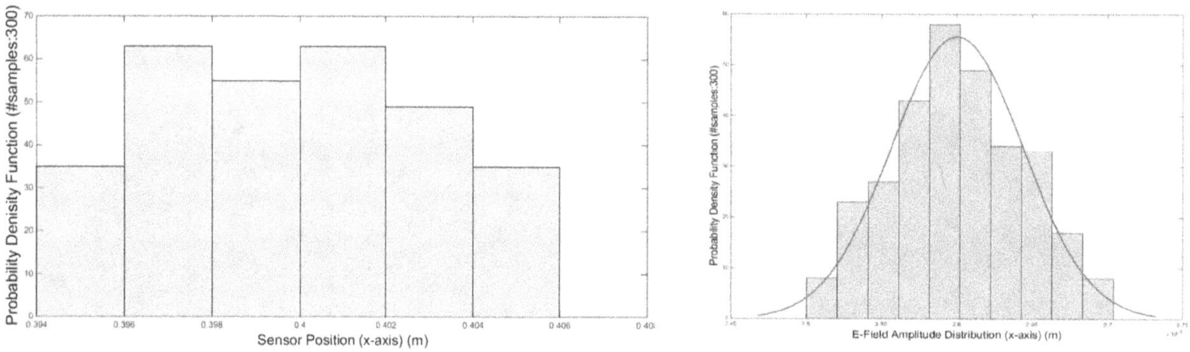

Figure 15. (a) Moving sensor position with Rayleigh distribution (b) Corresponding E-Field Distribution

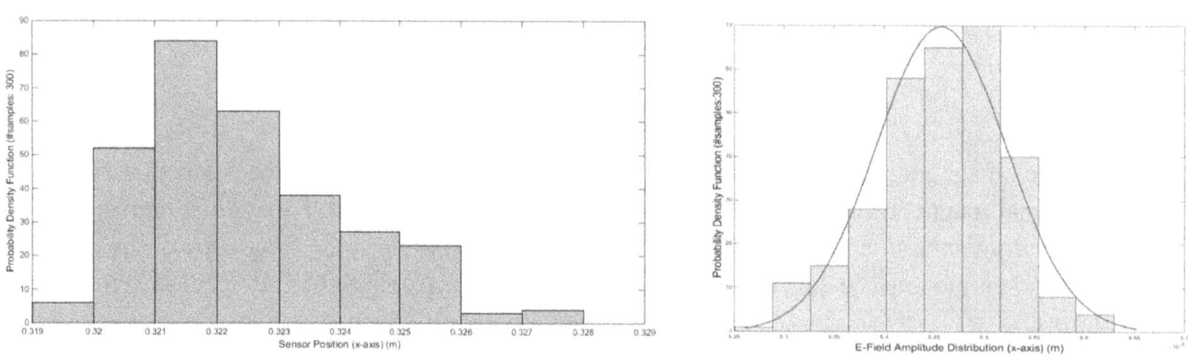

TEST FACILITIES DECOUPLING CONSIDERATIONS

Due to the extremely low frequency of the AC electric and magnetic field measurements, the measured fields are inevitably coupled to the measurement facility (e.g. anechoic tent, MCF). Thus, the method for test facility removal is a crucial issue in the framework of electromagnetic cleanliness, since it will

Figure 16. (a) Moving sensor position with Normal distribution, (b) Corresponding E-Field Distribution

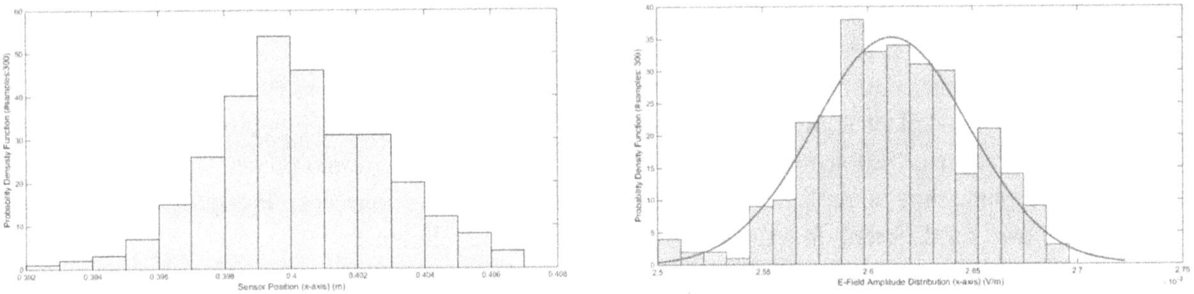

Figure 17. Indicative Image Dipole position and orientation from the anechoic chamber/ Tent's conductive surfaces when the DUT is placed at the center of the anechoic chamber/ Tent.

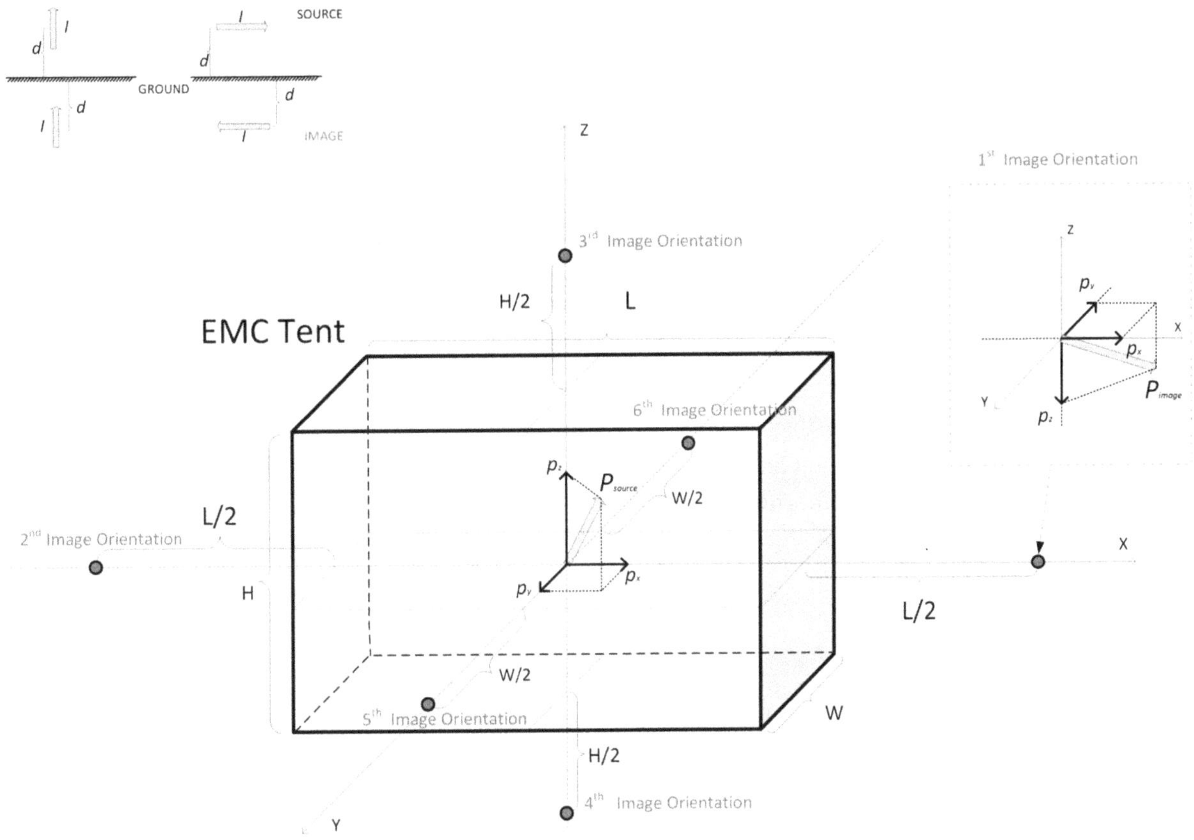

enable decoupling of the derived models from the facility that the measurements were performed and allow in-flight estimation of the EM fields in space.

The predicted models that are associated to the measurement test facility may be decoupled from it by considering the contribution of the image charges which in the majority of test cases aren't negligible. By taking into consideration the characteristics of the test set up, the contribution of the image sources can be defined in order to allow estimations in space.

Electric Field

In cases that an electric dipole source is located above a conducting plane, consideration of image theory (method of image sources) is essential. According to this theory, the electric field at an observation point above the ground plane may be calculated as a superposition of the electric fields generated by two sources: (i) the original electric dipole source and (ii) its mirror image located below the ground plane. The position of the image dipole source is symmetrical relative to the ground plane. Thus, in principle, the conducting plane may be replaced by the mirror image(s) of the source(s). In Figure 17 the image sources are shown for different directions of the electric dipole moment.

In the case of a single electric dipole, the total measured electric field can be calculated via:

$$E^{tot}\left(r_{obs}\right) = E^{s}\left(r_{obs}, x, y, z\right) + E^{i}\left(r_{obs}, x', y', z'\right) \tag{30}$$

and the electric moment of the image electric dipole may be expressed as:

$$\begin{cases} p_{\parallel}^{\,i} = -p_{\parallel}^{\,s} \\ p_{\perp}^{\,i} = p_{\perp}^{\,s} \end{cases} \tag{31}$$

Consequently, the two separate contributions can be dissociated and the AC electric signature of a unit can be decoupled from the measuring facility by removing the electric field contribution of the image dipole, as in Chatzineofytou et al., (n.d.), where measurements of an electric dipole above a conductive surface are employed in order to validate this method.

Furthermore, the anechoic/semi-anechoic chamber or tent (measurement facility) is comprised of multiple conductive planes/surfaces so this technique will be expanded in order to account at least 1st order images for all the surfaces that couple and play a role in the measurement results. This way the dominant effect of the measurement facility on the measurements, and consequently on each EUT model, can be removed yielding a more accurate EUT model, better suited for estimations in space.

Magnetic Field

The magnetic field at an observation point above a conducting ground plane may be calculated as a superposition of the magnetic fields generated by two sources: (i) the original magnetic dipole source $\mathbf{B}^{s}$ and (ii) its mirror image located below the ground plane $\mathbf{B}^{i}$, as shown in Figure 18. It should be noted that although the position of the image dipole source is symmetrical relative to the ground plane (as in

Figure 18. Magnetic dipole sources above a conducting plane and their mirror image sources.

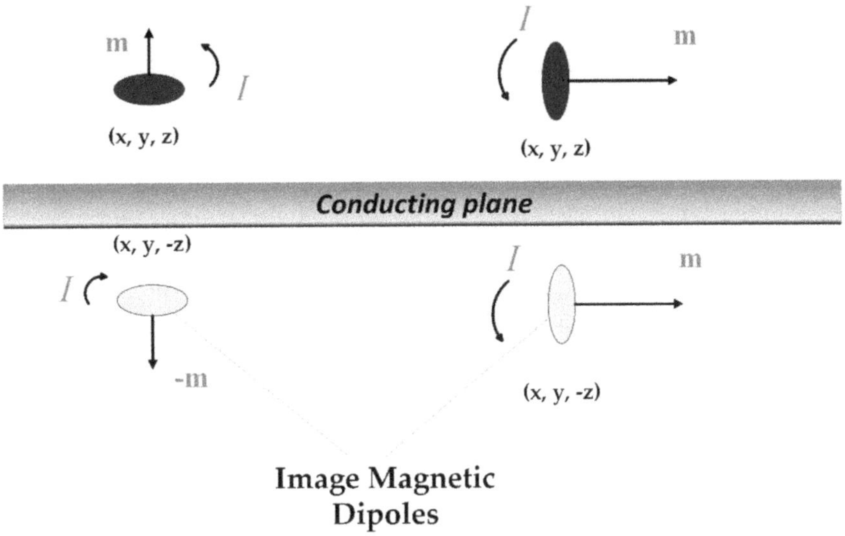

the electric dipole case), the magnetic moment of the image source exhibits the opposite orientation in contrast to the electric dipole case.

The magnetic field measurements were conducted in the MCF facility shown in Figure 19. The facility includes a conducting turntable that contributes to the measured low frequency AC magnetic field.

In the case of a single magnetic dipole (loop), the total measured magnetic field can be calculated via:

Figure 19. Magnetic Dipole Source in TAS-I Magnetic Coil Facility during validation measurements.

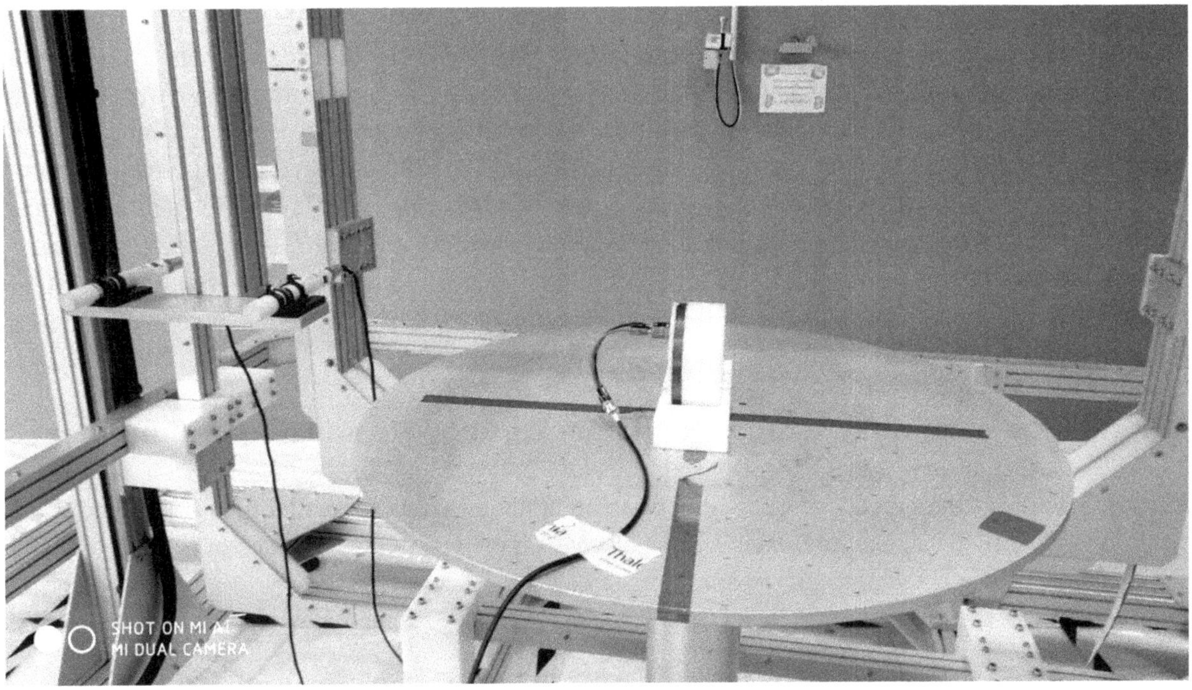

$$\mathbf{B}^{tot}\left(\mathbf{r}_{obs}\right)=\mathbf{B}^{s}\left(\mathbf{r}_{obs},x,y,z\right)+\mathbf{B}^{i}\left(\mathbf{r}_{obs},x',y',z'\right) \tag{32}$$

and the magnetic moments of the image magnetic dipole may be expressed:

$$\begin{cases} m_{\parallel}^{\;i}=m_{\parallel}^{\;s} \\ m_{\perp}^{\;i}=-m_{\perp}^{\;s} \end{cases} \tag{33}$$

Consequently, the two separate contributions can be dissociated and the AC magnetic signature of a unit can be decoupled from the measuring facility by removing the magnetic field contribution of the image dipole, as in Chatzineofytou et al., (n.d.), where measurements of a magnetic dipole loop are employed in order to validate this method.

CONTRIBUTION OF SPACECRAFT STRUCTURE IN MODEL'S SYSTEM LEVEL PREDICTIONS

Image Theory Approximations

Spacecraft internal space is limited and possible DUT location can be near the spacecraft's conducted metallic surfaces. In that case, modelling electromagnetic behavior of the DUT should consider the contribution of the image charges which in the majority of test cases they aren't negligible. Same considerations as in the previous section can be applied also in the case of system level contribution estimation.

Oversized Cavity Theory and Applications

The oversized cavity theory is very useful to assess electromagnetic fields inside an enclosure such as a satellite module or a launcher fairing that is much bigger than the wavelength at the frequencies of interest and geometrically complex. This theory is effectively the reverberation or stirred-mode chambers theory and is part of statistical electromagnetics. According to certain conditions concerning the size and complexity of the cavity, the field is statistically homogenous and isotropic and follows known distributions.

For a given calculated E the actual $E_{dBV/m}$ may be between $E_{dBV/m}+6dB\ and -12dB$ if one excludes the 0.1% of cases respectively higher and lower than this range. With this approximation and for the worst case of model ELF radiated emissions estimation, a level of 6 dB should be considered as a standard addition.

Contribution of Internal Sources Radiation Through Residual Apertures

According to the Bethe's theory of diffraction by small holes, an aperture smaller than λ/4 excited by an electromagnetic field can be represented by one electric dipole normal to the plane of the aperture and one magnetic dipole in the plane of the aperture.

In free space, the total power radiated by an electric dipole p_e, respectively by a magnetic dipole p_m is given by:

$$P = \frac{1}{12\pi}\frac{\omega^4 \mu_0 |p_e|^2}{c} \tag{34}$$

$$P = \frac{1}{12\pi}\frac{\omega^4 \varepsilon_0 |p_m|^2}{c} \tag{35}$$

Where $p_m = \mu 0_m$ and |m| is in A·m2

The aperture electric and magnetic dipoles are related to the internal incident field by:

$$p_e = 2\varepsilon_0 a_e E_{sc} \tag{36}$$

$$p_m = 2\mu_0 a_m \mathrm{H}_{sc} \tag{37}$$

Where:

- E_{sc} and H_{sc}, are the fields that exists on the inside of the cavity (spacecraft) at the location of the aperture if the aperture was closed (i.e shielded, short - circuited); the plane of the aperture is assumed perfectly conducting such that H_{sc} Is tangential and the E_{sc} in normal to the plane of the aperture;
- αe is the electric polarizability of the aperture.
- αm is the magnetic polarizability of their total power through the aperture.

With the convention that it has adopted, the dipoles do not need to be imaged in the conducting plane (and therefore doubled) as this factor 2 is already included in their definition. Those dipoles radiate half of their total power through the aperture.

Electromagnetic Shielding Approximations

Spacecraft internal space is limited and possible unit location can be near the spacecraft's conducted metallic surfaces. Normally, this results in an electromagnetic decoupling between the inside and the outside of the enclosure. By considering the characteristics of the spacecraft wall, the electric and magnetic fields generated by all the sources inside the spacecraft may be assessed at points of interest outside the enclosure by modeling the shielding effectiveness.

The classical transmission model of shielding (Paul, 2006) considers that the field is attenuated by a combination of absorption losses "A", depending on the number of skin depths through the wall, reflection losses "R" depending on the ratio between the wave and barrier impedances (i.e. mismatch) and multiple reflection term "M" which accounts for re-reflection effects and only plays a role if $t<\delta$, where t is the thickness of the wall and δ is the skin depth.

$$SEd_B = Ad_B + Rd_B + Md_B \tag{38}$$

This simple model assumes that the coupling between the screen currents and the source of the incident field is negligible, which is not guaranteed in practice.

The **absorption losses** can be approximated:

$$A_{dB} \cong 20\log_{10}(e^{t/\delta}) \tag{39}$$

where t is the thickness of the material, the skin depth is $\delta = 1/\sqrt{\pi f \mu \sigma}$,f is the frequency of the field, μ is the magnetic permeability of the shielding material and σ is its conductivity.

In order to estimate the **reflection losses,** the distinction between electric and magnetic field sources is essential. The intrinsic impedance of the material can be expressed:

$$\hat{Z}_i = \sqrt{\frac{j\omega\mu}{\sigma + j\omega\varepsilon}} \tag{40}$$

where ε is the electric permittivity of the material and $\omega=2\pi f$. For metal surfaces ($\sigma \gg \omega\varepsilon$) the above equation reduces to:

$$\hat{Z}_m = \sqrt{\frac{j\omega\mu}{\sigma}} = \frac{1+j}{\sigma\delta} \tag{41}$$

The barrier impedance of a metal wall is equal to $\hat{Z}_m$ if $t \gg \delta$. A more general formula of the barrier impedance of a metal wall is given by:

$$\hat{Z}_b = \frac{\hat{Z}_m}{1 - e^{-t/\delta}} \tag{42}$$

The **multiple reflection losses** can be approximated:

$$M_{dB} \cong 20\log_{10}\left|1 - e^{-2t/\delta} e^{-j2t/\delta}\right| \tag{43}$$

which can be neglected ($M_{dB} \cong 0$) for shields that are constructed of good conductors and whose thicknesses are much greater than the skin depth ($t \gg \delta$). However, for barrier thicknesses that are thin compared with a skin depth ($t \ll \delta$), the multiple reflection factor is negative in dB. In this case, multiple reflections reduce the shielding effectiveness of the barrier.

Electric Dipole Sources

The wave impedance of an elemental electric (Hertzian) dipole source can be expressed:

$$\hat{Z}_w = \frac{\hat{E}_\theta}{\hat{H}_\varphi} = n_0 \frac{j/\beta_0 r + 1/(\beta_0 r)^2 - j/(\beta_0 r)^3}{j/\beta_0 r + 1/(\beta_0 r)^2} \tag{44}$$

where $\beta_0 = \omega\sqrt{\mu_0 \varepsilon_0}$ and r is the distance between the source and the shielding material. By additionally considering the near electric field behavior approximation ($\beta 0_r \ll 1$) for the low frequency range under investigation (up to 250 KHz) the above equation reduces to:

$$\hat{Z}_{w_e} \cong n_0 \left(-j \frac{1}{\beta_0 r} \right) \tag{45}$$

In the near field of the electric dipole, the wave impedance is greater than the intrinsic impedance of the medium. Therefore, the electric dipole is referred to as a high impedance source. The magnitude of the wave impedance for an electric field dipole source is:

$$\left| \hat{Z}_w \right|_e = \frac{1}{2\pi f \varepsilon_0 r} \tag{46}$$

As an approximation, the shielding effectiveness can be expressed as the product of a reflection term, an absorption term and a multiple reflection term by substituting the wave impedance $\hat{Z}_{w_e}$ in the respective equations. Therefore, the reflection term can be expressed:

$$R_{dB} = 20\log_{10} \left| \frac{\left(\hat{Z}_{w_e} + \hat{Z}_i \right)^2}{4\hat{Z}_{w_e} \hat{Z}_i} \right| \tag{47}$$

The absorption loss term is unaffected by the type of source. The multiple reflection term can be expressed:

$$M_{dB} = 20\log_{10} \left| 1 - \left(\frac{\hat{Z}_{w_e} - \hat{Z}_i}{\hat{Z}_{w_e} + \hat{Z}_i} \right)^2 e^{-2t/\delta} e^{-j2t/\delta} \right| \tag{48}$$

If the barrier is a good conductor, $\hat{Z}_i$ can be replaced by $\hat{Z}_m$ (or $\hat{Z}_b$) in the above equations.

Magnetic Dipole Sources

Similarly, to the electric dipole case, the wave impedance for the elemental magnetic dipole source (loop) is:

$$\hat{Z}_w = \frac{\hat{E}_\varphi}{\hat{H}_\theta} = -n_0 \frac{j / \beta_0 r + 1 / (\beta_0 r)^2}{j / \beta_0 r + 1 / (\beta_0 r)^2 - j / (\beta_0 r)^3} \tag{49}$$

By additionally considering the near magnetic field behavior approximation ($\beta 0_r \ll 1$) for the low frequency range under investigation (up to 250 KHz) the above equation reduces to:

$$\hat{Z}_{w_m} \cong -jn_0\beta_0 r \tag{50}$$

In the near field of the magnetic dipole, the wave impedance is less than the intrinsic impedance of the medium. The magnetic dipole is referred to as a low impedance source. The magnitude of the wave impedance for the magnetic field dipole source is

$$\left|\hat{Z}_w\right|_m = 2\pi f \mu_0 r \tag{51}$$

The shielding effectiveness can be calculated as the product of a reflection term, an absorption term and a multiple reflection term by substituting the wave impedance $\hat{Z}_{w_m}$ in the respective equations.

FUTURE RESEARCH DIRECTIONS

Although several measurement techniques are already included in the standard procedure of EMC testing, complementary methods that are presented in this chapter have to be evaluated and possibly adopted in future releases of the aforementioned standards since they cover frequency ranges that are not discussed until now. Electromagnetic cleanliness is a key parameter for the success of the majority of the space missions. For that reason, future development towards this direction is essential and shall be the scope of the research of all space focused teams and engineers.

CONCLUSION

In this chapter, the main issues regarding ELF electric and magnetic field measurement and modeling techniques for a spacecraft unit are briefly discussed. A modeling methodology is reviewed and analyzed focusing on the open issues that need further exploration and development. Space engineering and aircraft design procedures cope with stringent requirements regarding electromagnetic compatibility and interference since electromagnetic cleanliness is a critical aspect in the majority of the space missions. Searching and developing new methodologies in this field seems crucial. This work highlights the importance of this issue and urge the researchers and engineers to include in future development all the contributing additional phenomena, such as harness topologies and solar arrays that cannot be treated as electric dipoles.

ACKNOWLEDGMENT

Author thank Thales Alenia Space (TAS -I) and European Space Agency (ESA) for the permission to publish Table 2, Figure 6, Figure 8 and Figure 19.

REFERENCES

Antonucci, F., Armano, M., Audley, H., Auger, G., Benedetti, M., Binetruy, P., Boatella, C., Bogenstahl, J., Bortoluzzi, D., Bosetti, P., Caleno, M., Cavalleri, A., Cesa, M., Chmeissani, M., Ciani, G., Conchillo, A., Congedo, G., Cristofolini, I., & Cruise, M., … Zweifel, P. (2011). LISA Pathfinder: Mission and status. *Classical and Quantum Gravity*. Advance online publication. doi:10.1088/0264-9381/28/9/094001

Baklezos, A. T., Nikolopoulos, C. D., & Capsalis, C. N. (2017). An equivalent dipole method with novel measurement positioning for modeling electric emissions in space missions. *Electromagnetics*, *37*(7), 1–15. doi:10.1080/02726343.2017.1376902

Baklezos, A. T., Nikolopoulos, C. D., Katsouris, A. G., Koutantos, G. I., & Capsalis, C. N. (2016). Electromagnetic Emission Modeling in Case of Shielded Cabling with Respect to the Ground Dielectric Properties. *IEEE Transactions on Electromagnetic Compatibility*, *58*(6), 1694–1700. Advance online publication. doi:10.1109/TEMC.2016.2588583

Baklezos, A. T., Nikolopoulos, C. D., Sigalas, I. R., & Capsalis, C. N. (2020, May 1). Measurement and modeling of spacewire radiation for electromagnetic compatibility assessment. *I2MTC 2020 - International Instrumentation and Measurement Technology Conference, Proceedings*. 10.1109/I2MTC43012.2020.9128502

Bayle, O., Lorenzoni, L., Blancquaert, T., Langlois, S., Walloschek, T., & Portigliotti, S. (2016). *Exomars entry descent and landing demonstrator mission and design overview*. Nasa Solar System.

Benkhoff, J., van Casteren, J., Hayakawa, H., Fujimoto, M., Laakso, H., Novara, M., Ferri, P., Middleton, H. R., & Ziethe, R. (2010). BepiColombo-Comprehensive exploration of Mercury: Mission overview and science goals. *Planetary and Space Science*, *58*(1-2), 2–20. Advance online publication. doi:10.1016/j.pss.2009.09.020

Brownlee, J. (2011). Clever Algorithms - Nature-Inspired Programming Recipes. Search.

Capsalis, C. N., Nikolopoulos, C. D., Spantideas, S. T., Baklezos, A. T., Chatzineofytou, E. G., Koutantos, G. I., Boschetti, D., Marziali, I., Nicoletto, M., Tsatalas, S., Mehlem, K., & Junge, A. (2019). EMC Assessment for Pre-Verification of THOR Mission Electromagnetic Cleanliness Approach. *Proceedings of 2019 ESA Workshop on Aerospace EMC, Aerospace EMC 2019*. 10.23919/AeroEMC.2019.8788958

Chatzineofytou, E. G., Spantideas, S. T., Nikolopoulos, C. D., Baklezos, A. T., Marziali, I., Nicoletto, M., Boschetti, D., & Capsalis, C. N. (n.d.). *Decoupling of ground plane effect on low frequency magnetic and electric field measurements & modeling*. Academic Press.

Dawar, D., & Ludwig, S. A. (2015). Differential evolution with dither and annealed scale factor. *IEEE SSCI 2014 - 2014 IEEE Symposium Series on Computational Intelligence - SDE 2014: 2014 IEEE Symposium on Differential Evolution, Proceedings*. 10.1109/SDE.2014.7031528

Dragoi, E. N., Curteanu, S., & Vlad, D. (2012). Differential evolution applications in electromagnetics. *EPE 2012 - Proceedings of the 2012 International Conference and Exposition on Electrical and Power Engineering*. 10.1109/ICEPE.2012.6463801

Drinkwater, M. R., Floberghagen, R., Haagmans, R., Muzi, D., & Popescu, A. (2003). GOCE: ESA's first earth explorer core mission. *Space Science Reviews*, *108*(1/2), 419–432. Advance online publication. doi:10.1023/A:1026104216284

ECSS-E-ST-20-07C rev-1 Space engineering Electromagnetic compatibility, ECSS-E-ST-20-07C rev-1 (2012).

Escoubet, C. P., & Schmidt, R. (2000). Cluster II: Plasma measurements in three dimensions. *Advances in Space Research*, *25*(7-8), 1305–1314. Advance online publication. doi:10.1016/S0273-1177(99)00639-0

Jackson, J. D. (2003). Electrodynamics, Classical. Digital Encyclopedia of Applied Physics. doi:10.1002/3527600434.eap109

Koutantos, G. I., Nikolopoulos, C. D., Baklezos, A. T., & Capsalis, C. N. (2017). On the Modeling of ELF Electric Fields for Space Mission Equipment. *IEEE Transactions on Electromagnetic Compatibility*, *59*(5), 1–8. doi:10.1109/TEMC.2017.2685199

Lapeña, E., Herranz, J. L., Gómez-Carpintero, F., & Rodríguez, M. (2017). The LEO PCDU EVO - A Modular and Flexible Concept for Low to Medium Power LEO & Scientific Missions. *E3S Web of Conferences*. doi:10.1051/e3sconf/20171618009

Müller, D., Marsden, R. G., St. Cyr, O. C., & Gilbert, H. R. (2013). Solar Orbiter: Exploring the Sun-Heliosphere Connection. *Solar Physics*. Advance online publication. doi:10.100711207-012-0085-7

Nikolopoulos, C. D., Baklezos, A. T., & Capsalis, C. N. (2019). Measuring Transient and Steady State Electric Field Emissions of Space Equipment for EMC and Cleanliness Purposes. *ICHVE 2018 - 2018 IEEE International Conference on High Voltage Engineering and Application*. 10.1109/ICHVE.2018.8642200

Nikolopoulos, C. D., Baklezos, A. T., Tsatalas, S., & Capsalis, C. N. (2019). Verification of Radiated Emissions Modeling for SpaceWire/ LVDS Links Routed on CFRP Ground. *IEEE Transactions on Aerospace and Electronic Systems*, 1–1. doi:10.1109/taes.2019.2914540

Obiekezie, C., Thomas, D. W., Nothofer, A., Greedy, S., Arnaut, L. R., & Sewell, P. (2013). Extended scheme using equivalent dipoles for characterizing edge currents along a finite ground plane. *Applied Computational Electromagnetics Society Journal*.

Obiekezie, C. S., Thomas, D. W. P., Nothofer, A., Greedy, S., Arnaut, L. R., & Sewell, P. (2014). Complex locations of equivalent dipoles for improved characterization of radiated emissions. *IEEE Transactions on Electromagnetic Compatibility*, *56*(5), 1087–1094. Advance online publication. doi:10.1109/TEMC.2014.2313406

Oliveri, G., Rocca, P., & Massa, A. (2012). Differential evolution as applied to electromagnetics: Advances, comparisons, and applications. *Proceedings of 6th European Conference on Antennas and Propagation, EuCAP 2012*. 10.1109/EuCAP.2012.6206056

Paul, C. R. (2006). Introduction to Electromagnetic Compatibility: Second Edition. doi:10.1002/0471758159

Scheeres, D. J., Marzari, F., Tomasella, L., & Vanzani, V. (1998). ROSETTA mission: Satellite orbits around a cometary nucleus. *Planetary and Space Science*, *46*(6-7), 649–671. Advance online publication. doi:10.1016/S0032-0633(97)00200-6

Storn, R., & Price, K. (1997). Differential Evolution - A Simple and Efficient Heuristic for Global Optimization over Continuous Spaces. *Journal of Global Optimization*, *11*(4), 341–359. Advance online publication. doi:10.1023/A:1008202821328

Vaivads, A., Retinò, A., Soucek, J., Khotyaintsev, Y. V., Valentini, F., Escoubet, C. P., Alexandrova, O., André, M., Bale, S. D., Balikhin, M., Burgess, D., Camporeale, E., Caprioli, D., Chen, C. H. K., Clacey, E., Cully, C. M., De Keyser, J., Eastwood, J. P., & Fazakerley, A. N. … Wimmer-Schweingruber, R. F. (2016). Turbulence Heating ObserveR - Satellite mission proposal. *Journal of Plasma Physics*. Advance online publication. doi:10.1017/S0022377816000775

KEY TERMS AND DEFINITIONS

Differential Evolution Algorithm (DEA): Is a method that optimizes a problem by iteratively trying to improve a candidate solution with regard to a given measure of quality.

Electric Cleanliness: The generated electric field from the spacecraft is required to not exceed specific thresholds at the location of the electric sensors and therefore, disturb their measurements.

Electric Dipole Model (EDM): A method that optimizes a problem by iteratively trying to improve a candidate solution with regard to a given measure of quality. Such methods are commonly known as metaheuristics as they make few or no assumptions about the problem being optimized and can search very large spaces of candidate solutions.

Electrical Ground Support Equipment (EGSE): Is an integrated suite of electrical satellite testing solutions to make sure that the satellites you are launching work perfectly.

Electromagnetic Compatibility (EMC): The branch of electrical engineering concerned with the unintentional generation, propagation and reception of electromagnetic energy which may cause unwanted effects such as electromagnetic interference (EMI) or even physical damage in operational equipment.

Electromagnetic Interference (EMI): Also called radio-frequency interference (RFI) when in the radio frequency spectrum, is a disturbance generated by an external source that affects an electrical circuit by electromagnetic induction, electrostatic coupling, or conduction.

Equipment or Device Under Test (EUT or DUT): Any instrument or unit related to the space mission that produces an electric field. These units can be inside (e.g., Power Conditioning and Distribution Unit, RF Switch, Battery, etc.) or outside the main body of the spacecraft (e.g., Solar Arrays, Boom, etc.) and need to be tested and electrically characterized.

European Space Agency (ESA): Is an intergovernmental organization of 22 member states dedicated to the exploration of space. Established in 1975 and headquartered in Paris.

Extremely Low Frequency (ELF): Is the ITU designation for electromagnetic radiation (radio waves) with frequencies from 3 to 30 Hz, and corresponding wavelengths of 100,000 to 10,000 kilometers, respectively.

Induced Magnetization: Magnetic properties of a material that purely depend on the presence of an applied external magnetic field and directly vanish in its absence.

Magnetic Cleanliness: The generated magnetic field from the spacecraft is required to not exceed specific thresholds at the location of the magnetic sensors and therefore, disturb their measurements.

Magnetic Sensors: Instruments that are designed to measure magnetic fields (fluxgate magnetometers, search coil magnetometers, etc.). For instance, the fluxgate magnetometers have a magnetically susceptible core wrapped by two coils and measure the induced electric currents from a background magnetic field.

Magnetic Signature: The magnetic field generated by an DUT typically depends on the distribution of magnetic material inside the unit, as well as the electronics and the current paths that are incorporated in its design. Therefore, each unit is expected to emit an individual distinctive magnetic field.

Magnetic Source: Any source that generates a magnetic field. These sources typically include magnetic materials (e.g., iron, steel, etc.) and flowing electric currents.

Multi-Frequency Electric Dipole Model (MFEDM): A method using electric dipoles to predict the radiated emission of an DUT.

Multiple Magnetic Dipole Modeling: The representation of a unit's generated magnetic field with a dipole model able to recreate and reproduce its magnetic signature.

Remnant Magnetization: Magnetic properties of a material that are always present, even in the absence of external fields (e.g., ferromagnetic materials).

Turbulence Heating Observer (THOR): A plasma mission, aiming at understanding the physical mechanisms underlying the energy dissipation of turbulent fluctuations in plasmas.

Chapter 2
SpaceWire:
An Overview, Measurements, and Modelling for EMC Assessment

Anargyros T. Baklezos
https://orcid.org/0000-0002-0532-6216
National Technical University of Athens, Greece

Christos N. Capsalis
National Technical University of Athens, Greece

ABSTRACT

SpaceWire is a point-to-point bit shipping protocol for high-speed data communication links and networks providing equipment compatibility and seamless component reusability. It has found great application in many space missions reducing the development cost, offering architectural flexibility, and improving reliability. This chapter delves into the standard describing the SpaceWire, focusing on the lower levels that play a key role in the electromagnetic behavior of the system and concern cable assemblies, shielding, bonding, and grounding. Findings regarding emissions affecting spacecraft components are presented as well as other EMC issues that have an impact on the system performance. Recent developments on the modelling of the cable of the system with a focus on radiated emissions of SpW systems are also presented and discussed.

INTRODUCTION

SpaceWire is a standard for high-speed data communication links and networks primarily intended for use on spacecraft systems. SpaceWire was initially defined in the European Cooperation for Space Standardization (ECSS) standard ECSS-E50-12A in 2003. The SpaceWire standard was authored by Steve Parkes, the University of Dundee with contributions from many individuals within the SpaceWire Working Group from the European Space Agency (ESA), European Space Industry, Academia, and NASA. It was replaced by ECSS-E-ST-50-12C (ECSS Secretariat, 2008) to ensure the standard was following the new ECSS numbering system. There are no actual technical content differences between

DOI: 10.4018/978-1-7998-4879-0.ch002

ECSS-E-50-12A and ECSS-ST-E-50-12C and any information relating to the 'old' ECSS-E-50-12A is still applicable to newer ECSS-ST-E-50-12C ("ECSS-E50-12A vs ECSS-E-ST-50-12C, what's the difference? | STAR-Dundee", 2017; "SpaceWire - The Standard", 2017).

The Standard specifies the physical interconnection media and data communication protocols to enable the reliable transmission of data at high-speeds, between 2 Mb/s and 400 Mb/s, among units. SpaceWire links are full-duplex, point-to-point, and serial data communication links. Since the SpaceWire standard was first published by the European Cooperation for Space Standardization, it has been adopted by ESA, National Aeronautics and Space Administration (NASA), Japan Aerospace Exploration Agency (JAXA), and Roscosmos State Corporation for Space Activities (ROSCOSMOS) for many missions and is being widely used on scientific, Earth observation, commercial and other spacecraft. Gaia, ExoMars, Bepi-Colombo, GOES-R, Lunar Reconnaissance Orbiter, ASTRO-H, and James Webb Space Telescope are among high-profile missions which use SpaceWire (Parkes, 2012).

The increasing need for spacecraft on-board data handling applications slowly led to the development of SpaceWire technology. The Standard provides a formal basis for the exploitation of SpaceWire in a wide range of on-board processing systems. For example, the integration and testing of complex on-board subsystems with ground support equipment connecting directly into the onboard data-handling subsystem. SpaceWire aims to offer equipment compatibility and reusability for components and subsystems. This means directly connecting a component made for one subsystem to another and operating without issues and readily use systems in one mission that were developed for another mission, thus reducing the cost of development, offering architectural flexibility, improving reliability, and most importantly increasing the amount of scientific work that can be achieved within a limited budget (ECSS Secretariat, 2008).

Since the publication of the ECSS-E-ST-50-12C standard (ECSS Secretariat, 2008), the engineering and scientific community have applied the guidelines that it dictates in many missions. Experience of practical applications highlighted the best practices as well as common problems. Different applications called for different designs, some of them were hard to implement while maintaining conformity with the standard. This fact led the European Cooperation for Space Standardization in 2012 to start the long process of updating the Standard including the experience of the past and the needs of the future. Working Group ECSS-E-ST-50-12C Rev.1 WG prepared Revision 1 of the standard (ECSS Secretariat, 2019), an effort concluded in May 2019 with the publication of the revised standard. The revision is an effort to overcome many of the earlier versions' shortcomings and limitations. Additionally, since the Standard's publication in 2008, the ESCC Detail Specification No. 3902/003 was updated and ESCC Detail Specification No. 3902/004 was published. Revision 1 of ECSS-E-ST-50-12C is expected to reflect these changes in the normative framework.

Taking into consideration the widespread use of the SpaceWire, personnel active in the space mission field is bound to come across some implementation of a system incorporating SpaceWire. The main objective of the chapter is to discuss the SpaceWire standard ECSS-E-ST-50-12C with regard to Electromagnetic Compatibility and Immunity issues. The authors aim to highlight the levels of the standard that define the electromagnetic compatibility (EMC) behavior of the system. This can be beneficial for the engineer or scientist either from a system modelling or a radiated emissions measurement standpoint. Students interested in space missions can also benefit from the concentrated information found in this chapter. Additionally, the chapter has been enhanced with various recent modelling methodologies focusing on different perspectives of the system which are also presented and discussed.

Background

Systems, space or not, are an arrangement of units interconnected to each other, constantly exchanging information of some form. Before there was a standard for interconnecting units, spacecraft and equipment manufacturers used their proprietary communication interfaces. As a result, several different communication links were used on a spacecraft, increasing the cost and extending the time required for spacecraft integration and test. There was a clear need for a standard on-board communication link that would simplify spacecraft development. (Parkes, 2012)

The work on what later became SpaceWire started in 1992. There was considerable interest in high-performance digital signal processing systems at that time. However, the single-chip devices could not satisfy the demands. This led to the investigation of the use of parallel processing and in return gave place to research concerning the interconnecting network of the individual processing elements. The Inmos Transputer, a microprocessor designed for parallel processing was studied, and the serial communication links being developed for the T9000 Transputer were identified as being an attractive solution for spacecraft onboard networking. This serial link technology was subsequently published as IEEE 1355-1995. Several radiation-tolerant devices were developed using it and the IEEE 1355-1995 standard was used on some space missions. However, there were many problems with this standard, which had to be resolved if this technology was to continue to be used for ESA spacecraft. The University of Dundee received a contract from ESA to examine and solve these problems. The result of this contract was the SpaceWire standard, a brief description of which follows (Parkes, 2012).

SpaceWire links are point-to-point data links that connect a SpaceWire node to another node or a switch. Information can be transferred in both directions of the link at the same time. Links are full-duplex, bidirectional able to operate at rates between 2 Mbps and 400 Mbps.

The SpaceWire Standard ECSS-E- ST-50-12C (ECSS Secretariat, 2008) uses Low Voltage Differential Signaling (LVDS), defined in ANSI/TIA/EIA-644, Electrical Characteristics of Low Voltage Differential Signaling Interface Circuits (Telecommunications Industry Association, 2001), as the physical level due to its capability of low power, low noise, and high-speed point-to-point communications.

Information is transmitted over a link as a serial bit stream using two signals, Data (D) and Strobe (S). Data and Strobe signals are driven across the link using LVDS and two wires for each signal according to the standard TIA-644-A; SpaceWire cable contains one signal pair for each signal (D or S) for each direction, so four screened twisted pairs in total.

SpaceWire sends the clock signal along with the serial data for bit synchronization. To reduce the maximum clock to data skew requirements, the clock signal is encoded into a Strobe signal so that an XOR function on the Data and Strobe signal recovers the clock signal. A stream of elementary characters is used to transmit Data. Character synchronization is only performed at the start of the link. Loss of character synchronization is detected from parity errors, which then leads to a restart of the link to recover the character's synchronization.

Link start, link operation, transmission of data, and error recovery are handled by the link state-machine, transparent to the user application, in the SpaceWire interface.

The SpaceWire (SpW) protocol is comprised of six levels as depicted in Figure 1:

- **Physical level**: defines connectors, cables, cable assemblies, etc.
- **Signal level**: defines encoding, voltage levels, noise margins, and signaling rates.
- **Character level**: defines the data and control characters used to manage data across a link.

- **Exchange level**: defines the protocol for link initialization, flow control, link error detection, and link error recovery.
- **Packet level**: manages data packetization and transmission over an SpW link.
- **Network level**: defines the structure of a SpaceWire network and source to destination node data transfers. Also defines link errors and network level error management.

The SpaceWire physical level uses Data-Strobe (DS) Encoded LVDS to communicate serial, full-duplex, bidirectional data. ANSI/TIA/EIA-644 defines the electrical signal levels only; it does not define a protocol. Instead, the protocol is defined in the SpaceWire Standard specification ECSS-S-ST-50-12C (ECSS Secretariat, 2008). LVDS is a method to transmit bi-directionally over differential media using a low voltage signal swing using a driver and a receiver. SpaceWire uses Data Strobe (DS) encoding to send information over the LVDS links. The data value is transmitted directly and the Strobe signal changes state whenever the Data remains unaltered from one data bit interval to the next. The XOR function on the Data and Strobe signals result in the clock signal. There is a slight delay between the edges of D and S and the recovered clock. Data is Non-Return to Zero, (NRZ) so a 100MHz clock equates to a data rate of 200Mbps.

Figure 1. SpaceWire ECSS-E-ST-50-12C Levels

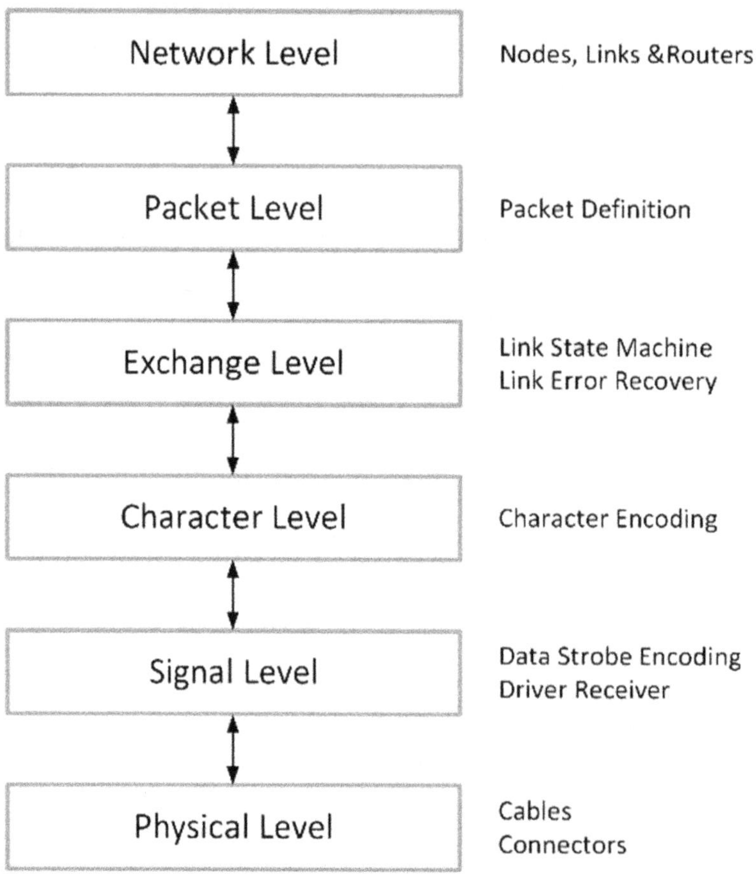

Two categories of characters are defined in ECSS-E-ST-50-12C standard, data and control characters. These characters are further defined as either link characters or normal characters. The Exchange level does not pass a link character to the Packet Level, meaning that a link character does not get entered into the memory space of a SpaceWire node. Flow control token (FCT), escape characters (ESC) as well as NULL control code (ESC followed by FCT) and the Time code characters (ESC followed by data character) are link characters. A Normal Character ends with an EOP or EEP and passes through an SpW network at a packet level. Data characters contain a parity bit, data-control flag, and an 8-bit data value transmitted LSB to MSB. The even or odd parity bit is calculated by adding the number of ones that are contained in the previous 8-bits data. Meaning if the number of 1's in bits added together is even, the data character is said to have even parity. There are two Control Characters that are formed using control characters and data characters, these are the NULL and Time Code Characters. NULL characters are used during the initialization sequence and are also used to keep the LVDS SpW lines active when no actual data is sent. Distributing system time is accomplished using Time Codes. The use of time codes is optional.

The exchange level defines the mechanisms for link initialization, link flow control, link error detection, and link error recovery. The state machine causes an SpW link to continuously attempt initialization as long as the link is active. Disconnection errors are detected after 850ns without Data and Strobe transitions. A parity error is detected when the first bit of the next character is received and link recovery starts. Initialization is executed upon the initial start of the link or reset when an error is detected (Parity Error, Escape Error, etc), or a link disconnection occurs. Under normal operation, a SpaceWire host node manages data by sending tokens to a destination node on the other end of the link. Each token indicates to the destination node that the host has 8 bytes of available buffer space. The destination node has a corresponding counter that records each token that is received. The destination node can send data to the host, according to the tokens of available buffer space at the host side it has received.

SpaceWire packets have a very flexible format. Each packet is composed of a Destination Address, a Cargo, and an End of Packet Marker (EOP) that signals the end of Cargo transmission. The Destination Address is a number of bytes of data characters describing the destination path of the packet. The Cargo consists of the transmission data, there is no requirement regarding the format or a limitation to the data size. Supported SpaceWire data rates are 2Mbps up to 400 Mbps. Nodes, Links, and Routers form a Network, which enables Packets to travel from a source to a destination end-point. SpaceWire nodes can be routers, sensors, memory units, processors, and telemetry subsystems. Nodes represent the source and destination of SpW packets and provide the interface to the application systems. The SpaceWire Standard supports various routing methods such as Wormhole Routing, Path Addressing, Logical Addressing, and Group Adaptive Routing for networking. Topologies like Distributed and Centralized network offer the designer a great deal of freedom. Distributed networks require look-up tables to be configured and provide lower harness mass, good throughput performance and increased reliability through redundant paths. A redundant network design ensures that the system remains functional if one or more of the system's components fail. Centralized networks require higher harness mass and offer high throughput performance and simple configuration. With centralized networks, there is a heavy reliance on the central router decreasing the system's fault tolerance. Other topologies are also possible, each with each own consideration.

THE SpaceWire STANDARD

This chapter discusses the SpaceWire standard from an EMC perspective. While a short overview of the whole standard is already presented, the authors aim to provide insight into the levels of the standard that mainly affect the Electromagnetic Emissions of the corresponding system, in particular the physical, signal and character levels. This is necessary to enable for example emissions modelling or interference minimization. To accomplish that, a more thorough presentation of the corresponding levels is mandatory.

Physical Level

The physical level of the SpaceWire standard covers cables, connectors, cable assemblies, and printed circuit board tracks (Figure 2). SpaceWire was developed to meet the EMC specifications of typical spacecraft. Some details concerning the construction of the cable and the connectors are useful to be provided at this point to more clearly present the physical level structure

Figure 2. Physical layer components, termination resistor Z = 100Ω

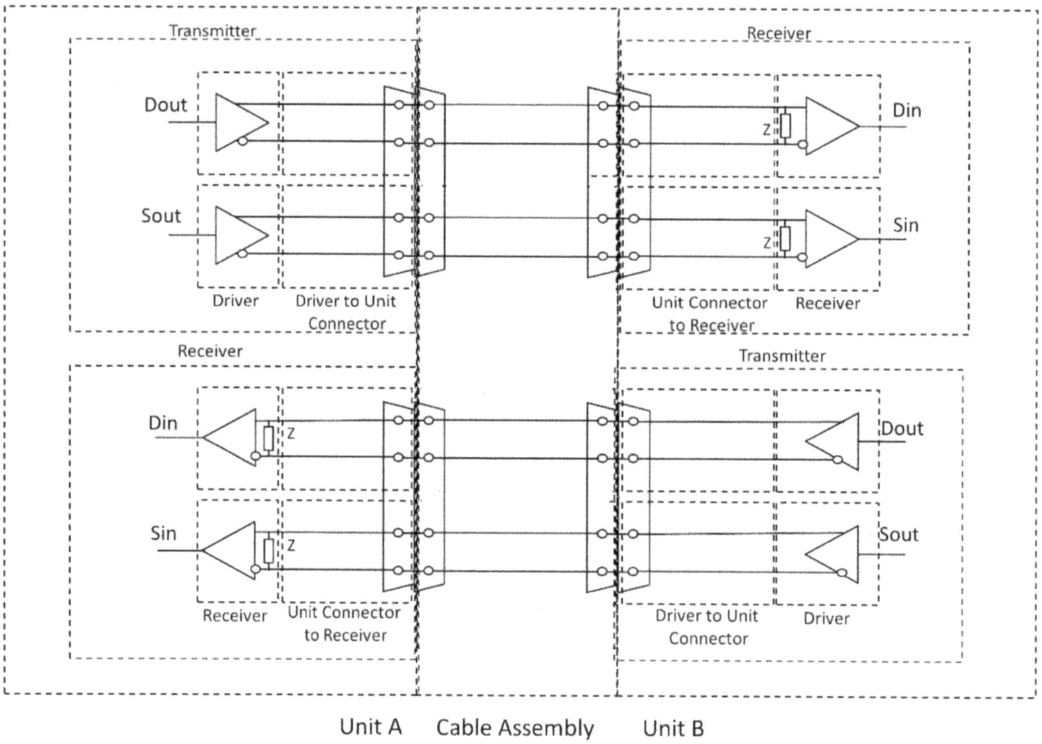

Cables

Typical SpaceWire cable is constructed according to ESCC 3902/003. The SpaceWire cable comprises four twisted pair wires with a separate shield around each twisted pair and an overall shield as depicted in Figure 3 (ECSS Secretariat, 2008).

Inner Conductors

Each signal wire is 28 AWG, constructed from seven strands of 36 AWG silver-coated, high-strength copper alloy. The thickness of the silver coating is minimum 2.0 μm. The maximum acceptable DC resistance of the inner conductor is 256 Ω/km. Each signal is insulated using white, expanded, microporous PTFE without any except those for processing and pigmentation. (ECSS Secretariat, 2008; ECSS Secretariat, 2003 and ECSS Secretariat, 2014a). In ECSS Secretariat, (2014a) two identically structured variants exist, with the exception of the diameter of the signal wire for the inner conductor. Variant 2 is 26AWG, hence all dimensions are a little larger. Since ECSS-E-ST-50-12C precedes ECSS Secretariat, (2014a) there is no mention of Variant 2 in the Standard, so in this chapter, Variant 1 (28 AWG) is always considered. It should be clarified that Variant 1 of Secretariat, (2003) is identical to Secretariat, (2014a).

Twisted Pair

The characteristic impedance of each differential signal pair is (100 ± 6) Ω differential impedance with a skew value between each signal in each differential signal pair less than 0.1 ns/m. Regarding the lay-length, it is stated in ECSS-E-ST-50-12C the lay-length of the two insulated conductors comprising a differential signal pair must not be less than 12 times and not more than 16 times the outside diameter of the unshielded twisted pair, yielding a lay-length of 24-32mm corresponding to 32-41 twists/m (ECSS Secretariat, 2014a). Also, Fillers, made of expanded microporous PTFE with only those additives for processing, are used with the differential signal pairs to ensure a smooth and uniform diameter under the shielding to contribute to a uniform impedance over the cable. Fillers have a diameter of 1.0 mm. Also, in the matter of inner shield, the ECSS-E-ST-50-12C (2008) states that each differential signal pair is shielded by a braided shield. The braided shield type is of push-back type and provides not less than 90% coverage. The shield wire size is 40 AWG. All strands used in the manufacture of the braided shield are silver-coated, soft or annealed oxygen-free high conductivity copper with a minimum thickness of silver 2.5 μm. The protective sheath for the shielded differential signal pairs is a white layer of extruded fluorpolymer PFA with a nominal wall thickness of 0.15 mm and only those additives for processing and pigmentation (ECSS Secretariat, 2008).

Complete Cable

The complete cable has a diameter of 7 mm maximum and a maximum weight of 80 g/m. It is constructed by four sets of differential signal pairs, twisted together not less than 12 times and not more than 16 times the outside diameter of two shielded and jacketed differential signal pairs, yielding a 55-77mm lay-length (ECSS Secretariat, 2014a). A filler of a diameter of 1.0 mm, is used in the center of the four differential signal pairs to ensure a smooth and uniform diameter under the shielding to contribute to a uniform impedance over the cable. The filler material as used for the complete cable is microporous

Figure 3. SpaceWire cable transverse view (ECSS Secretariat, 2008)

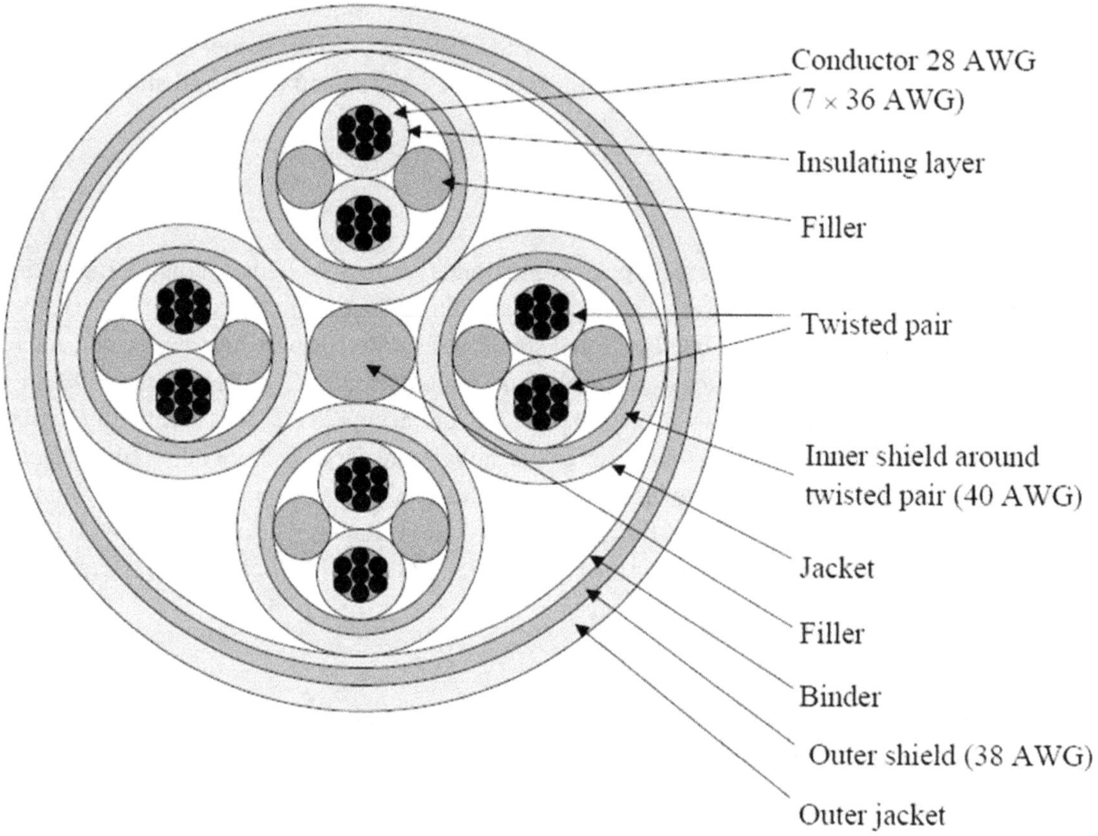

PTFE with only those additives for processing. A binder, of the same material as the filler, is applied over the four differential signal pairs and central filler to keep the signal pairs and filler together in a fixed position, wrapped with a 50% maximum overlap (ECSS Secretariat, 2008).

The set of four jacketed and screened differential signal pairs are shielded by an outer braided shield of push-back type that provides not less than 90% coverage. The shield wire size is 38 AWG. All strands used in the manufacture of the braided shield are silver-coated, soft or annealed oxygen-free high conductivity copper with a minimum thickness of silver of 2.5 μm. It should be noted that the twisted pair shields do not make contact with one another nor with the outer shield. The outermost jacket over the four twisted screened and jacketed differential signal pairs is a white layer of extruded Fluoropolymer PFA with only those additives for processing and pigmentation and no identifying marking since pressure applied to the cable during the marking process could adversely affect the electrical properties of the cable The nominal wall thickness of the jacket for the shielded differential signal pair is 0.25 mm (ECSS Secretariat, 2008).

The skew between the parts of the differential signal (intra-pair skew) in one differential signal pair is allowed to be a maximum of 0.1 ns/m. The skew between one differential signal pair and each other differential signal pair (inter-pair skew) within the cable must be less than the maximum of 0.15 ns/m (ECSS Secretariat, 2008).

The maximum length of the cable assembly is set to 10 m to ensure that the end to end skew and jitter introduced by the cable assembly does not exceed the maximum budget for the cable. Longer length cables can be used at slow data signaling rates provided that the signal attenuation and system jitter and skew limits are not violated at the operating data signaling rate (ECSS Secretariat, 2008). Detailed information can be found in clause 6 of the ECSS-E-ST-50-12C standard.

Shield Details

The coverage factor $K(\%)$ of the braided shield is calculated with regard to the number of strands N, the serving pitch P, the shield stand diameter $d$in mm, the angle $a$of the shield with the cable axis in degrees, the effective diameter d of the core under the shield in mm and the number of carriers C is calculated from equation 1 with the aid of equations 2 and 3 (ECSS Secretariat, 2014a):

$$K = \left(2F - F^2\right) \tag{1}$$

$$F = \frac{N \cdot P \cdot d}{\sin a} \tag{2}$$

$$\tan a = 2\pi \frac{\left(D \cdot 2d\right) \cdot P}{C} \tag{3}$$

The shielding effectiveness must conform to the limits for Variant 01 as depicted in Figure 4 (ECSS Secretariat, 2014a)

Connectors

The SpaceWire connector is a nine contact micro-miniature D-type with solder contacts, as defined in ESCC 3401/071, or crimp contacts. Connector contact identification as viewed from the rear of the receptacle or the front of the plug is depicted in Figure 5 (ECSS Secretariat, 2008).

The inner shield connection is connected to the inner shield of the SpaceWire cable. This inner shield of the SpaceWire cable is to be connected to signal ground according to the EMC requirements of the mission via a parallel resistor and capacitor (ECSS Secretariat, 2008).

In case of connections to a PCB, flying lead connectors are used. These connectors must have all the leads cropped to the same short length (less than 25 mm) and the wires comprising the differential signal pairs must be twisted together to minimize the discontinuity in impedance caused by the connector (ECSS Secretariat, 2008).

PCB mounting right-angled connectors use is generally discouraged. In case they are used, signal path length compensation is mandatory to be performed by adjusting the length of tracks on the PCB. The signals connected to the top row must be given correspondingly shorter PCB track lengths than tracks going to the bottom row since the topmost row of pins on the right-angled connect have longer leads than

Figure 4. Minimum Shielding Effectiveness (ECSS Secretariat, 2014a)

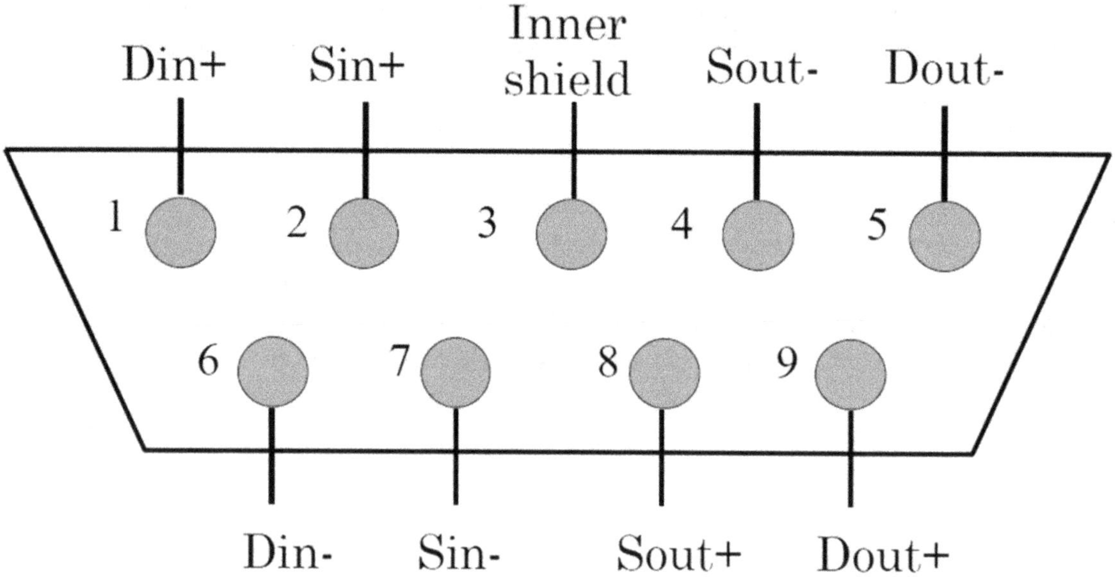

the bottom row. Also, in this case, track length compensation is mandatory to be performed at the connector end of the PCB tracks to maintain the differential signal across the PCB (ECSS Secretariat, 2008).

Cable Assemblies

SpaceWire cable assemblies are made from a length of SpaceWire cable of up to 10 m terminated at each end by nine-pin micro-miniature D-type connectors. Longer length cables are allowed at slow data signaling rates provided that the signal attenuation, system jitter and skew limits are not violated at the operating data signaling rate. The cable signal wires cross over to achieve a transmit-to-receive interconnection, e.g. Dout+ is connected to Din+ (ECSS Secretariat, 2008).

The SpaceWire cable assembly includes an outer-shield which is 360º terminated to the connector backshell at each end of the cable. The individual shields of the differential signal pairs carrying the output signals Dout+, Dout- and Sout+ and Sout- are connected together and to the pin 3 of the connector as depicted in Figure 6 and tabulated in Table 1 respectively. This way two of the differential pairs are connected at one end of the cable and the remaining two at the other end (ECSS Secretariat, 2008).

A metal shell is bonded to the main body of the connector via a low impedance connection (less than 1 Ω) to provide the necessary shielding to the connector. The connector shell is bonded to the outer shield of the cable via a low impedance connection (less than 1 Ω) (ECSS Secretariat, 2008).

Printed Circuit Board Tracks

SpaceWire can also be run over printed circuit boards (PCBs) or backplanes. Only point to point connections are supported on a PCB or backplane, not multi-drop bus structures. Bus type structures are built from point to point connections between nodes on the backplane. High-speed SpaceWire signals have a bandwidth of over 1 GHz (frequency of the signal edges); the PCB layout needs to be designed

Figure 5. Identification of SpaceWire connector pins (ECSS Secretariat, 2008)

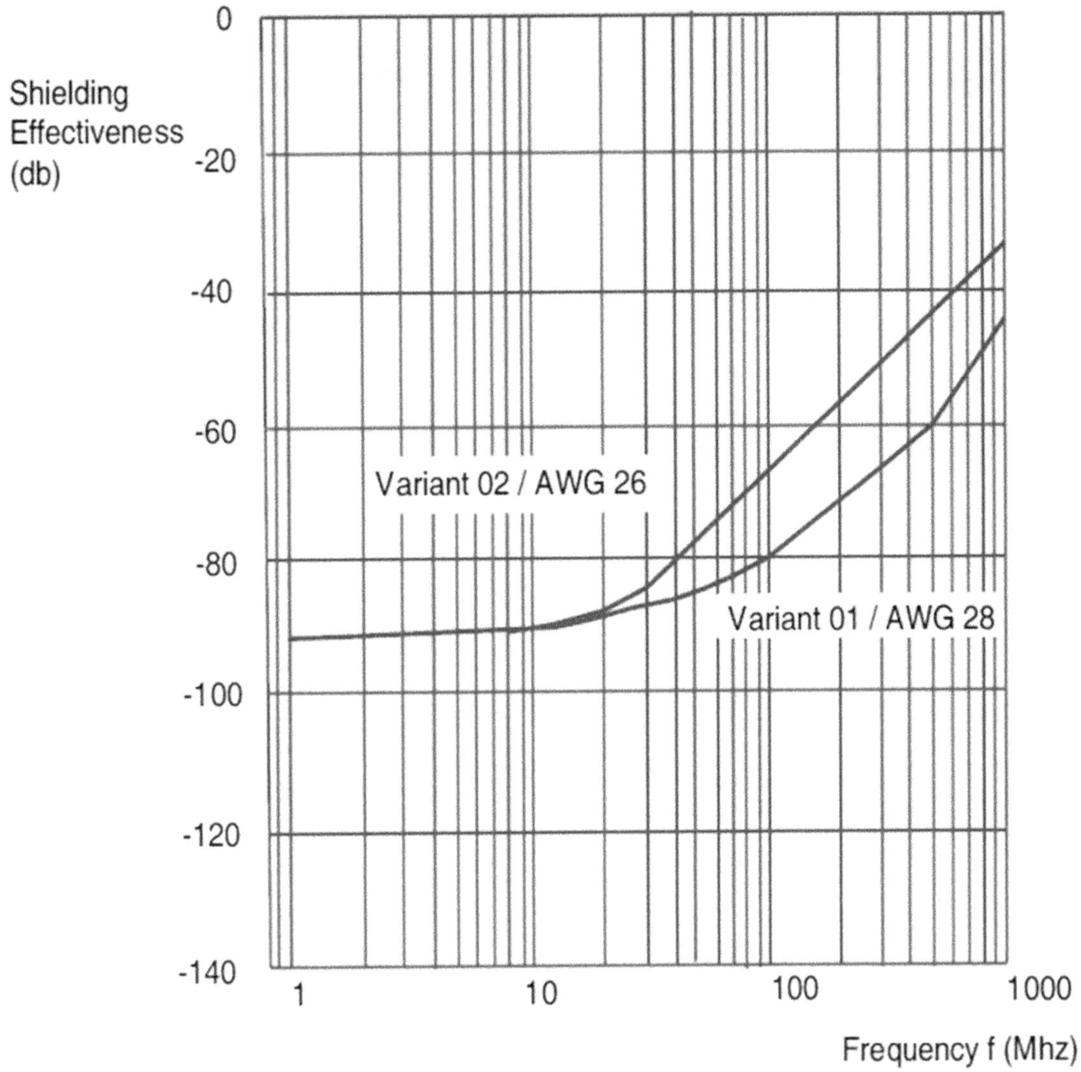

with some precautions. Among them, the PCB tracks must have (100 ± 6) Ω differential impedance, track pairs need to be well separated from other tracks, elimination of right-angle turns, minimal use of vias, each signal of the differential pair to be tracked identically and kept to the same length (difference in tracks less than 5 mm and less than 5% of the track length) to minimize skew of the differential signal, Data and Strobe signal pairs to be of the same length (less than 5 mm and less than 5% of the track length) to minimize the skew introduced between the Data and Strobe (ECSS Secretariat, 2008).

Signal Level

The signal level part of the SpaceWire standard covers signal voltage levels, noise margins, and signal encoding.

Figure 6. Complete assembly of SpaceWire cable (ECSS Secretariat, 2008)

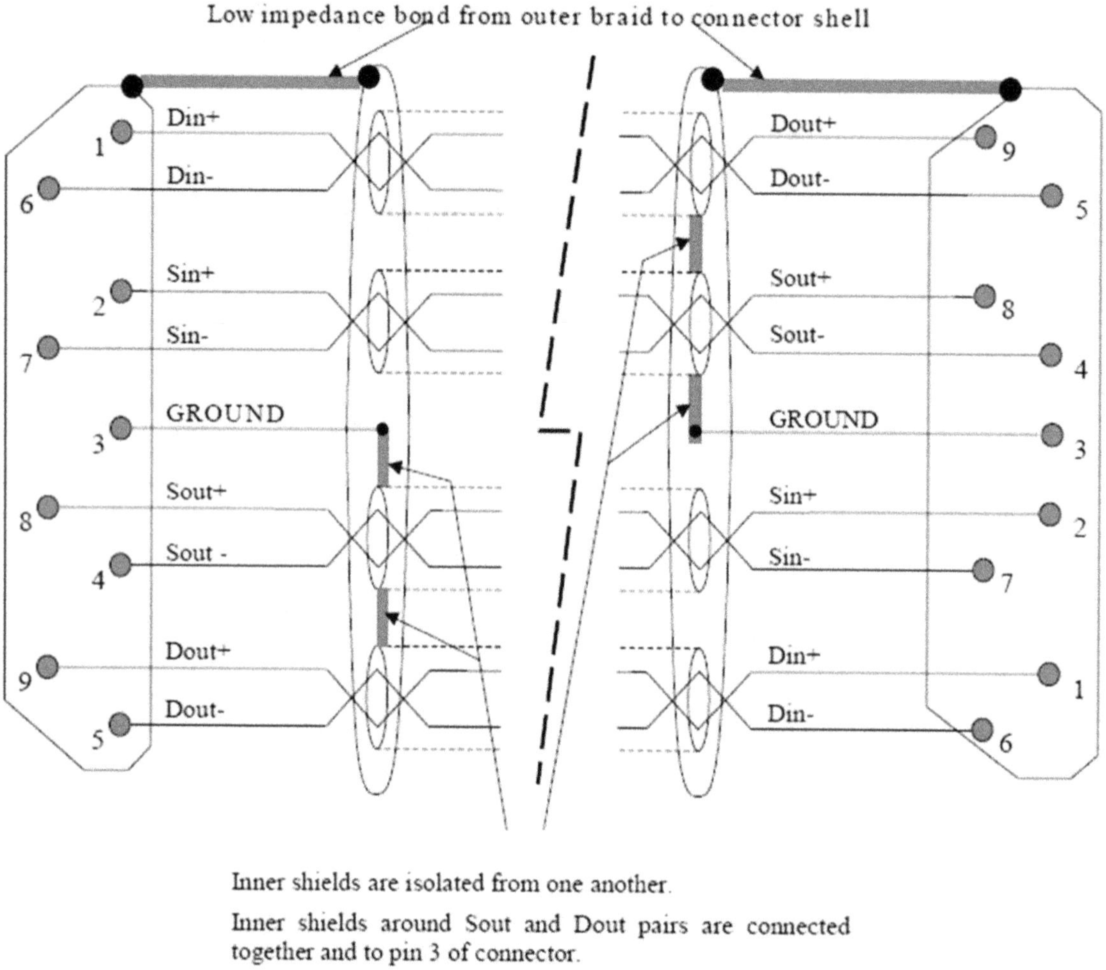

Low Voltage Differential Signaling (LVDS) as defined in ANSI/TIA/EIA-644 is specified as the signaling technique for SpaceWire. LVDS uses balanced signals to provide very high-speed interconnection using a low voltage swing (350 mV typical). The balanced or differential signaling provides an adequate noise margin to enable the use of low voltages in practical systems. The low voltage swing results in relatively low power consumption at high speed. LVDS is appropriate for connections between chips on a board, boards in a unit, and unit to unit interconnections over distances of 10 m or more (ECSS Secretariat, 2008). The LVDS signaling levels are illustrated in Figures 7 and 8.

A typical LVDS driver and a receiver are shown in Figure 9 connected by a medium (cable or PCB traces) of 100-ohm differential impedance. The LVDS driver uses current mode logic. A constant current source of around 3.5 mA provides the current that flows out of the driver, across the transmission medium, through the 100-ohm termination resistance and back to the driver via the transmission medium. Two pairs of transistor switches in the driver control the direction of the current flow through the termination resistor. When the driver transistors marked "+" are turned on and those marked "-" are turned off, current flows as indicated by the arrows on the diagram creating a positive voltage across the

Table 1. End to end signal wire connections (ECSS Secretariat, 2008)

Signal at A end	Pin at A end		Pin at B end	Signal at B end
A-Din+	1	Connection	9	B-Dout+
A-Din-	6	Connection	5	B-Dout
A-Sin+	2	Connection	8	B-Sout+
A-Din-	7	Connection	4	B-Sout-
A-(Drain of pairs 5,9 and 4,8)	3	No Connection	3	B-(Drain of pairs 5,9 and 4,8)
A-Sout+	8	Connection	2	B-Sin+
A-Sout-	4	Connection	7	B-Sin-
A-Dout+	9	Connection	1	B-Din+
A-Dout-	5	Connection	6	B-Din-
A-Shield	Shell	Connection	Shell	B-Shield

termination resistor. When the two driver transistors, marked "-", are turned on and those marked "+" are turned off, current flows in the opposite direction producing a negative voltage across the termination resistor (ECSS Secretariat, 2008).

LVDS has several features that make it very attractive for data signaling (ECSS Secretariat, 2008):

- Near constant total drive current (+3.5 mA for logic 1 and -3.5 mA for logic 0) which decreases switching noise on power supplies.
- High immunity to the ground potential difference between driver and receiver, LVDS can tolerate at least ±1 V ground difference.
- High immunity to induced noise because of differential signaling, normally using twisted-pair cable.
- Low EM emission because small equal and opposite currents create small electromagnetic fields that tend to cancel one another out.
- Failsafe operation
- Not dependent upon particular device supply voltage(s).

Figure 7. LVDS Voltage Swing across 100Ω termination resistor (ECSS Secretariat, 2008)

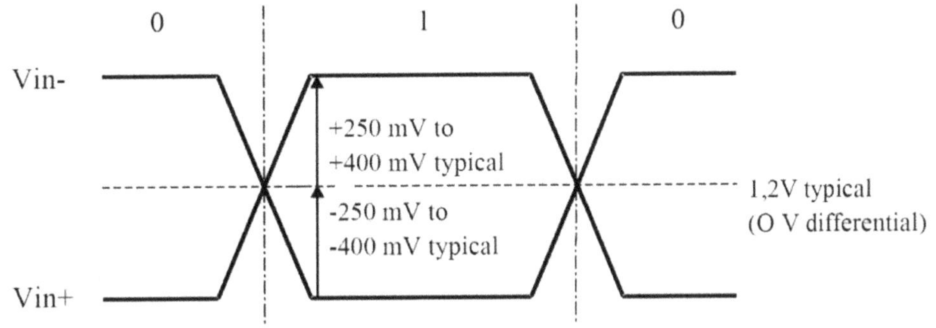

Figure 8. LVDS Receiver Input Thresholds (ECSS Secretariat, 2008)

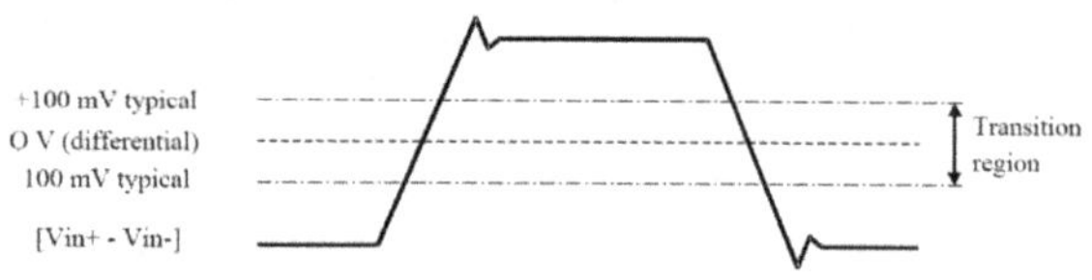

- Simple 100-ohm termination at the receiver.
- Power consumption is typically 50 mW per driver/receiver pair for LVDS compared to 120 mW for ECL or PECL signals used in IEEE Standard 1355-1995.

Figure 9. LVDS operation (ECSS Secretariat, 2008)

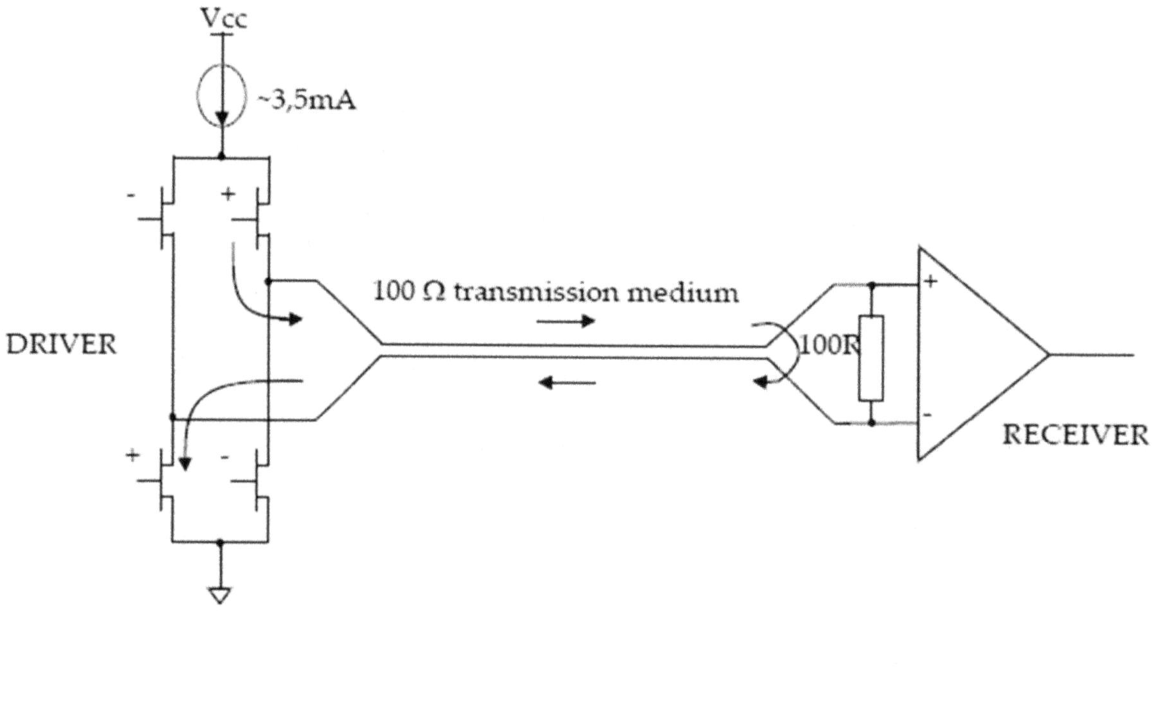

Short Description of the Failsafe Operation of LVDS

In order to prevent the system from operating under erroneous conditions and prevent undesired behavior, the following must be incorporated into the design of the driver and the receiver (ECSS Secretariat, 2008):

- When the driver is not powered its output should be high impedance (greater than 100 kΩ).
- When the receiver is not powered its input should be high impedance (i.e. greater than 100 kΩ).
- Assuming a noise threshold of 10 mV is not exceeded at the receiver input, the receiver outputs must be locked to a high state (inactive) and not oscillate when:

- The driver is not powered, disabled or not connected to the receiver
- Receiver inputs are shorted or open circuit, in case of the cable or a wire in the cable is disconnected

Signal Coding

SpaceWire uses Data-Strobe (DS) encoding. This coding scheme encodes the transmission clock with the data into Data and Strobe signals so that the clock is recovered, i.e. bit synchronization, by the execution of an XOR function on the Data and Strobe lines. The Data signal follows the data bitstream, i.e. is high when the data bit is 1 and low when the data bit is 0, and the Strobe signal changes state whenever the Data does not change state from one bit to the next (ECSS Secretariat, 2008). This coding scheme is illustrated in Figure 10.

Figure 10. Data-Strobe (DS) encoding (ECSS Secretariat, 2008)

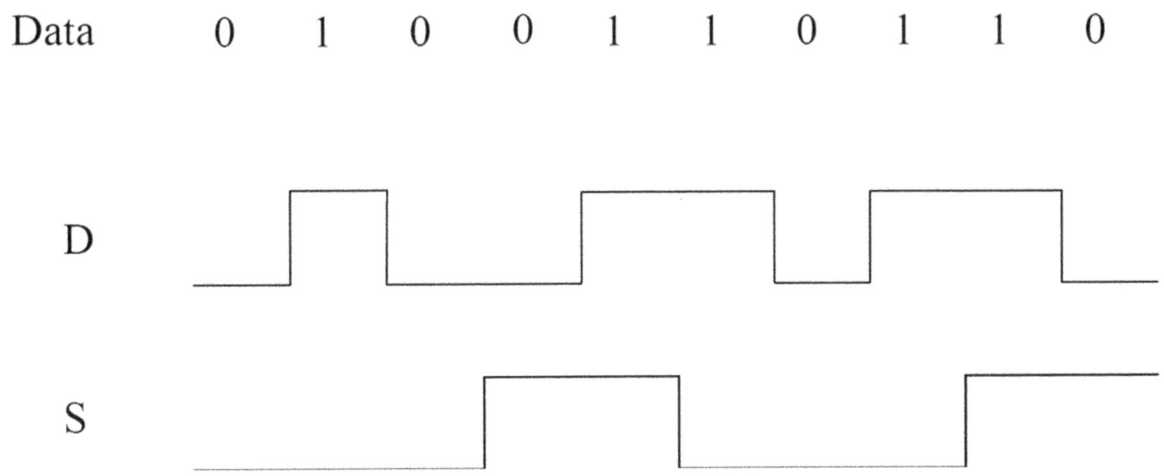

DS encoding improves the skew tolerance to almost 1-bit time, as opposed to 0.5-bit time for simple data and clock encoding. A link utilizes two pairs of differential signals, one pair transmitting Data and Strobe signals in one direction and the other pair transmitting Data and Strobe in the opposite direction, totaling eight wires for each bi-directional link. (ECSS Secretariat, 2008).

Data Signaling Rate

The minimum data signaling rate at which a SpaceWire link is allowed to operate is 2 Mb/s. On the contrary, the maximum data signal rate is the highest data signaling rate at which a SpaceWire link can operate and is defined by consideration of signal skew and jitter. The link in one direction can operate at a different data signaling rate to the same link in the opposite direction. Links within a system can operate at different data signaling rates. A SpaceWire link can operate at any data signaling rate between the minimum data signaling rate and the maximum possible data signaling rate. After a reset or disconnect the SpaceWire link transmitter initially commences operating at a data signaling rate of (10±1) Mb/s and continues at that rate until commanded otherwise (ECSS Secretariat, 2008).

Skew and Jitter Overview

Each system may achieve a different maximum data signaling rate. The maximum rate depends on several factors such as cable length, driver-receiver technology, and encoder-decoder design. Skew and jitter place limits on the maximum rate of the system. Figure 11 depicts the usual contributors to skew and jitter. A specific data rate value can be achieved when the total system skew and jitter added to the minimum edge separation of the receiver is lower than the bit period of the specific rate (ECSS Secretariat, 2008).

Figure 11. Skew and jitter contributors (ECSS Secretariat, 2008)

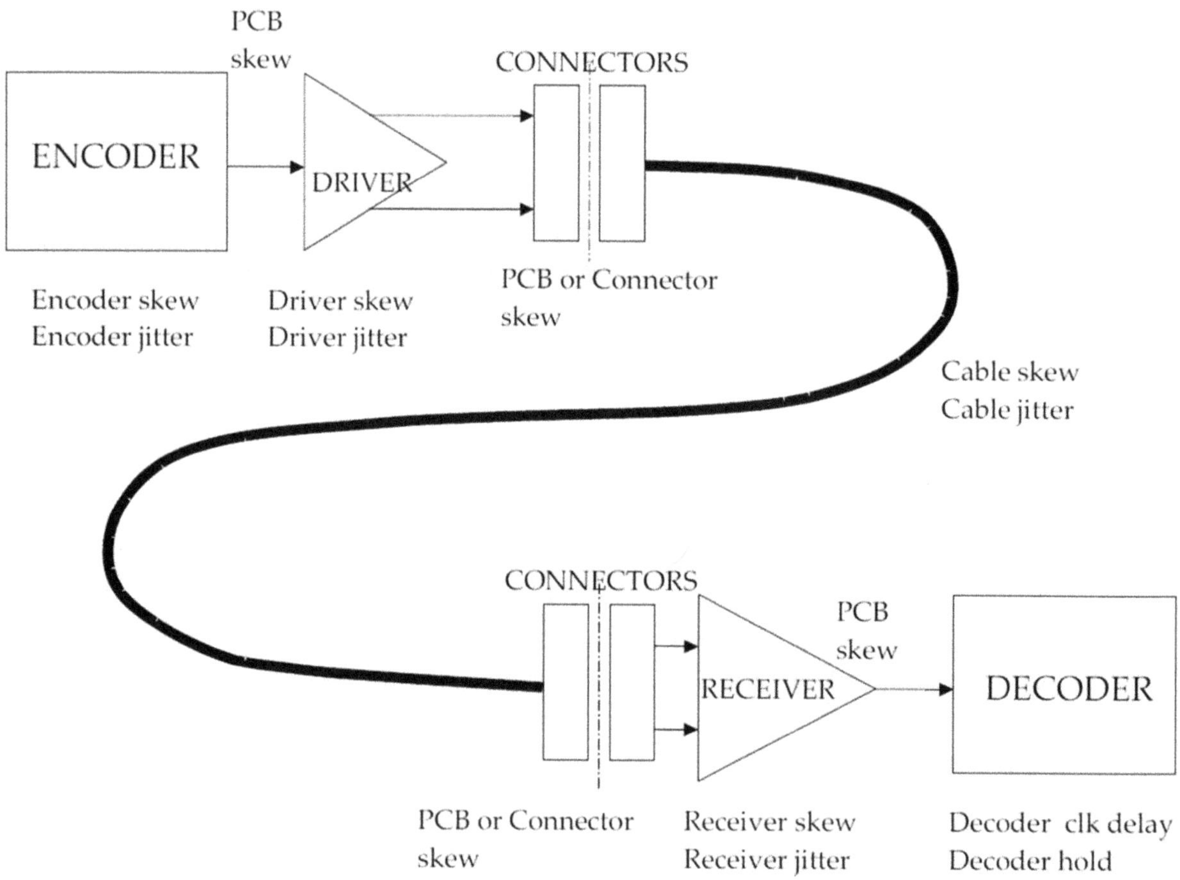

Character Level

SpaceWire supports only two types of characters: data and control characters as illustrated in Figure 13.

Data Characters

Data characters are 10-bit characters, they contain one parity bit, one data-control flag, and the eight-bits of data transmitted with the least-significant bit first. The parity bit covers the previous eight-bits of a data character or the two bits of a control character, the current parity bit and the current data-control flag as depicted in Figure 12. The parity bit is set to produce odd parity so that the total number of 1's in the field covered is an odd number. The data-control flag is set to zero to indicate that the current character is a data character (ECSS Secretariat, 2008).

Figure 12. Parity Coverage (ECSS Secretariat, 2008)

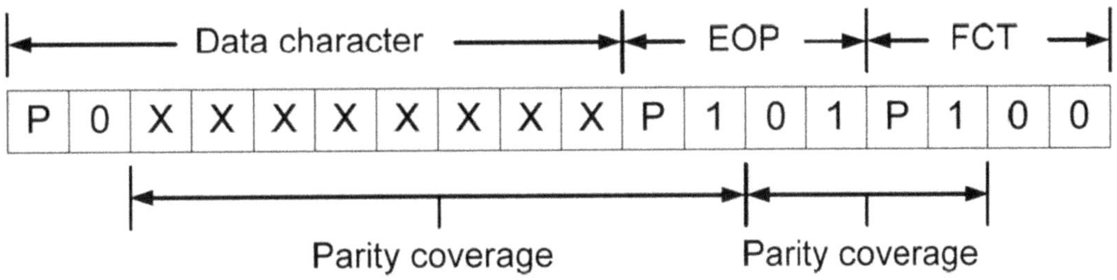

Control Characters and Codes

Control characters contain a two-bit control code. Each control character is formed from a parity bit, a data-control flag, and the two-bit control code. The data-control flag is set to one to indicate that the current character is a control character. Parity coverage is similar to that for a data character. One of the four possible control characters is the escape code (ESC). This can be used to form longer control codes. Two of these codes are the NULL code and the Time-Code, as depicted in Figure 13(ECSS Secretariat, 2008).

- NULL control code is formed from ESC followed by the flow control token (FCT). NULL is transmitted, whenever a link is not sending data or control tokens, to keep the link active and to support link disconnect detection. The parity bit (P) in the middle of the control code is zero.
- The Time-Code is used to support the distribution of system time across a network. A Time-Code is formed by ESC followed by a single data-character.

Character Synchronization

Character synchronization is performed only once when a link is started or re-started following a link disconnection. After reset or link error the Data and Strobe signals are set to zero. When the transmitter is enabled after reset the first bit that is sent is a parity bit set to zero so that the first transition is on the Strobe line as depicted in Figure 14 (ECSS Secretariat, 2008).

With all the provided information regarding SpaceWire Standard as well as the systems built conforming to it, the authors believe that readers have all the necessary data to start exploring the SpaceWire ecosystem, at least from an EMC point of view. Exchange level and higher, according to Figure 1, are out

Figure 13. Control characters, codes, and data characters (ECSS Secretariat, 2008)

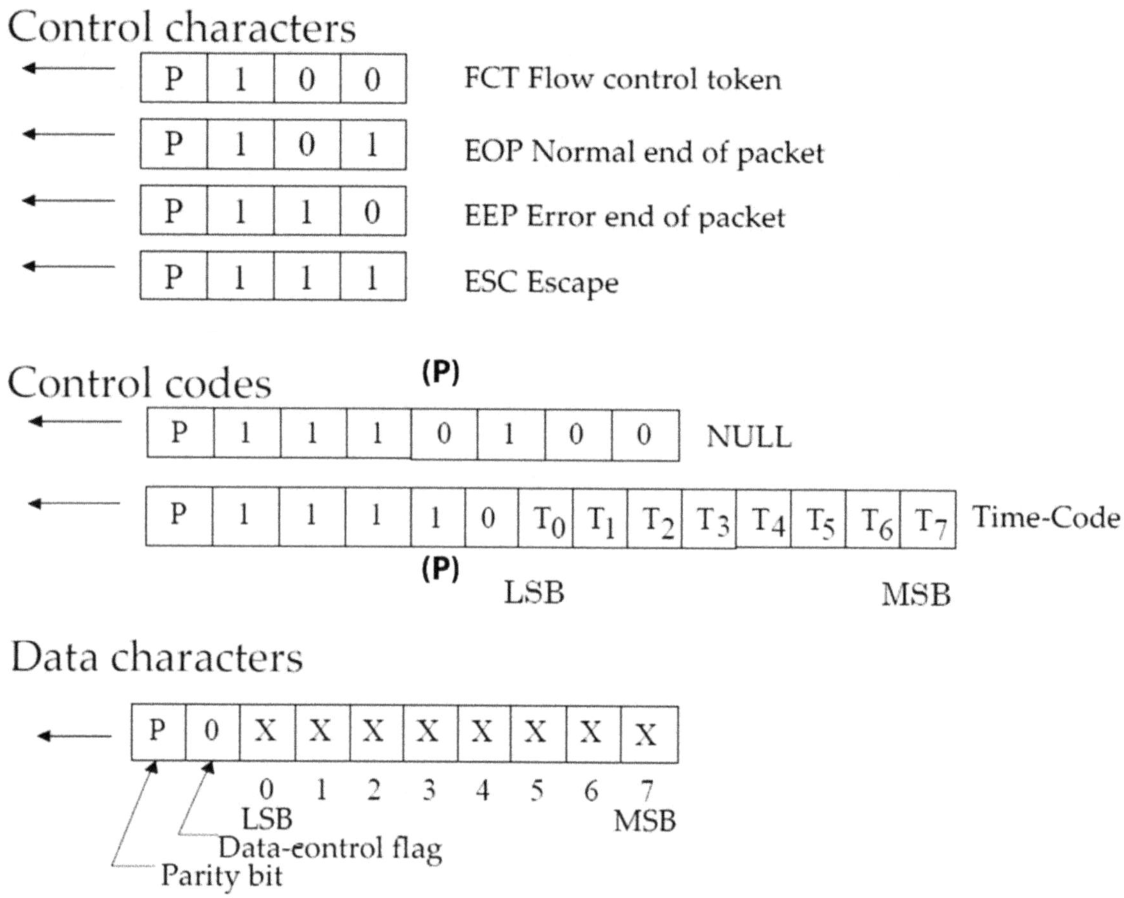

of the main scope of this chapter and they will not be analyzed further. Instead, some grounding strategies are presented and discussed. In addition to this, SpaceWire emissions, as well as some considerations to minimize them are discussed later in this chapter.

SpaceWire Revision 1

Since its first publication in 2003, the SpaceWire standard was not updated for more than 10 years. The European Cooperation for Space Standardization started the update process in 2012 to include the experience of the years of its application. Working Group ECSS-E-ST-50-12C Rev.1 WG prepared Revision 1 of the standard. Revision. 1 (ECSS Secretariat, 2019) was published in May 2019. It includes several changes as well as new additions, many of which are out of the scope of this chapter. In this section, the authors will try to include the most important updates with a focus on the cables and assemblies.

The inter-pair skew is limited to a maximum of 0.1 ns/m. It may be more than 0,1 ns/m, provided that when incorporated in a cable assembly the overall Data-Strobe skew budget requirement for the SpaceWire link is met. Also, the max skew allowed to be introduced by the connectors at both ends of the cable assembly is 0,07 ns (ECSS Secretariat, 2019).

Figure 14. Link Start-Up Bit Sequence of SpaceWire (ECSS Secretariat, 2008)

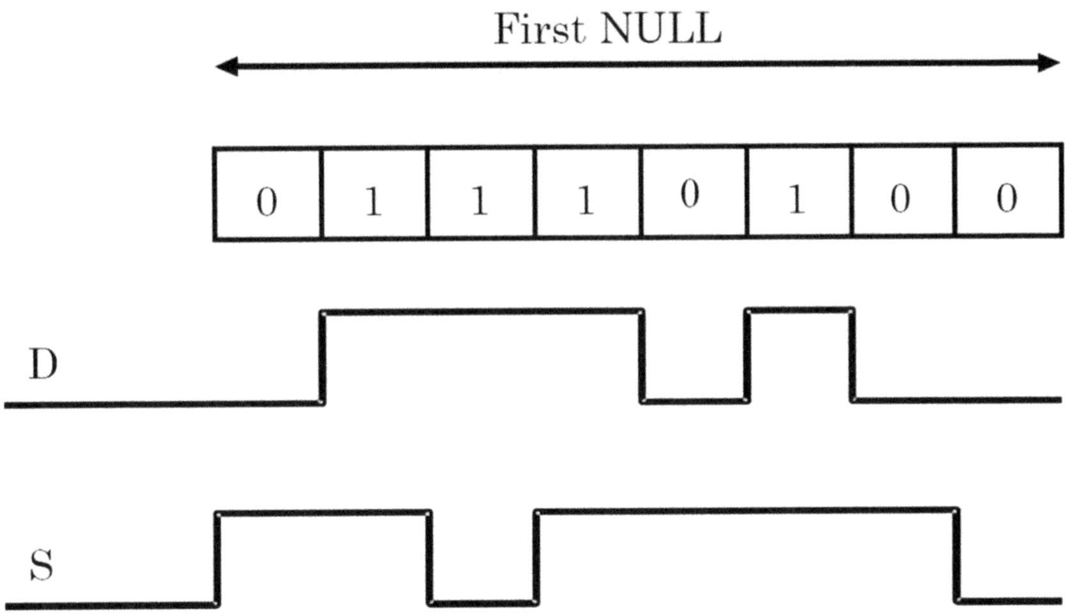

The insertion loss maximum limits for each differential pair in the SpW cable is calculated from equation 4

$$a_{cable}\left(\frac{dB}{m}\right) = k1 \cdot \sqrt{f} + k2 \cdot f + \frac{k3}{\sqrt{f}} \tag{4}$$

where f (MHz) is the Nyquist frequency applicable at the receiver (e.g. 100 MHz for a 200 Mbps link or 200 MHz for a 400 Mbps link), and k1, k2, k3 are constants related to differential pairs wire gauge according to Table 2. Figure 15 depicts the values for the insertion losses limits of the differential pairs of the cable for 28 AWG (Variant 1) and 26 AWG (Variant 2) wire gauges. Calculated insertion losses limits concern the cable only, the additional losses due to the connectors of the assembly are not taken into account (ECSS Secretariat, 2019).

Table 2. Constants for differential pair wire gauge

SpW Differential Pair Wire Gauge	K1	K2	K3
28 AWG (Variant 1)	4,5e-2	4,0e-4	5e-4
26 AWG (Variant 2)	3,5e-2	1,8e-4	5e-4

Specifications for the maximum levels of Power Sum Near End Crosstalk (PSNEXT) and Power Sum Equal Level Far End Crosstalk (PSELFEXT) are also included in the revision 1 as tabulated in Table 3 and

Figure 15. Differential pair insertion loss maximum limits a (ECSS Secretariat, 2019)

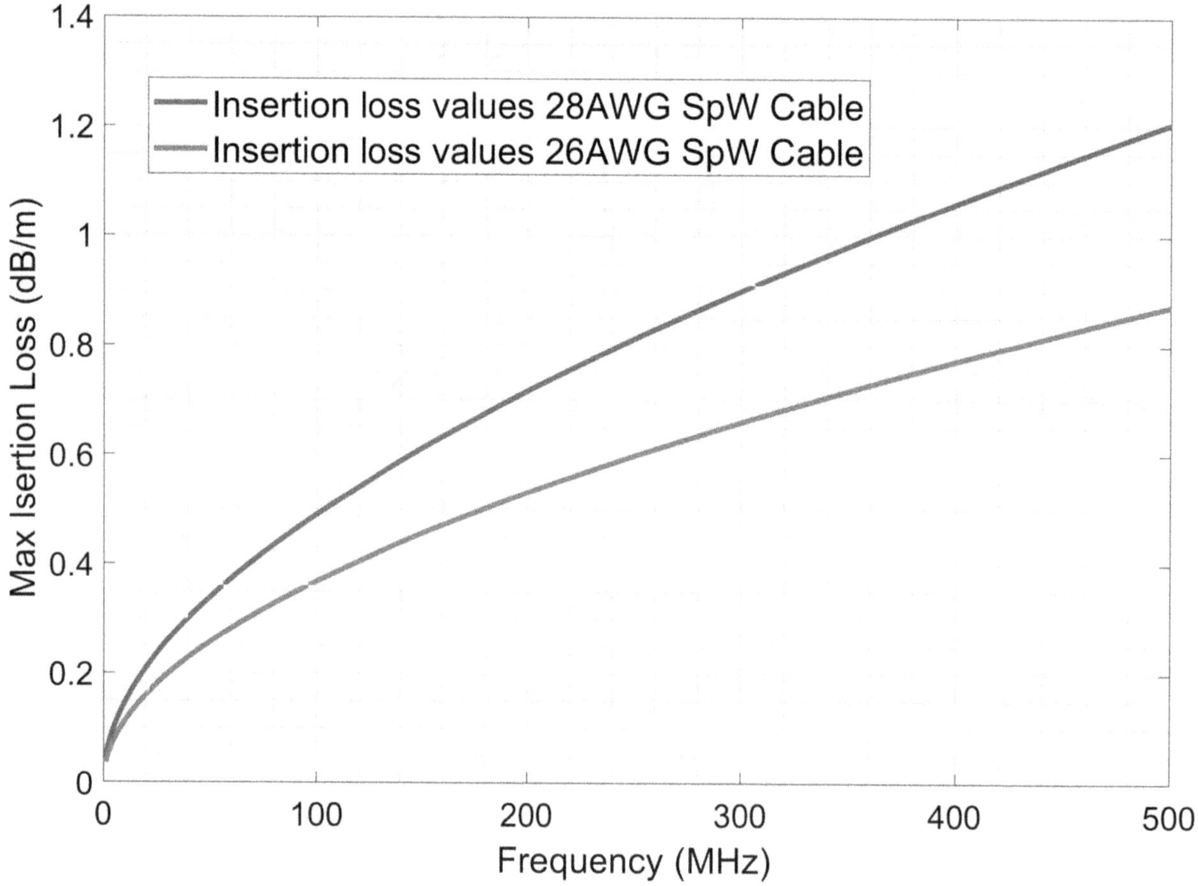

Table 4 (ECSS Secretariat, 2019). Values are for the cable only without considering the connectors of the assembly. In fact, this performance can only be achieved with EMI and crosstalk optimized connectors.

Table 3. Cable PSNEXT specification

Frequency (MHz)	PSNEXT (dB)
100	73.0
500	63.5
1000	39.0

Connectors

In revision 1 of the standard, the connectors for the SpaceWire cable assembly are either type A or type B. The type A connector is a microminiature Dtype connector with nine crimp or solder contacts, as

Table 4. Cable PSELFEXT specification

Frequency (MHz)	PSELFEXT (dB)
100	65.0
500	50.5
1000	41.5

defined in ESCC 3401/029 and ESCC 3401/071. In addition to type A connectors, other connectors (type B) are allowed, provided that they meet the requirements. Type B connectors must have the following characteristics (ECSS Secretariat, 2019):

- contact pairs with differential impedance of (100 ±6) Ω,
- compatible with ESCC 3902/003 cables or balanced cables in ESCC 3902/002,
- compatible with the space environment as defined in ESCC 3401.

A connector with female contacts must include a conductive gasket for EMI improvement. Connectors with male contacts will be used on cable assemblies. The outer shield of the SpaceWire cable must be circularly terminated to the connector body via an EMI backshell. The electrical resistance of the connection between the outer shield of the cable and the body of the connector with male contacts must be less than 10 mΩ at DC. The inner shields of the SpaceWire cable must be connected to the connector body by an impedance of less than 10 mΩ at DC. Any pigtails used to connect the inner shields to the backshell must be less than 10 mm in length. When type A connectors with female contacts are used to connect two cables, pin 3 must be connected to the body of the connector. Pin 3 of type A connectors with female contacts is connected to Circuit Ground also when used with type A cables to prevent charge accumulation (ECSS Secretariat, 2019).

In the case of Type B connector, the assignment of signals to connector contacts must be defined by the supplier but with the limitation that the two signals in each pair of conductors (Din+ and Din-, Sin+ and Sin-, Dout+ and Dout-, Sout+ and Sout-) must be assigned to adjacent pins of the connector. No connector pin will be connected to the cable shield. In the case where an outer shield is present, it must be circularly connected to the connector body. The inner shields must be circularly connected to the connector body when possible (ECSS Secretariat, 2019).

Cable Assemblies

Cable assembly type AL (type A Legacy) is the cable assembly specified in the current issue of the SpaceWire standard (ECSS Secretariat, 2008), prior to revision 1. Cable assembly type AL, after revision 1 was issued, is not recommended for use in new designs (ECSS Secretariat, 2019).

Acceptable cable assemblies in the standard revision 1 consist of

- two identical connectors of type A with male contacts joined by a length of cable,
- two connectors of Type B joined by a length of cable, or

- a length of cable in between one Type A connector with male contacts at one end, and one Type B connector at the other end.

The insertion loss through each differential pair in a cable assembly must be less than 7 dB at frequencies up to 1.5 times the data signalling rate (i.e. for a 200 Mbps signal, the frequency at which the insertion loss is measured is 300 MHz) . (ECSS Secretariat, 2019).

Cable assembly type A will use type A connectors and cable with an outer shield. Pin 3 of the connector must be left unconnected. The individual shields of each of the four differential signal pairs will be bonded to the connector shell via a low impedance connection of less than 10 mΩ. A metal backshell will be used for each connector to create, together with the outer shield, a continuous conductive barrier and to facilitate the termination of the inner shields. The outer shield of the cable must be 360° terminated to the connector backshell via a low impedance connection. The connection impedance between the main body of the connector and the backshell must be less than 10 mΩ. The differential far end and near end crosstalk (FEXT and NEXT) between any two pairs in the SpaceWire cable assembly must be less than 20 dB up to 1 GHz. (ECSS Secretariat, 2019).

Cable assembly type B will be a cable assembly with an outer shield and matched impedance type B connectors as defined by the supplier. The cable in Cable assembly type B may be another form of cable with the following properties:

- differential impedance of (100 ± 6) Ω,
- compatibility with ESCC 3902/003 cables or balanced cables in ESCC 3902/002,
- compatibility with the space environment as defined in ESCC 3401.

The individual shields of each of the four differential signal pairs must be bonded to the connector shell via a low impedance connection (less than 10 mΩ). A metal backshell must be used for each connector to create, together with the outer shield, a continuous conductive barrier and to facilitate the termination of the inner shields. The outer shield of the cable must be 360° terminated to the connector backshell via a low impedance connection. The main body of the connector must be connected to the backshell via a low impedance connection (less than 10 mΩ). The differential far end and near end crosstalk (FEXT and NEXT) between any two pairs in the SpaceWire cable assembly must be less than 50 dB up to 1 GHz including the contribution of the cables and of the connectors. (ECSS Secretariat, 2019).

Bonding and Grounding

The grounding strategy is a key factor of the spacecraft design. A ground path characterized by low-inductance and low-resistance is needed to provide a return path to fault currents as well as electromagnetic interference (EMI) induced currents. Ground paths traditionally were constructed of Aluminum, but Space Agencies need lightweight structures so increasingly, lighter composite materials such as CFRP (Carbon-Fiber-Reinforced polymer), are selected to replace the standard aluminum panels as the spacecraft haul.

Bonding methodologies are distinguished to direct and indirect. In the case of direct bonding, the junctions are permanently fixed together without additional joints. This way a low impedance path is formed, improving the electromagnetic compatibility (EMC) performance. Therefore, direct bonding is preferred, but the indirect bonding cannot be avoided in installations with units installed on partially or completely non- conductive ground planes like CFRP. The grounding network - usually composed of strips or foils of aluminum composing a "rail" structure to reduce mass, is a means to reduce the inductance of the ground connection and the common-mode current. This network is an additional item of the spacecraft composed of grounding rails routed under the cable bundles, or ground planes which are mounted on the face sheet. This solution, however, is not very flexible when the routing of cable bundles has to be adapted for late design modifications (Nicoletto, Boschetti, &Savi, 2014). Moreover, their successful exploitation must also take into account possible impacts on the electromagnetic compatibility performance of the onboard electronic systems. Consequently, electromagnetic interference issues have to be analyzed when such materials are used, using the radiated Electric Field (E-field) from high-speed digital lines as the parameter for this assessment.

Authors of Leininger et al. (2012) discuss the use of a network of round wires over a CFRP sheet as a replacement of the classical grounding rail system. They demonstrate that the electrical conductivity of carbon fibers can be used to simplify the grounding network. A low-resistance, low inductance network of metallic conductors is employed to divert fault currents from flowing through the CFRP structure. As an optimal alternative, the metallization of the CFRP face sheet is proposed, otherwise, ground conductors can be used. The use of rails, unless they are wide and track the entire harness, has no better results than the use of round wires, which can replace them and are easier to implement. In addition, to benefit from the ground plane properties of CFRP face sheet panels, low inductance connections of the electronic units to the panels need to be implemented (Leininger et al. 2012).

The authors of Nicoletto et al. (2014) compared the simulated radiated E-field for the three ground planes. The first is a 2.5:1 (length to width) aluminum plane. The second is a CFRP plane. The third is a modification of the second with a long aluminum strip (2.5x0.03x0.001 m) routed above the CFRP panel. This last case corresponds to a typical implementation of a grounding rail network. They concluded that the grounding plane made of aluminum, due to its intrinsic lower impedance, represents the best solution in terms of minimization of the radiated E-Field, in the whole frequency range analyzed. The plain CFRP plane induces higher levels of radiated E-field especially at lower frequencies (<200 MHz) compared to the grounding rail network. However, in all three cases, the radiated electric field levels are well inside the typical requirements applicable to Space equipment. Based on these results, CFRP panels may be used for common mode current return, avoiding the mass penalties induced by the implementation of a grounding network that tracks the entire harness (Nicoletto et al., 2014). This conclusion, regarding the suitability of CFRP for the ground plane, is also highlighted in Nikolopoulos et al., (2020).

Full Duplex Link Emissions Measurements

To acquire actual full-duplex link emissions from a SpaceWire link, the authors of Baklezos et al. (2020), proposed a measurement setup, based on military standard 461G (USA-DoD, 2015), as depicted in Figure 16 and Figure 17, The measurement setup imitates the typical harness routing in a satellite or spacecraft. A one-meter SpaceWire cable is placed over an aluminum (conductive) table. TELETEL's iSHAFT handles the signal generation and capturing. The iSHAFT Simulator/ Recorder is an advanced Data Front End with traffic generation capabilities that simulates SpaceWire instruments, enabling system integration tests before the availability of the flight models. It also provides a tool for the validation of satellite/ spacecraft on-board communication protocols and data networks implementing the SpaceWire protocol.

Authors in Baklezos et al. (2020) investigate the radiated emissions of the SpaceWire cable for the extreme values of the protocol-supported data rates. Link speeds of 10 and 400Mbps with payload sizes of 32 and 2048 bytes consist of 4 different possible modes of the full-duplex operation for the SpaceWire link. Table 5 tabulates the used measurement parameters of the EMI receiver according to the military standard 461G (USA-DoD, 2015). Authors (Baklezos et al., 2020) present the results of the measurements corresponding to these different modes of operation and report 6 dB as the estimated expanded uncertainty of the electric field amplitude measurement based on the EN 55016-4-2_2011, A2: 2018 Standard (tables D1-D6).

Figure 16. Radiated Emissions Setup Top-View (Baklezos et al., 2020)

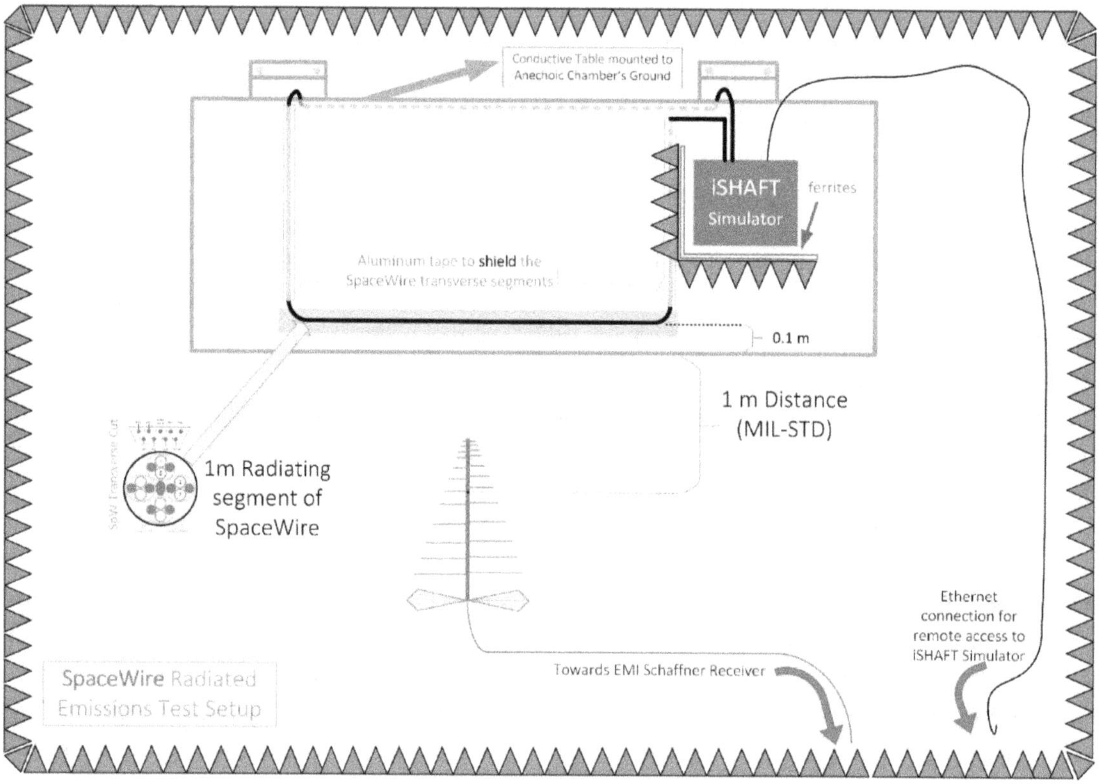

Figure 18 depicts the radiated emissions measurement results for both polarizations (Horizontal & Vertical) in the frequency range 30 MHz - 1 GHz. The operational mode of the SpaceWire link is 10Mbps, 32 Bytes payload size. Slightly higher levels of emissions are present in the Vertical Polarization along with more spectral spikes but overall, the spectrum is rather quiet (Baklezos et al., 2020).

The full-duplex SpaceWire link operating in the mode of 10 Mbps data rate and 2048-byte payload size was measured and the radiated emissions are depicted in Figure 19. Both polarizations are reported. The emissions present slightly lower maximum levels compared to the 32 bytes payload, other than that the link presents the same behavior in both polarizations (Baklezos et al., 2020).

Figure 20 depicts the results of radiated emissions measurements of the operational mode with a data rate set equal to 400 Mbps, i.e. the maximum supported by the protocol. The payload size was set to 32 Bytes. Both polarizations in the frequency range of 30 MHz to 1 GHz are measured. The link speed

Figure 17. Radiated Emissions Setup Side-View (Baklezos et al., 2020)

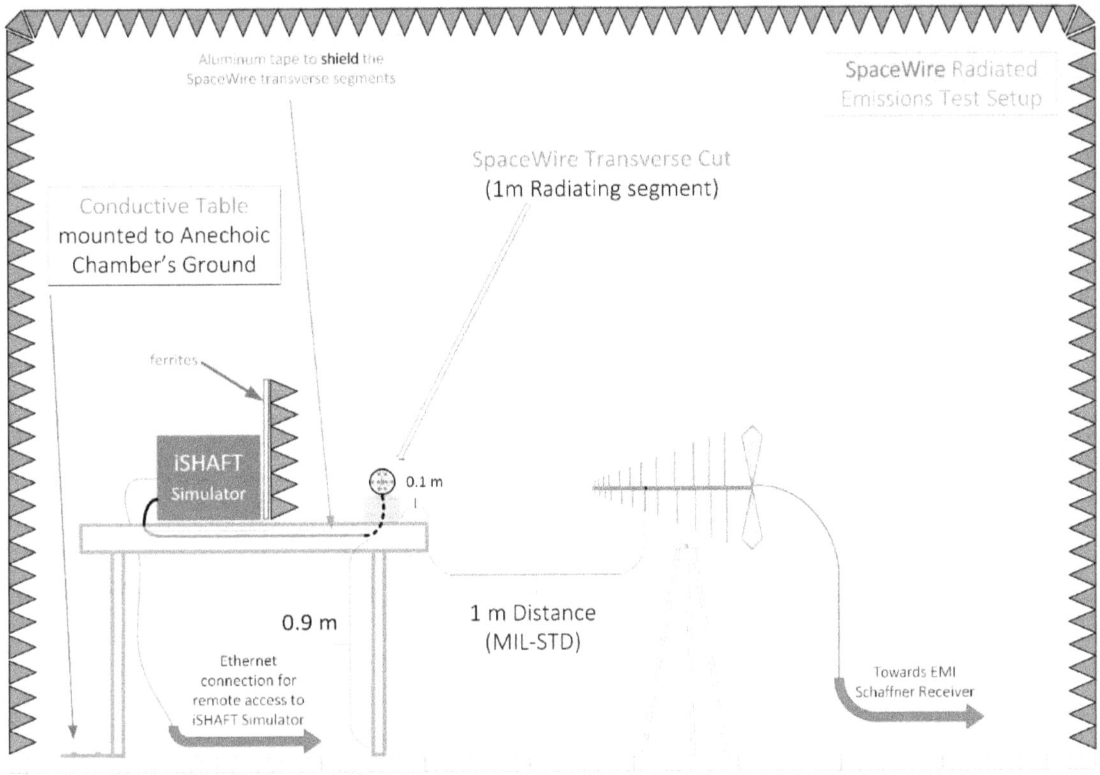

increase almost doubled the maximum emissions levels in both polarizations and the spectral spikes of the emissions are clearly distinguished. The contamination of the spectrum is occurring almost to the end of the measured frequency range (Baklezos et al., 2020).

Table 5. EMI Receiver Parameters

Step Frequency	IF Bandwidth	Detector	Frequency Sweep	Measure Time
100 kHz	120 kHz	Peak	30MHz -1 GHz	20ms

The last reported measured mode in Baklezos et al., (2020) is the one with the payload increased to 2048 bytes and the link speed set to 400Mbps. The measured radiated emissions for this mode are depicted in Figure 21. Data for both polarizations are included in the results for the frequency range 30 MHz - 1 GHz. More broadband emissions are present and they are smeared across the spectrum but they present slightly lower maximum levels when compared to the results of the smaller payload size measurement results. As the link speed increases it gets more and more difficult to isolate a quiet spectrum area (Baklezos et al., 2020).

Figure 18. Radiated Emissions Measurement Results (Mode: Data rate 10Mbps, payload 32 bytes, H/V Polarization) (Baklezos et al., 2020)

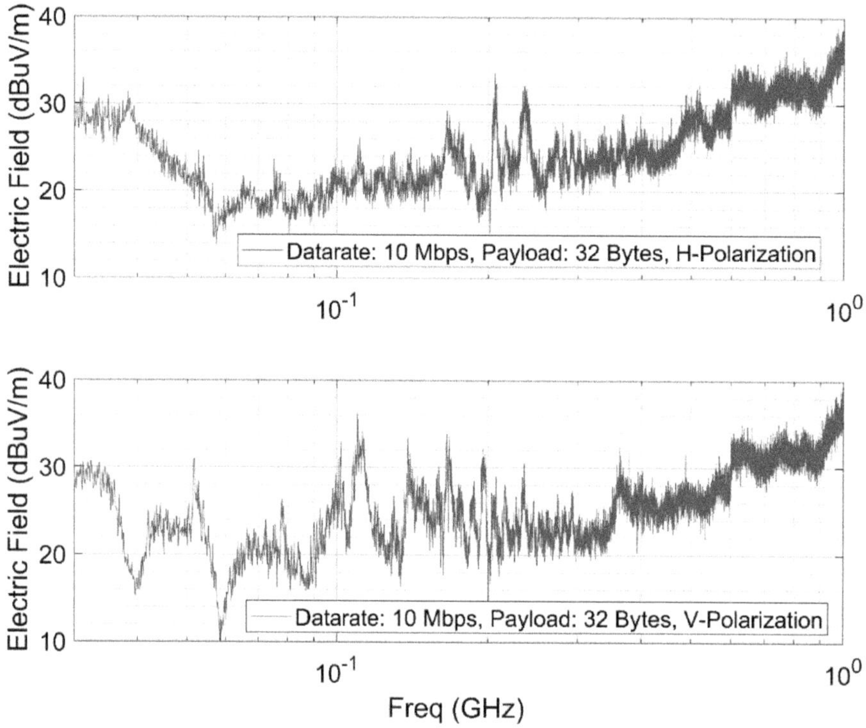

Figure 19. Radiated Emissions Measurement Results (Mode: Data rate 10Mbps, payload 2048 bytes, H/V Polarization) (Baklezos et al., 2020)

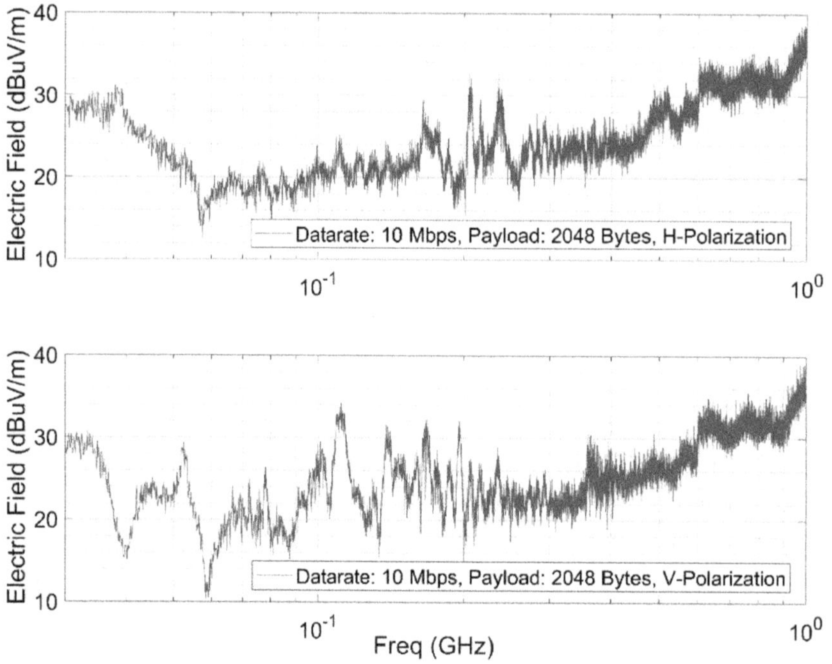

Figure 20. Radiated Emissions Measurement Results (Mode: Data rate 400Mbps, payload 32 bytes, H/V Polarization) (Baklezos et al., 2020)

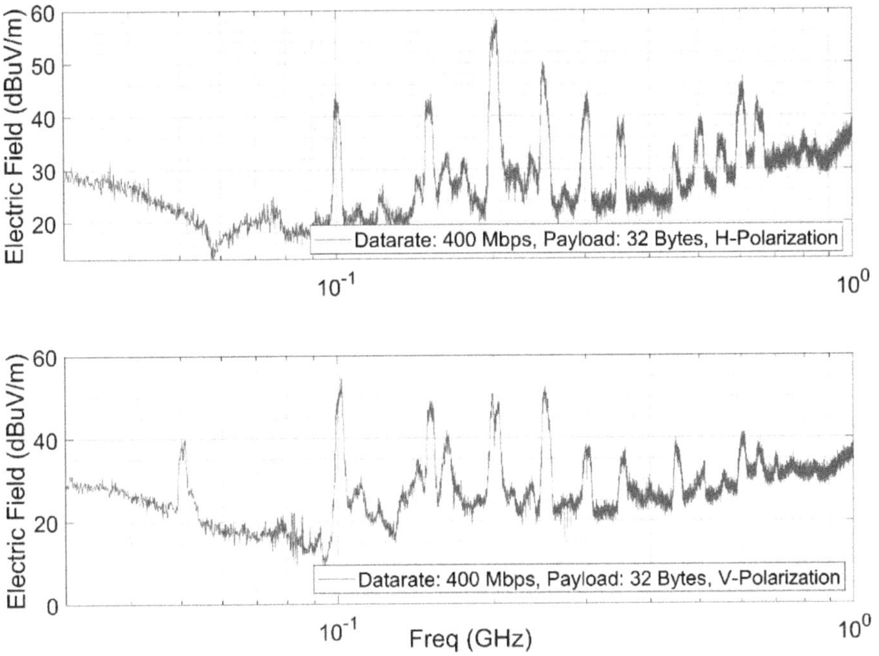

Figure 21. Radiated Emissions Measurement Results (Mode: Data rate 400Mbps, payload 2048 bytes, H/V Polarization) (Baklezos et al., 2020)

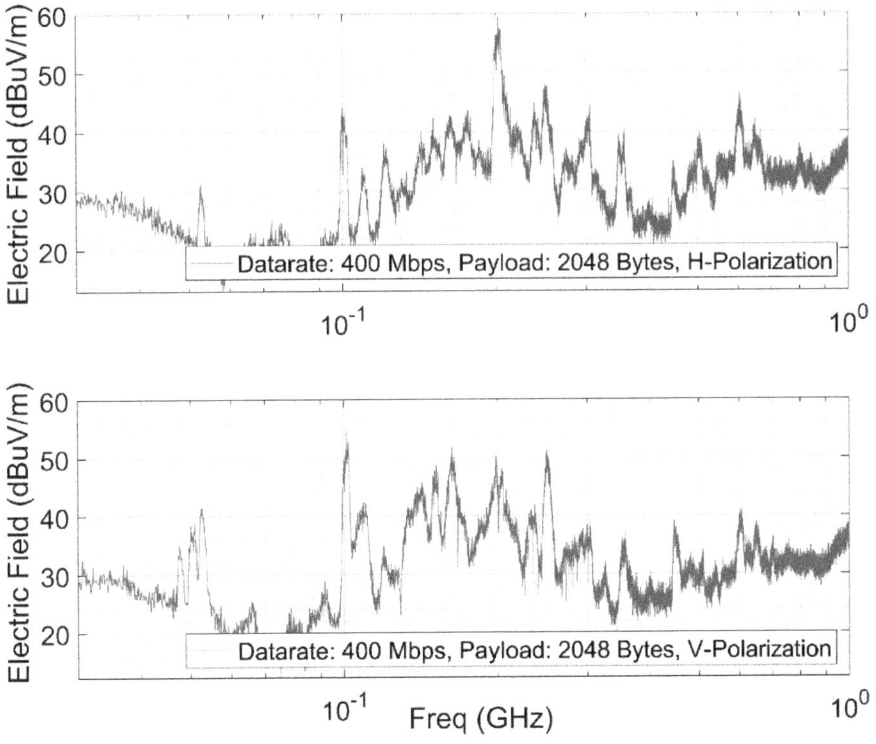

CABLE MODELS FOR EMC PURPOSES

Physical-Based Model

Authors of Treviso et al. (2019) developed and validated a physical-based model for frequencies up to 1GHz of the variant 01 of the SpaceWire cable terminated to two micro D-type connectors. This physical model consists of the SpW cable model and the connector model.

The cable model is built as a cascade of W-elements in HSPICE (Treviso et al., 2019), each element accounting for the various characteristics and effects. The first element is based on the per-unit-length parameters (as functions of frequency) of the untwisted versions of the wire pairs as estimated via a 2D solver based on the cable cross-section depicted in Figure 3 and considering all the shields solid. This element accounts for the dependence of the materials and phenomena like proximity effect, skin effect and shield transfer impedance. There is another element that accounts for the transition between the cross-sections of the cable and the connector. This transition consists of the cascade of two blocks with constant cross-sections. The two sections corresponding to the cable transitions have been characterized through their per-unit-length parameters computed via a 2D field solver. Two cascaded W-elements in HSPICE represent the resulting two sets of frequency-dependent per-unit-length matrices (Treviso et al., 2019).

CST MICROWAVE STUDIO is used to model the micro D-type connector (male) model in the frequency-domain. A vector-fitting approach (Treviso et al., 2019) is used to generate an equivalent lumped circuit based on the CST model results. The connector is electrically small for all frequencies under consideration, so the use of the lumped model is acceptable.

Overall, the cable and connector system model yields a netlist that can be employed by most commercial SPICE-like simulators. Authors (Treviso et al., 2019) point out that the circuit equivalent accounts for the frequency-dependence of materials and all interactions among the parts of the connector.

The physical-based model presented in (Treviso et al., 2019) is validated via Vector Network Analyzer (VNA) measurements for differential mode, common mode and mixed-mode scattering coefficients. The measurements are used to characterize the frequency domain response of the SpW physical layer. The aforementioned physical-based modeling of each of the components can replicate the measured scattering parameters.

Radiated Emissions Modelling

A different approach is to model the radiated behavior of the SpW cable based on emissions measurements as presented in Nikolopoulos et al. (2020). The core idea of the methodology is the representation of the cable by its shielding, ignoring the physical internal cable configuration, which then is treated as a travelling wave antenna. The concept was initially introduced in Baklezos et al. (2016). As a consequence, the radiated emissions are attributed to the current flowing on the shield. The estimation of this current on the cable shield across a frequency range of interest is achieved through a single standardized measurement, quite common in EMC (ECSS Secretariat, 2012). The methodology enables the prediction of EM emissions through the decomposition of the contributing phenomena. According to Baklezos et al. (2016), three contributing factors are considered: the current distribution $I(\omega)$, the dielectric properties of the ground and the reflected current at the cable and its image. The shield is considered a transmission line, and as such it may not be terminated properly. The model accounts for this possible improper termination of the shield with the addition of the reflected portion of the current distribution.

This is expressed with the complex reflection coefficient of the reflected current $\alpha = |\alpha| e^{j \cdot Arg(a)}$. (Nikolopoulos et al., 2020).

The source and image geometry are showcased in Figure 22 along with the designated angles needed for the calculation of the electric field. The ground plane in Figure 22 lies on the YZ plane while the top arrow indicates the current distribution (I) on the cable's outer shield. The bottom arrow indicates the image (I') of the current distribution. R_1 and R_2 are the distances of the measurements point P to the center of the cable length and its image, the distance from the measurement point to the origin is r and h is the height above the ground plane the cable is placed. The reflection coefficient Γ (image theory) is calculated from the equation 5:

$$\Gamma_{||} = \frac{\sin\psi - \sqrt{(\varepsilon_r - j\frac{\sigma}{\omega\varepsilon_0}) - \cos^2\psi}}{\sin\psi + \sqrt{(\varepsilon_r - j\frac{\sigma}{\omega\varepsilon_0}) - \cos^2\psi}} \tag{5}$$

Where μ is the relative permeability, εr is the relative permittivity and σ is the conductivity of the ground medium, εo is the permittivity of the vacuum, ω is the angular velocity corresponding to the frequency f and Ψ is the grazing angle (supplementary of the angle of incidence, between the incident wave and the ground plane) according to the formulation in Nikolopoulos et al. (2020).

The total electric field distribution is calculated by equation 6 considering the three independent contributing factors (Baklezos et al., 2016):

$$E_\gamma = \left[f(\gamma) + af'(\gamma)\right] I(\omega) \frac{j\omega\mu e^{-jkr}}{4\pi r}\left[e^{jkh\cos\theta} + \Gamma e^{-jkh\cos\theta}\right] \tag{6}$$

where $f(\gamma)$ and $f'(\gamma)$ are expressed by equations 7 and 8:

$$f(\gamma) = \sin\gamma \frac{1 - e^{-jkL(1-\cos\gamma)}}{jk(1-\cos\gamma)} \tag{7}$$

$$f'(\gamma) = \sin\gamma \frac{e^{jkL(1+\cos\gamma)} - 1}{jk(1+\cos\gamma)} \tag{8}$$

where k is the wavenumber, L is the length of the cable and γ is the angle between the ground plane and the line from the measurement point P to the origin O. The current distribution I(ω) and the complex α are calculated solving the inverse problem, iteratively for all frequencies, as presented in Figure 23, from the standard EMI/EMC measurement.

Figure 22. Electric field calculation over a ground plane (Nikolopoulos et al., 2020)

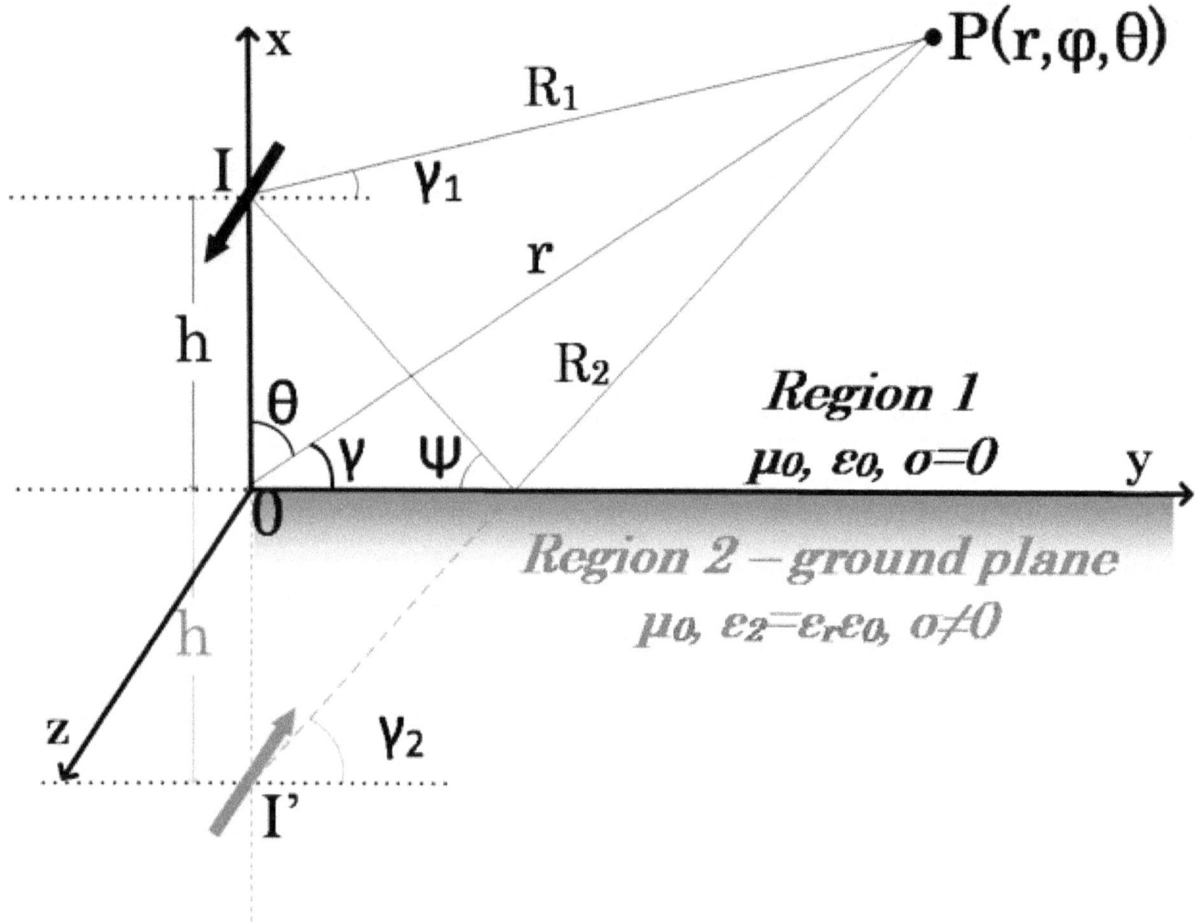

The initial step in the methodology is the measurement of the radiated emissions of the cable on a fixed height above a standard aluminum ground plane. From these experimental results, the proposed algorithm (Nikolopoulos et al., 2020) attempts to find an optimal model to match these measurements, which comprises of the solution values for the current distribution $I(\omega)$ and the complex α. These two independent optimized parameters synthesize the complete model of the cable. Given the current distribution and the complex α values that were calculated in the previous step, radiated emissions of the same system can be estimated, from equation 6, above any other ground (with known dielectric properties), placed at any height (optimized parameters of the extracted model) at a selected point P.

An iteration across the whole frequency range makes possible the prediction of electromagnetic emissions in a selected relative position and over a selected ground material (with known dielectric properties) completely ignoring the structural details of the cable. The model extracted with this methodology provides a safe estimation for the maximum emission levels over a wide frequency span (Baklezos et al. 2016).

In the aforementioned scheme, for each frequency, a genetic algorithm is used to match the estimated value of equation 6 with the measured value at point P, minimizing the objective function of equation 9 (Nikolopoulos et al., 2020).

Figure 23. Modeling methodology flowchart (Nikolopoulos et al., 2020)

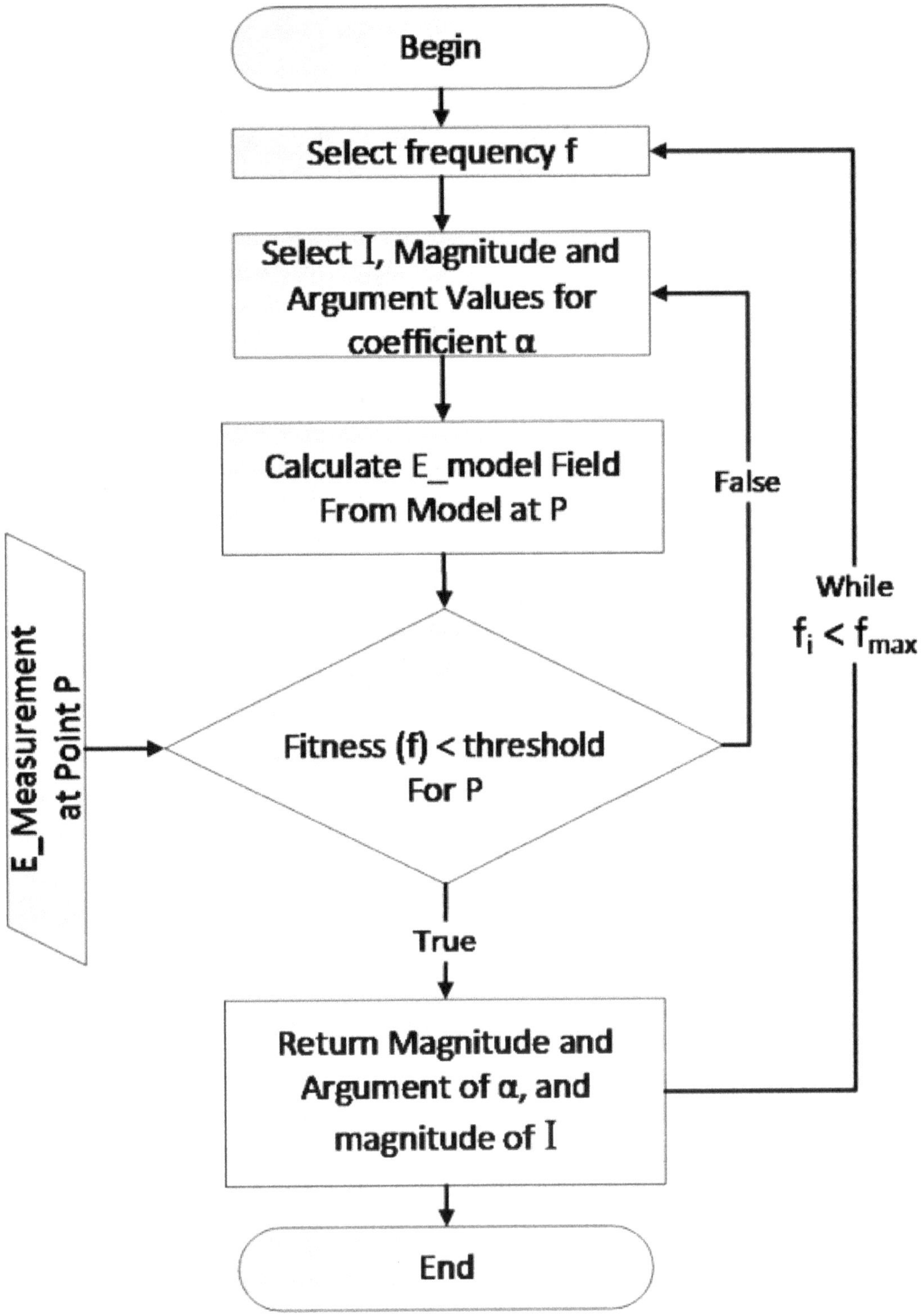

$$f_{fitness} = \left|1 - \frac{E_{\text{model}}}{E_{\text{measurement}}}\right|^2 \tag{9}$$

Authors in Nikolopoulos et al. (2020) applied the methodology to a 2.5 meters-long SpW cable of AWG 28 - constructed and wired according to ECSS-E-ST-50-12C, with one-meter transverse to the antenna direction. Two arbitrary function generators were used to feed the cable so the full-duplex operation was emulated. To simulate the operation of the SpW/LVDS link, a small PCB, acting as a complete LVDS interface, was implemented to drive the signal from the 50 Ω output of the generator to the 100 Ω SpW cable. This LVDS driver device was fed by the Agilent 81130A outputs with the Data and Strobe signals and provided LVDS output. Additionally, this driver had 100 Ω termination resistances to terminate signals fed from the other side of the SpW/LVDS link Authors employed two shielded enclosures to include the PCB of the LVDS driver on each side of the link along with their corresponding DC power sources to minimize the PCB emissions interference to the measurement and to increase the immunity of the circuitry of the driver.

Authors of Nikolopoulos et al. (2020) initially measured the radiated emissions of the cable routed over an aluminum ground for two pulse-rates 100 and 200 MHz. They report that all measurement results were denoised, modeling only emissions greater than a specific threshold noise level equal to 10 dBuV/m. The threshold was employed to avoid modeling noise values and to greatly reduce the modeling time since the time is linearly proportional to the number of frequency bins investigated. To detect this threshold level, authors benchmarked the environment noise with all equipment active excluding the cable, before the actual in the anechoic chamber. The maximum registered noise value was set as the threshold. This way only emissions originating from the cable are registered during the measurements (Nikolopoulos et al., 2020).

Measurements in Nikolopoulos et al. (2020) were conducted following ECSS-E-ST-20-07C Rev. 1: Space Engineering – Electromagnetic Compatibility (ECSS Secretariat, 2012) to maximize the reliability and the repeatability of the measurement results. However, some discrepancies were observed during the measurement campaign. A possible source of these deviations is that the SpW cable, as per ECSS-ST-50-12, had its inner shields terminated by pairs at pin 3 of the D-type connector. This is proved to be a not optimal setup and the recommendation is that all shields are terminated to the backshell and pin 3 at both ends of the connector. Another possible contributor to these deviations at emissions, beyond the normal EMC margin, can be traced to the 9-pin micro-miniature D-type connector specified in the SpaceWire standard that is not designed for balanced signals and causes emissions especially on higher frequencies.

The performance of the micro D-type connector is a known issue. To overcome that, Axon' Cable was selected to carry out the development of a new, compact impedance-matched SpaceWire connector under an ESA Technology Research Project. The findings of this project are presented in Enouf &Hemant, (2016). The authors presented a compact, impedance-matched connector with a desirable size coupled with significantly improved electrical and EMC performances, namely "MicroMach". A trade-off was made concerning the cable shield termination to the connector backshell (not full 360° screening) to minimize the size. Preliminary measurements are reported for bonding resistance between male & female contacts less than 5 mΩ, Near End and Far End Crosstalk (NEXT/FEXT) less than -50 dB up to 1 GHz, and Return loss less than -20 dB up to 1 GHz. The new MicroMach SpaceWire connector offers, at a

Figure 24. Field Emissions Comparative Results (Modelled vs Measured) (Pulse Rate 100MHz) (Nikolopoulos et al., 2020)

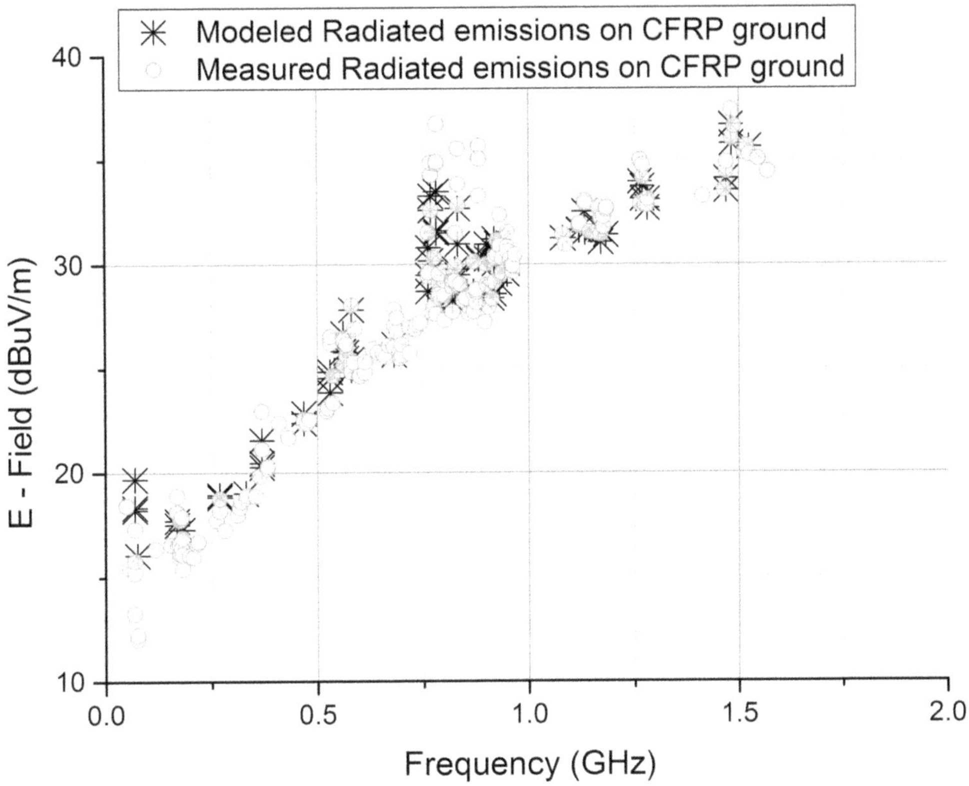

size marginally larger than the current micro-D, significantly improved performance in data rate, EMC and crosstalk compared to any of the current market solutions (Enouf &Hemant, 2016).

The proposed methodology treats the whole cable assembly is treated as a unit, not separating the connectors, which are normally one of the major causes of EMC emissions (Nikolopoulos et al., 2020). From the results of these measurements and according to the presented algorithm (Figure 23), the current distribution along with the parameter α for measurement point P are calculated. The dielectric properties of the ground plane materials that were used to calculate Γ in each case are tabulated in Table 6. The Genetic Algorithm (GA) procedure indicates that the magnitude and argument of complex coefficient α for the two ground planes optimized at the mean values of 0.726 and 76.890 for pulse-rate of 100 MHz and 0.712 and 62.020 for pulse-rate 200 MHz respectively (Nikolopoulos et al., 2020).

Table 6. Dielectric Characteristics of Ground Materials (Leininger et al., 2012)., (Akhtar et al., 2006).

Ground Materials	εr	σ (S/m)	μ
Aluminum	1	$38x10^6$	1
CFRP	2530	28.3	1

Figure 25. Field Emissions Comparative Results (Modelled vs Measured) (Pulse Rate 200MHz) (Nikolopoulos et al., 2020)

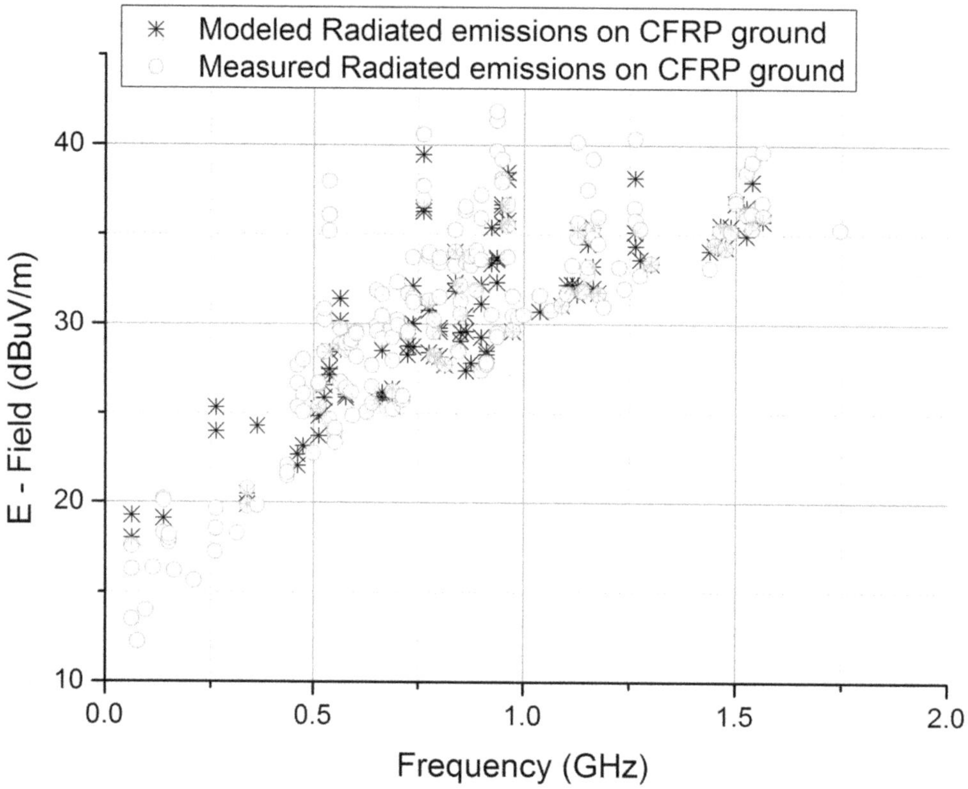

After the model parameters are extracted based on the measurements above the aluminum ground, the radiated emissions for the case of the cable above the CFRP ground plane were predicted using equation 6 and the Γ coefficient as adjusted accordingly from the dielectric characteristics of Table 6. The results of the modeled radiated emissions are depicted in Figure 24 for the 100 MHz pulse-rate and in Figure 25 for the 200 MHz pulse-rate against measurements. These figures indicate the good agreement of model and measurement results, a result quite satisfactory for a fast EMC pre-compliance method, especially when taking into account the high uncertainty (~ 5dB) involved in the measurements. Moreover, the capability of the methodology in providing a model that produces safe estimation regarding the radiated emissions level is found satisfactory especially over the wide frequency range. This fact is enhanced given the methodology requirement for only a single point measurement over a common ground material such as aluminum and the especially complex structure of the cable (Nikolopoulos et al., 2020).

Radiated Emissions Modified Modelling

Authors of Baklezos et al. (2020), adopted the modeling of Nikolopoulos et al. (2020) in the representation of the cable shield as a traveling wave antenna. Although the overall modeling rationale remains the same, work in Baklezos et al., (2020), differentiates from the one in Nikolopoulos et al. (2020) in the description of the current distribution. Instead of the distinct treatment for each frequency bin monochromatically which yields the current distribution, authors in Baklezos et al., (2020), employ a more realistic current description composed of pulses given the pulsed nature of the LVDS signal. They consider 6 pulses injected to the outer shield due to intra-pair and inter-pair skew of the DS encoding (Figure 26). These injected multiple pulses form a current distribution on the shield which in turn is the cause of the cable's radiation Additionally, the solving approach in this modeling methodology is enhanced considering the whole spectrum, instead of the previous approach: bin by bin. This way a single inverse problem is solved heuristically (via Differential Evolution - DE) as opposed to the numerous problems (depending on the frequency bins/frequency resolution) of the previous methodology. (Baklezos et al., 2020).

Figure 26. SpW Intra-pair and Inter-pair Skew effects (Baklezos et al., 2020).

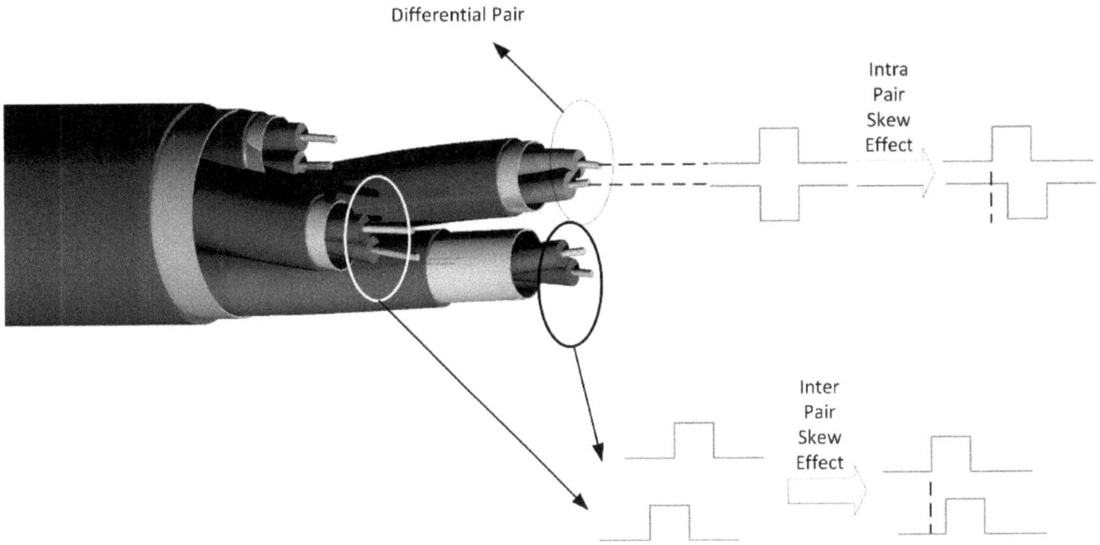

Equation 10 expresses the radiated emissions taking into account all the contributing phenomena such as the reflection on the ground via the coefficient Γ, the full-duplex transmission scheme via the terms on the first bracket, the cable's height -*h* and orientation via the angular coefficient γ).

$$E_{\gamma}(\omega)=\left[sin\gamma\frac{1-e^{-jkL(1-cos\gamma)}}{jk(1-cos\gamma)}+sin\gamma\frac{e^{jkL(1+cos\gamma)}-1}{jk(1+cos\gamma)}\right]\cdot I(\omega)\cdot\frac{j\omega\mu e^{-jkr}}{4\pi r}\cdot\left[e^{jkhcos\theta}+\Gamma e^{-jkhcos\theta}\right] \quad (10)$$

Authors in Baklezos et al. (2020) modeled the current distribution on the outer shield of a SpaceWire cable when the link operated in the 400 Mbps data-rate and 2048 bytes payload mode for the vertical polarization measurement (Figure 21). The flowchart of the exact methodology is depicted in Figure 27.

The current distribution spectrum is extracted from the measured electric field solving equation 10 and depicted in Figure 28. This current is considered the source of the radiation $I_{meas}(\omega)$ and expressed as the sum of the six (4 intra / 2 inter pair skew) pulses according to equation 11.

Figure 27. Algorithm flowchart (Baklezos et al., 2020)

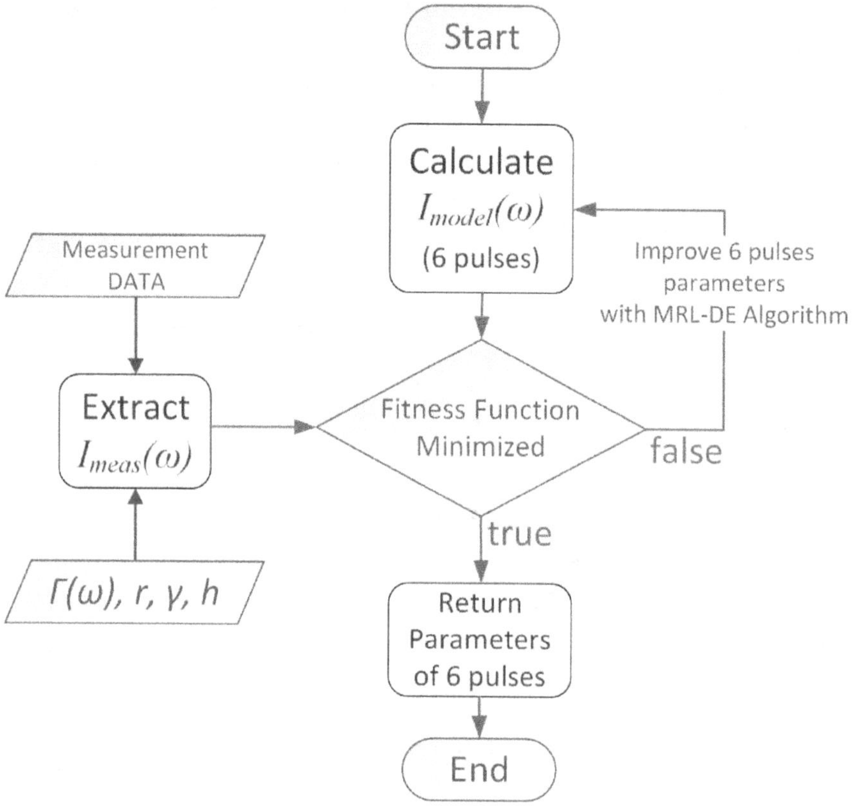

$$I(\omega)_{multipletrap} = \left| \sum_{i=1}^{N} \left\{ -j\frac{A}{2\pi n} e^{-\frac{jn\omega_0(\tau_i+\tau_{r_i})}{2}} \left(\frac{\sin\left(\frac{1}{2}n\omega_0\tau_{r_i}\right)}{\frac{1}{2}n\omega_0\tau_{r_i}} e^{\frac{jn\omega_0\tau_i}{2}} - \frac{\sin\left(\frac{1}{2}n\omega_0\tau_{f_i}\right)}{\frac{1}{2}n\omega_0\tau_{f_i}} e^{-\frac{jn\omega_0\tau_i}{2}} \right) e^{-j2\pi f t_{pi}} \right\} \right|$$

Table 7. Pulses Solution Parameters – Extracted Model (Baklezos et al., 2020)

# Pulse	Duration (s)	Trise (s)	Tfall (s)	DC Amplitude (A)	Occurrence Time (s)
1	2.73e-8	7.53e-9	5.32e-9	0.2	5e-6
2	0	1.22e-7	1.63e-7	0.21	4.9e-6
3	9.97e-22	2.5e-6	7.3e-8	0.57	5e-6
4	4.35e-11	4.17e-9	9.44e-9	0.47	4.9e-06
5	6.2e-8	2.56e-9	2.03e-8	0.097	5e-6
6	1.06e-9	5.41e-9	9.02e-9	0.59	4.97e-6

Figure 28. I_{meas} and I_{model} Comparison (Baklezos et al., 2020).

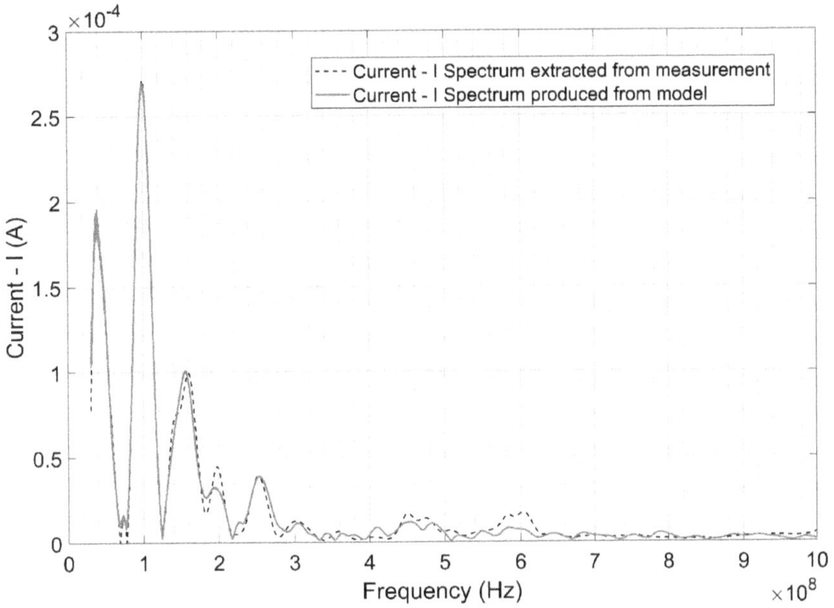

(11)

where $n=f \cdot T$ and $\omega 0\_2\pi/T$, T *is* the period of the pulse in seconds and f *is* the frequency in Hz and Ai the pulse amplitude in amperes. The parameters of the six pulses (amplitude – Ai, occurrence time - t_{p_i}, rise time - τ_{r_i}, rise time - τ_{f_i}, pulse width τi_j are extracted heuristically with the aid of the differential evolution algorithm and tabulated in Table 7. These parameters constitute the model representing the measured mode of operation of 400 Mbps and 2048 bytes payload size (Baklezos et al., 2020).

Next, authors in Baklezos et al., (2020) used this pulse model to reconstruct the current distribution $I_{model}(\omega)$ as depicted in Figure 28. The current $Im_{odel(}\omega)$ is in very good agreement with the one fed to the algorithm (the envelope of the current as calculated from the field measurements). Lastly, the model current distribution $Imo_{del(\omega}$) is used to estimate the electric field emissions with the aid of equation 10.

The reported results for the field values produced by the model prediction, as depicted in Figure 29, are in good agreement with the results of the measurement. Specifically, in the lower frequencies range (up to 300 MHz) an almost excellent agreement is clear. The maximum emission levels are near perfectly predicted by the model, which is critical for worst-case EMC methodologies. For the rest of the frequency range under study, there are deviations up to even 10dB. Overall, across the whole measured spectrum, over three-quarters of all field values lie within the 6dB margin of expanded uncertainty. The rest that exceeds this margin can be potentially attributed to other radiation phenomena that are currently not included in the model implemented by the authors (Baklezos et al., 2020). These sources of high-frequency emissions however are expected to have limited impact in lower data rate modes, so authors of Baklezos et al. (2020) are confident that the accuracy of the methodology will increase the other modes of operation as the spectral content will be limited to the lower frequency regions.

Figure 29. Electric field Comparison, predicted using I_{model}, Measured using measurements envelope (Baklezos et al., 2020).

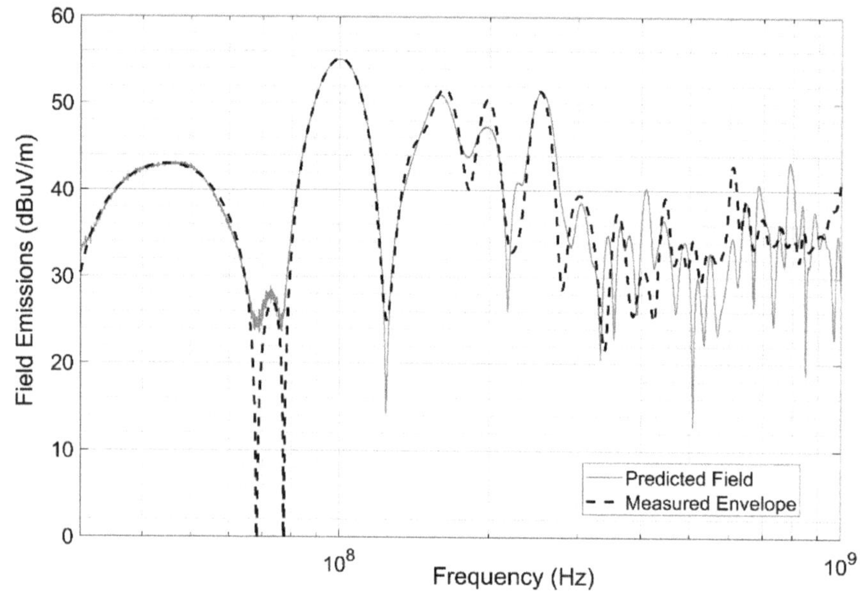

FUTURE RESEARCH DIRECTIONS

Space missions are an active field in engineering. The SpaceWire ecosystem is full of new challenges. An issue that is worth the focus is the research on the applicability and effectiveness of Galvanic Isolation of SpaceWire as it promises reduced sensitivity to common-mode noise compared to LVDS. Another relative issue worth exploring is DC balanced SpaceWire. A Direct-Current line-balanced interface provides the ability to isolate the physical layer to achieve common-mode voltage rejection or even the complete galvanic isolation. Moreover, it could minimize the number of conductors and transceivers by eliminating the need for a Strobe line. Another important direction is the dreaded 10 m link length limitation. While this length is sufficient for the actual missions, it hinders the efforts at the verification stages since it proves to be a limiting factor at system level measurements. So work is underway to surpass this limitation. At another front, in the SpaceWire ecosystem a new protocol is introduced, the SpaceFibre, which supports both electrical and fiber-optic cables. It was developed to cover the need for higher and higher data rates offering multi-Gigabit/s speeds. SpaceFibre was standardized in 2019 as ECSS-E-ST-50-11C. Another important aspect to consider is the use of SpaceWire in robotics systems in space. Current SpaceWire standard is not tailored to robotic use, for instance, the cables and connectors. Research that contributes to any of those areas will greatly affect the future of the field of space missions.

CONCLUSION

SpaceWire is a point-to-point bit shipping protocol using LVDS as the physical layer. Given the wide use of the SpaceWire, due to the standardization, it is near impossible to work or study in the space mission field without some level of familiarity with SpaceWire. The main objective of the chapter is to discuss the SpaceWire standard ECSS-E-ST-50-12C and ECSS-E-ST-50-12C Rev.1 from an EMC engineer's point of view, with a special focus on modeling methodologies aiming at the Electromagnetic Compatibility and Immunity aspects. This chapter describes the SpaceWire standard, focusing on the levels that play a major role in the electromagnetic behavior of the system and concern cable assemblies, shielding, bonding and grounding. Measured radiated emissions that can possibly affect spacecraft components are presented along with other EMC/EMI issues that may have an impact on the system performance. The levels of the standard that define the EMC behavior of the system were highlighted. Engineers or scientists regardless of their approach to the standard, theoretical or practical, can only gain from the information herein. Students interested in space missions can also benefit from the concentrated information found in this chapter. Finally, by including a section in this chapter that presents various radiated emissions measurement results and discussing recent modelling methodologies authors hope to provide the readers with a more complete-all-around-understanding of the system as well as some initial tools to perform preliminary assessments.

REFERENCES

Akhtar, M. J., Feher, L. E., & Thumm, M. (2006). A waveguide-based two-step approach for measuring complex permittivity tensor of uniaxial composite materials. *IEEE Transactions on Microwave Theory and Techniques*, *54*(5), 2011–2022. doi:10.1109/TMTT.2006.873623

Baklezos, A. T., Nikolopoulos, C. D., Capsalis, C. N., & Tsatalas, S. (2017, March). Effect of LVDS link speed and pattern length on spectrum measurements of a Spacewire harness. In *Antenna Technology: Small Antennas, Innovative Structures, and Applications (iWAT), 2017 International Workshop on* (pp. 38-41). IEEE. 10.1109/IWAT.2017.7915291

Baklezos, A. T., Nikolopoulos, C. D., Katsouris, A. G., Koutantos, G. I., & Capsalis, C. N. (2016). Electromagnetic emission modeling in case of shielded cabling with respect to the ground dielectric properties. *IEEE Transactions on Electromagnetic Compatibility*, *58*(6), 1694–1700. doi:10.1109/TEMC.2016.2588583

Baklezos, A. T., Nikolopoulos, C. D., Sigalas, I. R., & Capsalis, C. N. (2020). Measurement and Modeling of SpaceWire Radiation for Electromagnetic Compatibility Assessment. In *2020 IEEE International Instrumentation and Measurement Technology Conference (I2MTC)* (pp. 1-5). IEEE. 10.1109/I2MTC43012.2020.9128502

Cook, B., & Walker, P. (2007). Reducing Electromagnetic Emissions From SpaceWire. In *DAta Systems in Aerospace*. DASIA.

Duffy, A. P., Martin, A. J., Orlandi, A., Antonini, G., Benson, T. M., & Woolfson, M. S. (2006). Feature selective validation (FSV) for validation of computational electromagnetics (CEM). part I-the FSV method. *IEEE Transactions on Electromagnetic Compatibility*, *48*(3), 449–459. doi:10.1109/TEMC.2006.879358

ECSS-E50-12A vs ECSS-E-ST-50-12C, what's the difference? | STAR-Dundee. (2017). *Star-dundee.com*. Retrieved 15 June 2017, from https://www.star-dundee.com/knowledge-base/ecss-e50-12a-vs-ecss-e-st-50-12c-whats-difference

Enouf, K., & Hermant, S. (2016, October). Compact, impedance-matched SpaceWire connector development—"MicroMach SpaceWire": SpaceWire components, long paper. In SpaceWire Conference (SpaceWire), 2016 International (pp. 1-7). IEEE.

European Cooperation for Space Standardization (ECSS) Secretariat. (2008). *ECSS- E-ST-50-12C, Space Engineering – SpaceWire- Links, nodes, routers and networks, ESA-ESTEC Noordwijk, The Netherlands, Requirements & Standards Division European Cooperation for Space Standardization (ECSS) Secretariat (2003), ESCC Detail Specification No. 3902/003, CABLE, "SPACEWIRE", ROUND, QUAD USING SYMMETRIC CABLES, FLEXIBLE, -200 TO +180 °C, ESA-ESTEC.* Requirements & Standards Division.

European Cooperation for Space Standardization (ECSS) Secretariat. (2012). *ECSS-E-ST-20-07C Rev. 1, Space Engineering –Electromagnetic compatibility, ESA-ESTEC Noordwijk.* Requirements & Standards Division.

European Cooperation for Space Standardization (ECSS) Secretariat. (2014). *ESCC Detail Specification No. 3902/003, CABLE, "SPACEWIRE", ROUND, QUAD USING SYMMETRIC CABLES, FLEXIBLE, -200 TO +180 °C, ESA-ESTEC.* Requirements & Standards Division.

European Cooperation for Space Standardization (ECSS) Secretariat. (2019). *ECSS- E-ST-50-12C Rev.1, Space Engineering – SpaceWire- Links, nodes, routers and networks, ESA-ESTEC Noordwijk.* Requirements & Standards Division.

Karlsson, I. (2015, May). Common mode effects in high speed serial links. In *2015 IEEE 19th Workshop on Signal and Power Integrity (SPI)* (pp. 1-4). IEEE. 10.1109/SaPIW.2015.7237394

Leininger, M., Thurecht, F., Pfeiffer, E., & Ruddle, A. (2012, May). Advanced grounding methods in the presence of carbon fibre reinforced plastic structures. In *Aerospace EMC, 2012 Proceedings ESA Workshop on* (pp. 1-6). IEEE.

Lellouch, G., Mishra, A. K., & Inggs, M. (2016). Design of OFDM radar pulses using genetic algorithm based techniques. *IEEE Transactions on Aerospace and Electronic Systems*, *52*(4), 1953–1966. doi:10.1109/TAES.2016.140671

Nicoletto, M., Boschetti, D., & Savi, P. (2014, September). High Speed Digital Lines routed on non-metallic Spacecraft structures. In *Electromagnetic Compatibility (EMC Europe), 2014 International Symposium on* (pp. 785-789). IEEE. 10.1109/EMCEurope.2014.6931011

Nikolopoulos, C. D., Baklezos, A. T., Tsatalas, S., & Capsalis, C. N. (2020). Verification of Radiated Emissions Modeling for SpaceWire/LVDS Links Routed on CFRP Ground. *IEEE Transactions on Aerospace and Electronic Systems*, *56*(1), 393–402. doi:10.1109/TAES.2019.2914540

Orlandi, A., Duffy, A. P., Archambeault, B., Antonini, G., Coleby, D. E., & Connor, S. (2006). Feature selective validation (FSV) for validation of computational electromagnetics (CEM). part II-assessment of FSV performance. *IEEE Transactions on Electromagnetic Compatibility*, *48*(3), 460–467. doi:10.1109/TEMC.2006.879360

Parkes, S. (2012). SpaceWire User's Guide. STAR-Dundee Limited.

Roberge, V., Tarbouchi, M., & Labonté, G. (2018). Fast genetic algorithm path planner for fixed-wing military UAV using GPU. *IEEE Transactions on Aerospace and Electronic Systems*, *54*(5), 2105–2117. doi:10.1109/TAES.2018.2807558

SpaceWire - The Standard. (2017). *Spacewire.esa.int*. Retrieved 15 June 2017, from http://spacewire.esa.int/content/Standard/Standard.php

Telecommunications Industry Association. (2001). *ANSI/TIA/EIA-644, Electrical Characteristics of Low Voltage Differential Signaling (LVDS) Interface Circuits*. Standards and Technology Department.

Treviso, F., Trinchero, R., & Canavero, F. G. (2019, May). Validation of a Physical-Based Model for a Spacewire Cable. In *2019 ESA Workshop on Aerospace EMC (Aerospace EMC)* (pp. 1-6). IEEE. 10.23919/AeroEMC.2019.8788945

U.S.A. Department of Defence. (2015). *MIL-STD-461G - Requirements for The Control of Electromagnetic Interference Characteristics of Subsystems and Equipment*. Author.

KEY TERMS AND DEFINITIONS

AWG: American wire gauge.

Data Strobe Encoding (D-S): encoding scheme in which a sequence of data bits and clock is encoded as the original Data bit sequence, together with another bit sequence (Strobe) which changes state whenever the Data bit sequence does not. The clock can be recovered by performing the XOR function on the Data and Strobe.

Electromagnetic Compatibility (EMC): The ability of a system to function satisfactorily in its electromagnetic environment without introducing intolerable electromagnetic disturbances to anything in that environment.

Electromagnetic Immunity: The ability of a system to maintain its proper functionality under exposure to various sources of electromagnetic disturbance.

Jitter: Random errors in the timing of a signal.

Lay-Length: Number of twists per length unit expressed as the length between one complete turn of a single end in the cable.

Shielding: A method for the mitigation of electromagnetic disturbances which is based on the concept of creating an electrically continuous enclosure around the under-protection structure, mainly by the use of conductive materials.

Skew: Difference in time between the edges of two signals which should ideally be concurrent.

SpaceWire: A standard that specifies the physical interconnection media and data communication protocols to enable the reliable sending of data at high-speed (between 2 Mb/s and 400 Mb/s) from one unit to another. SpaceWire links are full-duplex, point-to-point, and serial data communication links.

Chapter 3
Advances in Electromagnetic Environmental Shielding for Aeronautics and Space Applications

Rafael Vargas-Bernal
https://orcid.org/0000-0003-4865-4575
Instituto Tecnológico Superior de Irapuato, Mexico

ABSTRACT

Electromagnetic environmental shielding is one of the main research topics for the development of aeronautical and space applications. Numerous research groups around the world study the problems that space systems and astronauts experience when these are subjected to space radiation. Despite the progress made so far, different proposals of advanced materials have been continuously proposed throughout the history of space career to protect space systems and astronauts against the solar particle events (SPE), cosmic rays galactic (GCRs), and proton-electron radiation (PERs). This chapter presents the recent advances made about space environmental shielding and that have been reported so far to visualize the future perspectives that this type of research must carry out so that future space voyage is completely reliable for space systems and astronauts. This research area is fully current, and its experimental success will depend on the work done by all space researchers and professionals.

INTRODUCTION

The space environment has been considering commonly as a vacuum; however, it is composed of innumerable constituents such as photons, particle radiation, meteorites, neutral and ionized molecular and atomic interactions, variable thermal environment, and orbital debris, which regularly produce a decrease of performance or even catastrophic failures to aerospace systems (Piscane, 2008). The natural space environment is composed of the neutral thermosphere, thermal environment, plasma, meteorites, and orbital debris, solar environment, ionizing radiation, magnetic field, gravitational field, and mesosphere.

DOI: 10.4018/978-1-7998-4879-0.ch003

In addition, the relative impact of the environmental effects of space have been determined according to the orbit in which the space system is located (Calders, 2018):

- Low Earth orbit (LEO) (Kleiman, 1995),
- Middle earth orbit (MEO),
- Geosynchronous equatorial orbit (GEO),
- International space station (ISS),
- Earth observation system (EOS),
- Tropical rainfall measuring mission (TRMM),
- Global positioning system (GPS), or
- The Defense satellite communications system (DSCS).

The effects considered in the aerospace environment are direct sunlight, gravity field, magnetic field, trapped radiation, solar particle events, galactic cosmic rays, orbital debris, meteorites, ionosphere, spacecraft electric charge, and neutral atmosphere (Piscane, 2008). According to the impact that these effects of the space environment have on a space application, NASA established a relative scale that ranges from zero to ten. This scale establishes that:

- the effect can be ignored (value of 0),
- the effect may or will cause upsets (value of 1 or 2, respectively),
- the effect may or will require design changes (value of 3 or 4, respectively),
- the effect may or will reduce mission effectiveness (value of 5 or 6, respectively),
- the effect may or will shorten the mission (value of 7 or 8, respectively),
- Moreover, the effect may or will deny the mission (value of 9 or 10, respectively).

The effects of the aerospace environment have had a significant impact on the performance of the following subsystems of a space shuttle, space stations, satellites, or aircraft: determination and control of altitude, avionics, electrical power systems, environmental control and life support, guided navigation and control, instrumentation, materials and/or structure, thermal control, as well as in telemetry, tracking, and communications.

Any system or living being sending to a space mission will be exposed to harmful radiation in the form of energetic particles from solar and galactic sources (Letfullin, 2019). The three main contributors to space radiation are galactic cosmic rays (GCRs), solar energy particles (SEPs), and bremsstrahlung X-rays (BXRs). Cosmic ray particles beyond the Earth's magnetosphere are composed of 85% protons, 14% alpha particles, and 1% heavier nuclei with energies in the range of billions of electron volts. SEPs are making up protons and electrons coming from the Sun. Heavy nuclei are composed of light elements such as lithium and beryllium. The charged particles within the solar system are releasing from the upper atmosphere of the sun. The solar wind is made up of electrons, protons, and alpha particles with energies ranging from 1.5 to 10 KeV (Letfullin, 2019). Also, these particles travel in the speed range from 400 to 750 Km/s. Bremsstrahlung X-rays are basing on photons emitted from an electron collision with a proton during their random thermal movement inside a hot gas, such as the sun's atmosphere. Besides, these can be generating through the braking of high-energy charged particles by their collision with spacecraft shielding material. Both charged particles and X-ray photons can destroy the electronic components of space systems and causing biological damage to DNA, possibly leading to the presence of cancer in astronauts (Geng, 2015. Traditionally, these systems and their occupants are protecting through some configuration of a massive shielding material that absorbs the energy of the particles that enter them. In the case of high energy galactic cosmic rays (GCRs), passive shields are too massive to be practical and will likely produce secondary radiation showers that may even be more damaging than the GCRs themselves (Kamsali, 2019; Warden, 2019). Active shielding involves very large magnetic and/or electric fields being adding to a spatial design to deflect energy particles that reduce or eliminate these effects (Tripathi, 2008). Numerous groups of researchers around the world are investigating possible solutions through multidisciplinary approaches to the development of these shielding systems (Spillantini, 2000; Jang, 2016; Kartashov, 2018; Rawal, 2018; Nenarokomov, 2019; Samwel, 2019; Shuvalov, 2020). Materials based on metals, polymers, ceramics, and composites can be used to develop electromagnetic shielding. Some examples of these materials are those used to block neutrons and gamma rays (Bel, 2019), only neutrons (Cataldo, 2019), and electromagnetic interference (Fayazbakhsh, 2012; Liu, 2019; Tishkevich, 2019; Wanasinghe, 2019; Zhou, 2019). The introduction of nanomaterials is a driving force for future generations of materials that will be using in future generations of shielding solutions.

The purpose of this chapter is to describe and analyze the evolution that the materials used to offer electromagnetic shielding have experienced since the first space system launched in 1957. This work is intending to serve two audiences: first, those studying undergraduate engineering courses and graduates related to the design and engineering of spacecraft, and second, those experts in the field of the space environment. The chapter begins by describing the problems associated with the space environment. Subsequently, the history of the different materials used to allow electromagnetic shielding is discussing. Next, the impact of these materials on the design of space systems is analyzing. The prospects of engineering materials that must be developing in future decades are discussing later. Finally, conclusions related to this work are given.

Background

The space environment defines the atmosphere that surrounds or circulates a body in space. The study of this atmosphere seeks to understand and direct the existing conditions that affect the design, development, and operation of the systems or devices sent to space, as well as the effects on living beings. The space environment involves naturally occurring phenomena such as atomic oxygen (AO) and atmospheric density, ionizing radiation, plasma, etc. and man-made factors such as orbital debris (Herr, 1994). The natural space environment is not static, multiple, and different events, with either relevant or irrelevant effects, can occur in an almost unpredictable manner. Therefore, the science and engineering associated with the space environment are constantly changing given this susceptibility to environmental factors. Understanding the effects of natural space environments and manufactured factors makes it possible to establish a management program that more effectively optimizes aspects of space missions such as risk, cost, quality, weight, verification, science, and technology.

Earth's atmosphere is a set of layers of gases commonly known as air, which surrounds planet Earth and is retaining by Earth's gravity (Herr, 1994). It protects life on Earth by creating pressure, which allows liquid water to exist on the Earth's surface, absorbs ultraviolet solar radiation, reheats the Earth's surface through heat retention, and reduces the extremes of temperature between day and night. Dry air contains by volume: 78.09% nitrogen, 20.95% oxygen, 0.93% argon, 0.04% carbon dioxide, and small amounts of other gases. Its mass is 5.15×10^{18} Kg, three-quarters of which are within about 11 Km of the surface. Besides, the air contains a variable amount of water vapor, the average values being: 1%

at sea level and 0.4% over the entire atmosphere. Its composition, temperature, and pressure vary with altitude. This becomes thinner and thinner with increasing altitude, without a defined boundary between the atmosphere and outer space. The Kármán line located at 100 Km altitude is frequently using as the boundary between the atmosphere and outer space. The atmospheric effects are noticeable during the re-entry into the atmosphere at an altitude of 120 Km.

Different layers can be distinguished in the atmosphere according to their temperature and composition, as illustrated in Figure 1 (Herr, 1994; Silverman, 1995a; Silverman, 1995b). Starting from the ground level moving upwards these layers are called the troposphere, stratosphere, mesosphere, thermosphere, and exosphere. The troposphere is the layer of Earth's atmosphere where humans live, and where a comfortable environment is generating when people are closer to ground level because the temperature decreases as altitude increases. The stratosphere is the layer of Earth's atmosphere where ozone is finding, which absorbs dangerous high-energy ultraviolet light from the Sun, and which is converting into heat. Unlike the troposphere, in the stratosphere, both wind turbulence and temperature are less aggressive at lower altitudes. The mesosphere is the layer of Earth's atmosphere where most meteors burn and where atmospheric pressure and temperature decrease as altitude increases, making the air non-breathable. The thermosphere is the layer of Earth's atmosphere where high-energy X-rays and ultraviolet radiation from the Sun are absorbed, and its temperature rises to hundreds or thousands of degrees Celsius. Besides, in this layer, the air is so thin that humans would freeze and where most satellites to orbit. Thus this layer is where the northern and southern lights occur. The exosphere is the layer of Earth's atmosphere that is considering the boundary of Earth's gas envelope and from which air gradually escapes and it gradually fades into what is calling deep space. In addition to these layers, there is another layer called the ionosphere located between the mesosphere and the thermosphere, where high-energy radiation from the sun collides with atoms and molecules releasing electrons and producing electrically charged ions.

Figure 1. Diagram showing the primary layers of the Earth's atmosphere

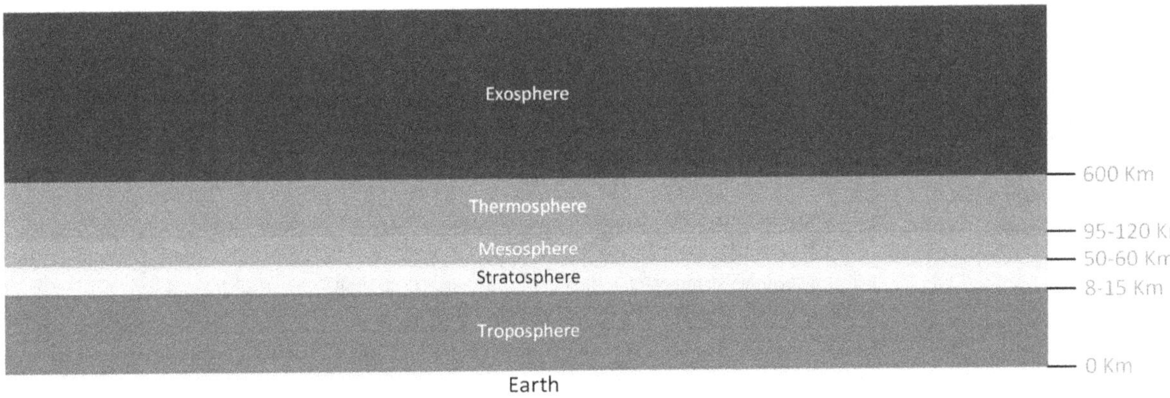

A low Earth orbit or LEO is defining as an Earth-centered orbit with an altitude of 2,000 km or less, with at least 11.25 periods per day, or an eccentricity less than 0.25. In this orbit, most of the space systems operate either controlled by Earth or autonomously. The previous definition assumes the use of circular orbits, but in reality (Hastings, 1996), the orbits are elliptical, and also, the Earth is not completely a sphere, so the reference altitude can vary greatly in different geographical positions on the planet. Space stations as well as most satellites operate on LEO, but human space flights have exceeded LEO since 1968, but the altitude record has not been exceeding since then.

Table 1 summarizes the main space environments. The effects of the space environment will influence the following subsystems of a system or device sent to space: avionics, electrical power, guidance, navigation and control and/or pointing, materials, optics, propulsion, structures, telemetry, tracking and communications, thermal control, and mission operations (Herr, 1994; Bedinger, 1996). A more detailed description can be reading in (Herr, 1994).

On the other hand, space weather determines the dynamic processes that occur in the atmosphere, ionosphere, and geomagnetic space field, and that affect or modify the behavior of devices or systems as well as living beings. The effects of space weather on Earth include ionospheric storms, temporary changes in ozone densities, and disturbance in radio communications to GPS signals and underwater positioning. Even scientists theorize about a connection between sunspot activity and glacial ages (Belk, 1997). Space weather can then be defined as the conditions and processes that occur in space and that can harm space systems and astronauts. Different processes occur in space weather such as changes in the solar wind (energy from the sun in the form of particles such as protons or electromagnetic radiation), changes in the interplanetary magnetic field, disturbances in the Earth's magnetic field, and coronal mass ejections (CME) (Rask, 2008). Therefore, experts decide the ideal time to travel to space, and which materials should be used for space systems, as well as astronauts and/or living beings that accompany them.

Radiation is a form of energy from rays, electromagnetic waves, and/or particles (Silverman, 1995a; Silverman, 1995b). This can be visible or felt, while other forms - such as X-rays and gamma rays - can only be observing with special equipment. Besides, it can produce negative effects on the systems

Table 1. Main space environments (Partially extracted from Herr, 1994).

Space Environment	Description
Neutral Thermosphere	This describes the variations in density found in space, its atmospheric components such as the presence of atomic oxygen, and the presence of winds found in the thermosphere.
Thermal Environment	This involves solar radiation (Albedo (percentage of radiation that the sun's surface reflects the radiation that falls from the sun's atmosphere) and outgoing longwave radiation (OLR) (electromagnetic radiation emitted from the Earth and its atmosphere) variations), radiative transfer (electromagnetic radiation diffused from the sun's atmosphere), and atmospheric transmittance (effectiveness in transmitting radiant energy).
Plasma	This implies three types of plasma: ionospheric (partially ionized plasma found in the ionosphere), auroral (magnetically trapped plasma) and magnetospheric (shocking solar wind plasma).
Meteoroids and Orbital Debris	This determines parameters of the flow of meteorites and orbital debris such as its size distribution, mass distribution, speed distribution and directionality.
Solar Environment	It associates the physics and dynamics of the sun, geomagnetic substorms, solar activity predictions, solar and geomagnetic indices, solar constant, and the solar spectrum.
Ionizing Radiation	It groups the trapped proton-electron radiation, galactic cosmic rays (GCRs), and solar particle events.
Magnetic Field	This relates to the natural magnetic field that represents the interaction of the solar wind with the charged particles emanating from the sun.
Gravitational Field	This corresponds to the influence of the natural gravitational field that explains the influence that a body exerts on the space around itself.
Mesosphere	This describes the variations in density found in space, its atmospheric composition, and the presence of winds found in the mesosphere.

regardless of their nature and properties. Space radiation is comprised of atoms in which its electrons have been removing from the atom - normally, only the nucleus of the atom remains - when it was accelerating in interstellar space at speeds approaching the speed of light.

Effects on space systems or devices arise from radiation, space debris, and meteorite impact, upper atmospheric drag, and electrostatic charge (Rask, 2008). Radiation in space mainly arises from three sources: 1) Van Allen radiation belts (as shown in Figure 2), 2) events of solar protons and solar energetic particles, and 3) galactic cosmic rays. A Van Allen radiation belt is defining as a region with charged particles from the solar wind, which are capturing by or held around the Earth by its magnetic field. Earth has two bells and sometimes others can appear temporarily. They are located from an altitude of 640 to 58,000 km above the surface at which the radiation levels of the region vary. Most of the particles they contain come from the solar wind and others from cosmic rays. When the solar wind is trapping, the magnetic field deflects the energy particles and protects the Earth's atmosphere from destruction (Piscane, 2008). They are located inside the magnetosphere. Bells regularly trap energetic electrons and protons, and rarely alpha particles. They jeopardize space systems when their components are not adequately shielding, and they remain for long periods in these regions of space.

In long-term missions, radiation absorbed by systems or devices can damage electronic components and solar cells (Dever, 2012). Besides, the effects of single events, such as the single event upsets (SEUs),

Figure 2. Van Allen radiation belts

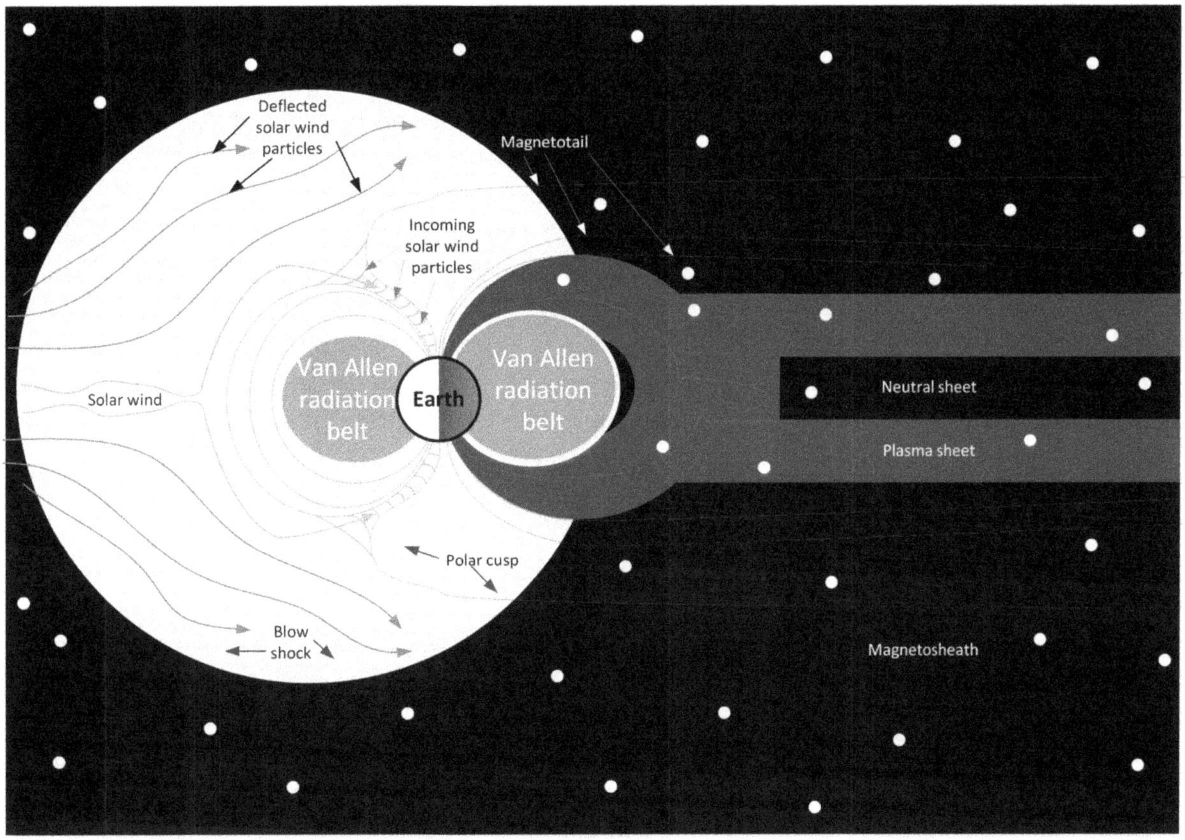

produce changes in the electrical state by ionizing simple particles such as ions, electrons, or photons by striking these with sensitive nodes of active electronic devices producing so-called soft errors or SEUs.

In the case of crewed missions, these normally avoid passing or are placing in the radiation belts, as is the case of slow transit systems or systems permanently installed such as the International Space Station (ISS) to reduce their harmful effects (Dever, 2012). When a solar energy event occurs, such as coronal mass ejections and solar flares, particles from radiation reach very high energies and touch the Earth in times as short as 30 minutes, although it usually takes a few hours. The particles that commonly cause radiation damage are protons and heavy ions, and these damage astronauts and electronic circuits.

Space debris and meteorites impact space systems or devices at high speeds, causing them mechanical or electrical damage (Belk, 1997). Space debris reaches average speeds of 10 km/s, while meteorites reach higher speeds of the order of 58 km/s or more. Mechanical and electrical damage causes anomalies such as loss of altitude control, which leads to adjustments on the stable operation achieved.

The geostationary orbit that surrounds the Earth contains plasma (Herr, 1994). The perturbations in the solar wind-heat this orbit during geomagnetic substorms. The hot plasma environment around the Earth produces an electrostatic charge on space devices or systems. The generated surface electrostatic charge contains hot electrons with energies in the range of kilo electron volts, and if discharges are present, these will surely cause anomalies in space devices and systems.

Cosmic rays are atomic nuclei and high-energy protons moving in space at the speed of light. They come from the sun, outside the solar system, and distant galaxies (Hastings, 1996). When these collide with Earth's atmosphere, it is possible to produce showers of secondary particles that sometimes reach the surfaces. 99% of cosmic rays come from atomic nuclei, while only 1% is producing by solitary electrons. For those from atomic nuclei, about 90% are simple protons, 9% alpha particles and 1% are heavy elements known as HZE (High (H) atomic number (Z) and energy (E)) ions.

Due to the various effects of the space environment, scientists and engineers from the first system sent to space began to devise space-shielding solutions that involve mechanical and electrical improvements. Before launching these systems into space, engineers evaluate environmental effects through radiation belt models, models of interaction with the space plasma, and atmospheric models to predict drag effects found in the lower orbits and during re-entry (Piscane, 2008).

An aurora is a display of natural light in the Earth's sky, primarily seen in the high latitude regions (around the Arctic and Antarctic). The auroras are perturbations in the plasma of the magnetosphere caused by the solar wind. These are very strong to alter the trajectories of charged particles in both solar wind and magnetospheric plasma. Thanks to the energy of the solar wind, particles such as electrons and protons, precipitate into the upper atmosphere called thermosphere and/or exosphere (James, 1994).

WHY DESIGN MATERIALS FOR ELECTROMAGNETIC SHIELDING FOR SPACE APPLICATIONS?

Space radiation penetrates habitats, spaceships, equipment, spacesuits, and can damage the astronauts or living beings that accompany them (Rawal, 2018). Space radiation penetrates habitats, spaceships, equipment, spacesuits, and can damage the astronauts or living beings that accompany them. In low Earth orbit (LEO), astronauts and space systems lose the natural shield against cosmic and solar radiation provided by Earth's atmosphere. In deep space, they also lose the armor provided by the Earth's magnetic field. Therefore, a priority for researchers and businesspersons is to offer alternatives for protection against space radiation to minimize physiological changes in living beings, as well as protection of the space systems that transport them or where they live while they are in space (Fayazbakhsh, 2012). Various space agencies monitor the radiation environment inside and outside a manned aircraft through the radiation dosimetry technique. This technique performs the process of monitoring, characterization, and quantification of the radiation environment. In addition to offering an assessment of the radiation biology support, this technique estimates the amount of radiation to which the crew is exposed during extra-vehicular activities, evaluates the radiation on any equipment transported in the aircraft, and performs computer modeling of the radiation on the crew. The equipment that performs this activity is a so-called dosimeter, and this is continuously using by astronauts on the International Space Station to monitor the levels of space radiation in their blood.

The effects and phenomena that space systems, astronauts, or living beings sent to a space mission described in the previous section have led numerous research groups to develop basic and applied scientific research to protect them through electromagnetic shielding (Silverman, 1995a; Silverman, 1995b). Protection against electromagnetic radiation is essential for humans to live and work safely in space. Conventional methods to protect space systems and their occupants from these forms of radiation involve some shielding topology based on massive or passive material to absorb energy from incoming particles. Since astronauts will be exposing to high-energy galactic cosmic rays, these massive materials become impractical that will also likely produce secondary radiation showers that could be even more damaging than the cosmic rays themselves. The composition, thickness, and properties of a material using in electromagnetic shielding affect its capability to shield radiation (Rawal, 2018). A more interesting alternative is the so-called active shield, whose operation consists of the use of magnetic and/or electric fields to deflect energy particles. The design of materials that produce a magnetic field that is strong enough to deflect particles from cosmic rays but weak enough not to harm astronauts is a challenge.

Radiation shielding requirements vary due to different types and levels of radiation (Kamsali, 2019). More penetrating ionizing radiation such as those produced by gamma rays and galactic cosmic rays, pass through aluminum but are stopping by thick and dense materials such as cement. Generally, better shields can block a range of radiation. Hydrogen and hydrogen-rich materials are ideal for radiation shielding because they do not break easily to form secondary radiation sources. For example, onboard the space station, polyethylene, which is a hydrogen-rich material, has reduced astronauts' exposure to space radiation in their sleeping rooms and the galley. The location of the International Space Station and the space shuttle in the low Earth orbit (LEO) reduces the requirements of electromagnetic shielding thanks to the protection of the Earth's atmosphere, which would not be so for a base on the surface of the Moon. In the design of a lunar module where astronauts could live on the Moon, electromagnetic shielding requirements would involve thick materials. This is because primary cosmic rays from high-energy protons and heavy ions can penetrate the rooms. Materials such as metal sheets or insulating layers of lunar water, or both would be required for this purpose.

Electromagnetic shielding materials must not interact with space radiation particles. When these interactions appear, nuclear by-products called secondary such as neutrons and other particles occur (Kamsali, 2019). The shielding material must contain these products generated in space systems, as these are more harmful than primary space radiation. Regularly, materials based on heavier chemical elements generate more secondary radiation than materials based on lighter chemical elements. Water being a hydrogen-rich molecule can absorb radiation, however, oxygen makes the molecule heavier than polyethylene, and therefore, more expensive to launch. Commonly, light shields greatly reduce the effects of incoming space radiation particles, but they cannot completely stop their entry.

An interesting alternative for the development of electromagnetic shielding was the introduction of electrostatic or passive radiation shielding, which generates positive and negative electrical charges that deflect incoming electrically charged space radiation (Letfullin, 2019). The lunar regolith, a powdered powder material found on the surface of the Moon has been proposing to shield human colonies. The shielding material, although partially solving the radiation problem, must be carefully designed so as not to increase the weight and cost of launching into space. Shielding materials with thicknesses of five to seven centimeters block only 30 to 35% of the radiation, so astronauts or space systems are exposing up to 70% to the radiation that passes these shields.

The ionizing radiation energy is expressing in the Gray (Gy) unit, which is equivalent to one joule of radiation absorbed per kilogram of organ or tissue weight (Letfullin, 2019). An older unit called rad is using to express the absorbed dose and its equivalence is one Gray is equal to 100 rad. To assess the harmful effects of radiation, the radiation dose is expressing as an equivalent dose whose unit is Sievert (Sv). Other books use as a unit the rem to measure the equivalent dose and its equivalence called Sv which is equal to 100 rem of radiation. Epidemiological studies on populations exposed to radiation have shown that doses above 100 mSv increase the risk of cancer and even these doses are reduced for childhood whose values are in the range of 50 to 100 mSv.

The time limit for exposure to space radiation for an astronaut is 100 days. To extend this time, space missions must be carried out within Earth's magnetic field. Furthermore, although many types of radiation occur in space flight, Heavy Particle Radiation (HZE) is one of the most dangerous types in terms of harm to humans (Wilson, 2004).

RECENT INNOVATIONS IN MATERIALS FOR SHIELDING AND HEALTH FOR ASTRONAUTS

Recently, space systems professionals and researchers have shown a growing interest in developing environmental space shielding materials based on polymeric matrices and micro and/or nanofillers of controlled shape and size. The advantage of using composite materials is to offer various degrees of freedom to design tailored shielding materials to counteract the ionizing radiation found in space. In the following paragraphs, some of the advances in materials that have been proposed for environmental space shielding for aerospace systems, as well as the health challenges that astronauts are facing are presented.

Materials for Space Systems

Shielding from space radiation is essential to protect from the effect of the Van Allen belts and radiation environments that include galactic cosmic rays (GCRs) and solar particle events (SPEs) to manned space flight (Harrison, 2008). In this section, protection strategies based on shielding materials against different radiation storms are analyzing. Depending on the type of space event to which the space system is subjecting, the protection strategy is directly related. In the case of solar energy particles consisting of low energy protons, the preferred solution is to add a shielding material to the surfaces to be protecting. For galactic cosmic rays being made up of ions from elements between hydrogen and nickel, the required protection varies greatly in the range of 10 MeV to millions of MeV. Low atomic number chemical element shielding materials can fragment galactic cosmic rays resulting in low numbers of secondary particles, especially neutrons that cause further biological damage. From a biological point of view, the preferred protection material is aluminum, which, even though it is thick, does not prevent cancer induction.

Chemical elements such as copper, tin, iodine, barium, iron, magnesium, chromium, aluminum, and titanium have also been investigating as shielding materials for X-ray radiation as substitutes for lead (Lin, 2000; Liu, 2004). However, these materials have not been able to offer good mechanical properties due to their low chemical reactivity with the polymer in the formation of composite materials. Lead dimethylmethacrylate ($Pb(MA)_2$) copolymerized with styrene and methacrylic acid was synthesizing to produce a composite material, which showed high X-ray absorption, high transparency in visible light, higher glass transition temperature, and higher refractive index.

In deep space, astronauts and space systems can suffer from mass reduction compromising their safety, so the choice of system design materials, as well as the crew profile, are the most important aspects of designing a space mission (Tripathi, 2003). Among the materials that have been proposed for radiation shielding are aluminum 2219, polyetherimide, polysulfone, polyethylene, lithium hydride, liquid methane, graphite nanofibers, and liquid hydrogen. High hydrogen content materials regularly offer better shielding effectiveness, but these do not have the necessary structural integrity for a vehicle or habitat. This is normally accomplished using organic polymers. Polysulfone and polyetherimide offer high-performance structural options. Fuels based on methane and liquid hydrogen offer global protection. Lithium hydride is used as a shielding material for nuclear reactors but is not suitable for other applications. Hydrogen-impregnated graphite nanofiber represents an alternative option for multifunctional space structures.

Composite materials based on lead oxide and natural rubber (albeit with some degree of toxicity) have shown good X-ray radiation absorption capacity for energy values greater than 88 KeV or for the range of 13 to 40 KeV (Liu, 2004). However, for the 40 to 88 KeV range, this absorption capacity is poor. As an alternative to this energy range, lanthanides have lower toxicity than lead and have better absorption for the 38.9 to 63.3 KeV range. A composite material based on gadolinium acrylate ($Gd(AA)_3$) and natural rubber can be synthesized to offer X-ray shielding in the range previously described.

Knowledge of the transport of heavy, high-energy ions into deep space in different materials provides insight into the risk of nuclear radiation to reduce both the physical doses and the biological effectiveness of radiation in human space missions (Guetersloh, 2006). Hydrogenous materials and light chemical elements are considered effective shielding materials against deterioration caused by galactic cosmic rays (GCRs). Polyethylene has been chosen by NASA as reference material for radiation testing of multifunctional composite materials under a beam of 1 GeV/amu ^{56}Fe ions.

A composite material made of low-density polyethylene and hollow borosilicate glass microspheres has been synthesizing to absorb radiation from heavy particles due to its high hydrogen content and low gravimetric density (Ashton-Patton, 2006). The purpose of this radiation shielding material is to extend the period outside of Earth's atmosphere without space systems and astronauts experiencing detrimental biological effects.

High-density polyethylene and boron nitride-based structural composite materials for spacecraft can be designed to absorb low-energy neutrons, as well as flame retardant properties, and these can be functionalized with alkoxysilane to achieve interfacial adhesion (Harrison, 2008). This composite material with a third of the density of aluminum and 2% by volume of boron nitride has radiation shielding capabilities like that of aluminum. An additional merit is the ability of the shield to emit high-energy gamma radiation and its use around nuclear reactors.

In the search to reduce mass and energy in space applications, low-power electronic devices were used, which led to problems of electromagnetic interference (EMI) (Fayazbakhsh, 2012). Electronic packaging is used to shield electronic devices against electromagnetic interference produced by themselves or by the environment around them. The materials used for this electromagnetic shielding must be carefully selected so that not only the electromagnetic but also the mechanical and thermal properties are satisfied. The best proposals that have been made are based on metallic or composite materials. A composite material based on nickel-coated carbon fibers with a filler percentage of 30 showed the best performance even on metallic materials such as aluminum and stainless steel.

Multifunctional composite-based structures are being incorporated into aircraft, thanks to their electronic, thermal, radiation-shielding, and load-bearing handling capabilities to reduce mass and volume (Jang, 2016). One of the multi-functional material alternatives used is carbon fibers and tungsten sheet laminates, which offer light-weight radiation shielding capabilities for sensitive devices with the selective region.

A composite material based on poly (methyl methacrylate) (PMMA) and colemanite $Ca_2B_6O_{11}\times5H_2O$ (CMT) has been proposed as a shielding material for neutron radiation and gamma rays (Bel, 2019). The polymer is polymerized using atom transfer radicals (ATRP). Incorporation of 40 wt% CMT has increased neutron and gamma-ray attenuation using a ^{239}Pu-Be neutron source and the Cs-197 radioisotope, respectively. The potential applications of the composite material are in low Earth orbit satellites and high-altitude airplanes.

A proposal for a composite material based on polyurethane and boron carbide (B_4C) at 20.9% by weight to provide shielding against thermal neutron radiation with a neutron attenuation index of 100 has been presented (Cataldo, 2019). In addition to the electromagnetic shielding advantage, this composite material shows an insignificant change in glass transition temperature thanks to an insignificant change in crosslinking density and the absence of a chain scission mechanism.

Astronauts

Active shielding techniques for radiation protection have the advantage of saving mass over passive shielding techniques (Geng, 2015). Active techniques are based on producing a resistant magnetic field to provide shielding for protons from events of solar particles (SPEs) and galactic cosmic ray particles (GCRs). This type of space environmental shielding is useful for the effective protection of astronauts on spacecraft.

Experience in studying the effects of space radiation has shown a disparity between research results in simulated space environments and the observed empirical effects found in human crews who have traveled into space (Chancellor, 2018). The current interest in trips to the Moon and Mars have stimulated to determine the possible impacts of space radiation not only from a medical point of view but the search for materials for the design of clothing capable of mitigating the exposure of interplanetary radiation. Classic studies of human health risks after exposure in space flight are primarily associated with bone behavior, human behavior, and nutrition. Besides, at the Earth level, the tests to astronauts are based on their exposure to mono-energy sources as well as to simple acute ions such as protons, lithium, carbon, oxygen, silicon, iron, etc., which do not replace the space environment complex. Other tests include applying to the astronaut's cumulative mission doses in one time, or sequential and rapid doses to experimental animals, but whose values do not agree as much in intensity as in temporality as it happens in space.

One of the studies developed to estimate the radiation levels to which astronauts are exposed is the ray-tracing method that uses geometry models with depth dose curves (Kartashov, 2018). The experiment was tested on the International Space Station in the period from 2004 to 2016. The results showed that the efficiency of the shield to the environmental space presents a radiation risk that depends on the depth of the organ in the body as well as the orientation of the body member to the shield.

In future space voyage plans such as to Mars, astronauts will be exposed to space for periods of at least 1 to 2 years to obtain information about their surface and geological characteristics, as well as environmental and atmospheric phenomena (Kamsali, 2019). During this period astronauts will be exposed to sun damage and cosmic radiation (HZE), and specifically on Mars, the situation is more dangerous due to a non-existent magnetosphere and relative marginal natural shielding of its atmosphere.

An electrostatic field can be used as a shield against high-energy cosmic rays to reduce astronauts' exposure to radiation (Letfullin, 2019). This electric field can be produced with multi-plate low-energy capacitors that surround the spacecraft habitat capsule using ultralight composite materials consisting of two or more layers. The results showed high effectiveness to the environmental shielding of the space thanks to the high volume of hydrogen of the used composite materials, and to its versatility of providing multifunctional materials.

FUTURE RESEARCH DIRECTIONS

Through the space missions conducted by NASA from the 1960s and 1970s to got to the Moon and the activities carried out in recent decades to go to Mars or beyond, a lot of information has been continuously exploited in the most recent missions. The time of flight of the spacecraft as well as the time that the international space station has been operational in space have allowed accumulating experience on the exposure of astronauts to space radiation. It is now important to develop reliability-based designs which are based on the biological and physical uncertainties of radiation shielding materials as well as increasing the level of confidence in the success of protecting astronauts from radiation hazards. As space explorations move away from protecting the low Earth orbit, the risks of space radiation need to be more fully understood. To achieve this goal, it is necessary to develop more modeling techniques that emulate the space environment, find appropriate biological substitutes to determine the effects on humans and establish flux data on astronauts from space radiation. The effects of space radiation need to be more extensively investigated in aspects such as genomic sequencing, as well as transcriptional, proteomic, and epigenomic studies at the cellular level. In the case of environmental space shielding materials, prospects are associated with the development of multi-functional composite based approaches to offer passive and/or active strategies. The use of nanomaterials (based on carbon such as carbon and graphene nanotubes, two-dimensional materials such as transition metal dichalcogenides (TMDs), MXenes, silicene, germanene, etc.), advanced materials and composite materials with high contents of hydrogen or light chemical elements they will continue to be the main trends in the coming decades to provide shielding for space radiation.

CONCLUSION

Until now, space professionals and researchers have learned a lot about space radiation to study its effects on space systems and astronauts as well as the possible materials associated with electromagnetic environmental shielding. This learning comes from the history of space flight as well as the health of space pioneers. In the case of astronauts, the number of cancer cases and degenerative diseases found in the passenger crews of the Apollo, Space Shuttle, and International Space Station is worrisome. In the case of environmental space shielding materials, the prospects involve offering passive and/or active strategies for the development of multi-functional composite materials-based approaches. The use of nanomaterials, advanced materials, and composite materials with high contents of hydrogen or light chemical elements will continue to be the main trends in the coming decades to offer shielding to space radiation.

ACKNOWLEDGMENT

The author acknowledges the support provided by CONRICYT of the National Council of Science and Technology (CONACYT) from Mexico for the access to scientific articles, as well as the support of the Thematic Network of Nanotechnology and Nanoscience and Thematic Network of Space Science and Technology of CONACYT, Mexico. Besides, he wants to thank your family for their support in carrying out this research.

REFERENCES

Ashton-Patton, M.M., Hall, M.M., & Shelby, J.E. (2006). Formation of Low Density Polyethylene/Hollow Glass Microspheres Composites. *Journal of Non-Crystalline Solids, 352*(6-7), 615-619. doi:.2005.11.058 doi:10.1016/j.jnoncrysol

Bedinger, K. L., Leach, R. D., & Alexander, M. B. (1996). *Spacecraft System Failures and Anomalies Attributed to the Natural Space Environment. NASA Reference Publication 1390, Marshall Space Flight Center (MSFC), National Aeronautics and Space Administration*. NASA.

Bel, T., Arslan, C., & Baydogan, N. (2019). Radiation Shielding Properties of Poly(Methyl Methacylate)/Colemanite Composite for the Use in Mixed Irradiation Fields of Neutrons and Gamma Rays. *Materials Chemistry and Physics, 221*, 58–67. doi:10.1016/j.matchemphys.2018.09.014

Belk, C. A., Robinson, J. H., Alexander, M. B., Cooke, W. J., & Pavelitz, S. D. (1997). Meteoroids and Orbital Debris: Effects on Spacecraft. NASA Reference Publication 1408, Marshall Space Flight Center (MSFC), National Aeronautics and Space Administration (NASA).

Bothmer, V., & Daglis, I. A. (2007). *Space Weather: Physics and Effects*. Springer. doi:10.1007/978-3-540-34578-7

Calders, S., Messios, N., Botek, E., De Donder, E., Kruglanski, M., Evans, H., & Rodgers, D. (2018, May). *Modeling the Space Environment and Its Effects on Spacecraft and Astronauts using SPENVIS*. Paper presented at 15th International Conference on Space Operations, Marseille, France. 10.2514/6.2018-2598

Cataldo, F., & Prata, M. (2019). Neutron Radiation Shielding with PUR Composites loaded with B_4C or Graphite. *Fullerenes and Carbon Nanostructures, 27*(7), 531–537. doi:10.1080/1536383X.2019.1604512

Chancellor, J. C., Blue, R. S., Cengel, K. A., Auñón-Chancellor, S. M., Rubins, K. H., Katzgraber, H. G., & Kennedy, A. R. (2018). Limitations in Predicting the Space Radiation Health Risk for Exploration Astronauts. *NPJ Microgravity*, *4*(1), 8. doi:10.103841526-018-0043-2 PMID:29644336

Dever, J., Banks, B., de Groh, K., & Miller, S. (2012). Degradation of Spacecraft Materials. In M. Kutz (Ed.), *Handbook of Environmental Degradation of Materials* (2nd ed., pp. 717–770). Elsevier. doi:10.1016/B978-1-4377-3455-3.00024-9

Fayazbakhsh, K., & Abedian, A. (2012). Materials Selection for Electronic Enclosures in Space Environment considering Electromagnetic Interference Effect. *Advances in Space Research*, *49*(3), 586–593. doi:10.1016/j.asr.2011.10.025

Geng, C., Tang, X., Gong, C., Guan, F., Johns, J., Shu, D., & Chen, D. (2015). A Monte-Carlo-based Radiation Safety Assessment for Astronauts in an Environment with Confined Magnetic Field Shielding. *Journal of Radiological Protection*, *35*(4), 777–788. doi:10.1088/0952-4746/35/4/777 PMID:26484984

Guetersloh, S., Zeitlin, C., Heilbronn, L., Miller, J., Komiyama, T., Fukumura, A., Iwata, Y., Murakami, T., & Bhattacharya, M. (2006). Polyethylene as a Radiation Shielding Standard in Stimulated Cosmic-Ray Environments. *Nuclear Instruments & Methods in Physics Research. Section B, Beam Interactions with Materials and Atoms*, *252*(2), 319–332. doi:10.1016/j.nimb.2006.08.019

Harrison, C., Weaver, S., Bertelsen, C., Burgett, E., Hertel, N., & Grulke, E. (2008). Polyethylene/Boron Nitride Composites for Space Radiation Shielding. *Journal of Applied Polymer Science*, *109*(4), 2529–2538. doi:10.1002/app.27949

Hastings, D., & Garrett, H. (1996). *Spacecraft-Environment Interactions*. Cambridge University Press. doi:10.1017/CBO9780511525032

Herr, J. L., & McCollum, M. B. (1994). Spacecraft Environments Interactions: Protecting Against the Effects of Spacecraft Charging. NASA Reference Publication 1354, Marshall Space Flight Center (MSFC), National Aeronautics and Space Administration (NASA).

James, B. F., Norton, O. W., & Alexander, M. B. (1994). The Natural Space Environment: Effects on Spacecraft. NASA Reference Publication 1350. Marshall Space Flight Center (MSFC), National Aeronautics and Space Administration (NASA).

Jang, T. S., Rhee, J., Hyun, B.-S., Seo, H.-S., Kim, T. Y., & Seo, J. K. (2016). A Study of Grid-Stiffened Multi-Functional Composite Structures with Radiation Spot Shielding. *Aerospace Science and Technology*, *54*, 330–339. doi:10.1016/j.ast.2016.05.002

Kamsali, N., Chakravarty, S. C., & Basuvaraj, P. K. (2019). Investigation of HZE Particle Fluxes as a Space Radiation Hazard for Future Mars Missions. *Heliyon*, *5*(12), e02972. doi:10.1016/j.heliyon.2019.e02972 PMID:31867459

Kartashov, D., & Shurshakov, V. (2018). Analysis of Space Radiation Exposure Level at Different Shielding Configurations by Ray-Tracing Dose Estimation Method. *Acta Astronautica*, *144*, 320–330. doi:10.1016/j.actaastro.2018.01.013

Kleiman, J. I., Iskanderova, Z. A., Pérez, F. J., & Tennyson, R. C. (1995). Protective Coatings for LEO Environments in Spacecraft Applications. *Surface and Coatings Technology*, *76-77*(Part 2), 827–834. doi:10.1016/0257-8972(95)02497-2

Letfullin, R. R., George, T. F., & Ramazanov, A. K. (2019). Multifunctional Cosmic-Ray Shielding of Spacecraft with Elements of Systems Engineering Design. *Journal of Spacecraft and Rockets*, *56*(5), 1312–1321. doi:10.2514/1.A34440

Lin, Q., Yang, B., Li, J., Meng, X., & Shen, J. (2000). Synthesis, Characterization and Property Studies of Pb^{2+}-containing Optical Resins. *Polymer*, *41*(23), 8305–8309. doi:10.1016/S0032-3861(00)00212-3

Liu, C., & Kang, X. (2019). Facile Fabrication of Conductive Silver Films on Carbon Fiber Fabrics via Two Components Spray Deposition Technique for Electromagnetic Interference Shielding. *Applied Surface Science*, *487*, 1245–1252. doi:10.1016/j.apsusc.2019.04.186

Liu, L., He, L., Yang, C., Zhang, W., Jin, R.-G., & Zhang, L.-Q. (2004). In situ Reaction and Radiation Protection Properties of $Gd(AA)_3$/NR Composites. *Macromolecular Rapid Communications*, *25*(12), 1197–1202. doi:10.1002/marc.200400077

Nenarokomov, A. V., Alifanov, O. M., Krainova, I. V., Titov, D. M., & Morzhukhina, A. V. (2019). Estimation of Environmental Influence on Spacecraft Materials Radiative Properties by Inverse Problems Technique. *Acta Astronautica*, *160*, 323–330. doi:10.1016/j.actaastro.2019.04.014

Piscane, V. L. (2008). The Space Environment and Its Effects on Space Systems. Reston, VA: American Institute of Aeronautics and Astronautics (AIAA). doi:10.2514/4.862533

Rask, J., Vercoutere, W., Navarro, B.J. & Krause, A. (2008). *Space Faring: The Radiation Challenge, An Interdisciplinary Guide on Radiation and Human Space Flight*. NASA.

Rawal, S. (2018). Materials and Structures Technology Insertion into Spacecraft Systems: Successes and Challenges. *Acta Astronautica*, *146*, 151–160. doi:10.1016/j.actaastro.2018.02.046

Samwel, S. W., El-Aziz, E. A., Garrett, H. B., Hady, A. A., Ibrahim, M., & Amin, M. Y. (2019). Space Radiation Impact on Smallsats during Maximum and Minimum Solar Activity. *Advances in Space Research*, *64*(1), 239–251. doi:10.1016/j.asr.2019.03.025

Shuvalov, V. A., Gorev, N. B., Tokmak, N. A., & Kuchugurnyi, Y. P. (2020). Drag on a Spacecraft produced by the Interaction of Its Magnetic Field with the Earth's Ionosphere. Physical Modelling. *Acta Astronautica*, *166*, 41–51. doi:10.1016/j.actaastro.2019.10.018

Silverman, E.M. (1995a). *Space Environmental Effects on Spacecraft: LEO Materials Selection Guide*. Langley Research Center, NASA Contractor Report 4661 Part 1.

Silverman, E.M. (1995b). *Space Environmental Effects on Spacecraft: LEO Materials Selection Guide*. Langley Research Center, NASA Contractor Report 4661 Part 2.

Spillantini, P., Taccetti, F., Papini, P., & Rossi, L. (2000). Radiation Shielding of Spacecraft in Manned Interplanetary Flights. *Nuclear Instruments & Methods in Physics Research. Section A, Accelerators, Spectrometers, Detectors and Associated Equipment*, *443*(2-3), 254–263. doi:10.1016/S0168-9002(99)01091-8 PMID:11543201

Tishkevich, D. I., Grabchikov, S. S., Lastovskii, S. B., Trukhanov, S. V., Vasin, D. S., Zubar, T. I., Kozlovskiy, A. L., Zdorovets, M. V., Sivakov, V. A., Muradyan, T. R., & Trukhanov, A. V. (2019). Function Composites Materials for Shielding Applications: Correlation between Phase Separation and Attenuation Properties. *Journal of Alloys and Compounds*, *771*, 238–245. doi:10.1016/j.jallcom.2018.08.209

Tripathi, R. K., Wilson, J. W., Cucinotta, F. A., Anderson, B. M., & Simonsen, L. C. (2003). Materials Trade Study for Lunar/Gateway Missions. *Advances in Space Research*, *31*(11), 2383–2388. doi:10.1016/S0273-1177(03)00551-9 PMID:14696588

Tripathi, R. K., Wilson, J. W., & Youngquist, R. C. (2008). Electrostatic Space Radiation Shielding. *Advances in Space Research*, *42*(6), 1043–1049. doi:10.1016/j.asr.2007.09.015

Wanasinghe, D., & Aslani, F. (2019). A Review on Recent Advancement of Electromagnetic Interference Shielding Novel Metallic Materials and Processes. *Composites. Part B, Engineering*, *176*, 107207. doi:10.1016/j.compositesb.2019.107207

Warden, D., & Bayazitoglu, Y. (2019). New Comparative Metric for Evaluating Spacecraft Radiation Shielding. *Journal of Spacecraft and Rockets*, *56*(4), 1024–1038. doi:10.2514/1.A34360

Wilson, J.W., Clowdsley, M.S., Cucinotta, F.A., Tripathi, R.K., Nealy, J.E. & De Angelis, G. (2004). Deep Space Environments for Human Exploration. *Advances in Space Research*, *34*(6), 1281-1287. doi: 101016/j.asr.2003.10.052

Zhou, Z.-H., Liang, Y., Huang, H.-D., Li, L., Yang, B., Li, M. Z., Yan, D.-X., Lei, J., & Li, Z.-M. (2019). Structuring Dense Three-Dimensional Sheet-like Skeleton Networks in Biomass-derived Carbon Aerogels for Efficient Electromagnetic Interference Shielding. *Carbon*, *152*, 316–324. doi:10.1016/j.carbon.2019.06.027

ADDITIONAL READING

Chancellor, J. C., Scott, G. B. I., & Sutton, J. P. (2014). Space Radiation: The Number one Risk to Astronaut Health beyond Low Earth Orbit. *Life (Chicago, Ill.)*, *4*(3), 491–510. doi:10.3390/life4030491 PMID:25370382

Chumakov, A. I., Vasil'ev, A. L., Kozlov, A. A., Kol'tsov, D. O., Krinitskii, A. V., Pechenkin, A. A., Tararaksin, A. S., & Yanenko, A. V. (2010). Single-Event-Effect Prediction for ICs in a Space Environment. *Russian Microelectronics*, *39*(2), 74–78. doi:10.1134/S1063739710020022

Huang, C.-L., Spence, H. E., & Kress, B. T. (2009). Assessing Access of Galactic Cosmic Rays at Moon's Orbit. *Geophysical Research Letters*, *36*(9), L09109. Advance online publication. doi:10.1029/2009GL037916

Kaul, R. K., Barghouty, A. F., & Dahche, H. M. (2004). Space Radiation Transport Properties of Polyethylene-based Composites. *Annals of the New York Academy of Sciences*, *1027*(1), 138–149. doi:10.1196/annals.1324.013 PMID:15644352

Shingledecker, C. N., & Herbst, E. (2018). A General Method for the Inclusion of Radiation Chemistry in Astrochemical Models. *Physical Chemistry Chemical Physics, 20*(8), 5359–5367. doi:10.1039/C7CP05901A PMID:29067367

Tribeault, S. A., Kang, J. H., Sauti, G., Park, C., Fay, C. C., & King, G. C. (2015). Nanomaterials for Radiation Shielding. *MRS Bulletin, 40*(10), 836–841. doi:10.1557/mrs.2015.225

Zheng, Y., Ganushkina, N. Y., Jiggens, P., Jun, I., Meier, M., Minow, J. I., O'Brien, T. P., Pitchford, D., Shprits, Y., Tobiska, W. K., Xapsos, M. A., Guild, T. B., Mazur, J. E., & Kuznetsova, M. M. (2019). Space Radiation and Plasma Effects on Satellites and Aviation: Quantities and Metrics for Tracking Performance of Space Weather Environment Models. *Space Weather, 17*(10), 1384–1403. doi:10.1029/2018SW002042 PMID:31894181

Zhou, D., Sun, Y., Zhang, B., Zhang, S., Sun, Y., Liang, J., Zhu, G., Jing, T., Yuan, B., Zhang, H., Zhang, M., Wang, W., & Zhao, L. (2018). Radiation Measured for Chinese Satellite SJ-10 Space Mission. *Journal of Geophysical Research. Space Physics, 123*(2), 1690–1700. doi:10.1002/2017JA024697

KEY TERMS AND DEFINITIONS

Cosmic Rays: Protons and high-energy atomic nuclei, originating from the sun and other distant galaxies, which move through space at almost the speed of light and produce showers of secondary particles (ionized particles (protons, electrons, muons, antiprotons, alpha particles, pions, positrons) and electromagnetic radiation or X-rays) that reach the surfaces.

Electromagnetic Shielding: Physical property that reduces the electromagnetic field in space or material by blocking the field with barriers made of conductive and/or magnetic materials.

Electronvolt (eV): Amount of kinetic energy gained (or lost) by a single electron accelerating from rest through an electric potential difference of one volt in a vacuum.

Gamma Rays: Penetrating electromagnetic radiation arising from the radioactive decay of atomic nuclei.

Geomagnetic Storm: Temporary disturbance of the Earth's magnetosphere caused by a solar wind shock wave and/or cloud of a magnetic field that interacts with the Earth's magnetic field.

Low Earth Orbit: Earth-centered orbit with an altitude of 2,000 km or less where most man-made objects are in outer space.

Solar Wind: Steam of charged particles or plasma based on electrons, protons, and alpha particles with kinetic energy between 0.5 and 10 keV released from the upper atmosphere of the sun.

Van Allen Belt: Zone of charged energy particles mainly originated from the solar wind that is capturing by and held around a planet by its magnetic field.

Chapter 4
Electromagnetic Compatibility Testing for Space Systems:
Specifications, Methodologies, Tailoring, and Recent Trends

Panagiotis K. Papastamatis
High Voltage Laboratory, National Technical University of Athens, Greece

Eleni P. Nicolopoulou
High Voltage Laboratory, National Technical University of Athens, Greece

Ioannis F. Gonos
High Voltage Laboratory, National Technical University of Athens, Greece

ABSTRACT

Electromagnetic compatibility has emerged in the last decades as one of the most important aspects of product design. Space equipment, in particular, due to their increased sensitivity requirements and a greater risk of failure, are required to comply with a variety of EMC requirements. The scope of this chapter is to provide a detailed overview of these requirements according to the two main standards applicable, MIL-STD461G and ECSS-E-ST-20-07C, describing the proposed methodologies, the associated testing instrumentation, and the tailoring capabilities. The limitations of these methodologies and relative research, aiming to validate or improve them, are also presented highlighting potential deficiencies. This chapter aims to serve as a compact guide to EMC testing of space equipment according to the requirements of two of the most active organizations of space engineering.

DOI: 10.4018/978-1-7998-4879-0.ch004

INTRODUCTION

Electromagnetic Compatibility (EMC) evaluation is always carried out according to specific guidelines. However, the large scale and variety of most space programs have led to varying compatibility requirements for components to be used in each program. As a result of this non-uniform approach to EMC requirements among the different programs, the requirements for specific equipment under test can vary significantly. Most of these guidelines are based on two major Standards: a) The MIL-STD-461 (Latest version: G) of the US Department of Defense and b) The ECSS-E-ST-20-07C Rev.1 by the European Cooperation for Space Standardization. The following section, including the different EMC requirements for evaluating the performance of equipment to be installed in a spacecraft, will highlight the differences between the requirements of these two Standards and how these differences can affect the product designing and manufacturing process.

BACKGROUND

Electromagnetic Compatibility has emerged in the last decades as one of the most important aspects of product design. The rapid progress in electronic and system engineering, the co-existence of complex and sensitive integrated and embedded systems with power electronics in a very limited amount of space and the overall increase in the variety of implemented technologies has placed the ability of the system to function correctly under the influence of various electromagnetic phenomena, as well as the requirement for the system to not interfere with the function of other systems in its vicinity, in the center of research and manufacturing attention. These two system characteristics are modernly defined as:

- Electromagnetic disturbances (emissions), which is defined as an electromagnetic phenomenon which can potentially degrade the performance of other equipment or system and can be manifested either as noise or as unwanted signals.
- Electromagnetic susceptibility (immunity), which is the evaluation of the system's ability to operate normally and without unacceptable performance degradation in the presence of electromagnetic disturbances.

These two aspects of electromagnetic compatibility are generally addressed by performing specific and targeted tests according to the relevant Standards, published by international bodies, committees or organizations. Each of these Standards studies, not only the applicable electromagnetic environment the equipment may be installed to, but also the specific functions of each type of equipment. This results in specifying different types of applicable tests, different levels and limits for these tests and performance criteria directly related to each function of the equipment.

The same general rationale in assessing the electromagnetic compatibility of a system is valid for Space Systems and equipment to be installed in such an environment (Satellites, Spacecraft, etc.). Many of the requirements for such systems are based on military and aeronautics Standards, applying however also methodologies that are often used for the evaluation of commercial products. As it is expected, the performance criteria and the defined limits are way stricter for the space environment. This is mainly because in the premises of a space project or mission, even the slightest unidentified electromagnetic compatibility problem, could result in the performance degradation of the entire system and potentially jeopardize the success of the entire project or mission. This increased sensitivity of the system requires detailed EMC analysis, including special mitigation techniques and measures to protect and shield the more sensitive equipment from interference propagating throughout the system. This analysis is often referred to as intra-system EMC analysis and is crucial for each space mission or project, where the consequences of failure can be catastrophic contrary to commercial applications where the consequences are mainly financially, and most of the time limited. The different approaches to this intra-system analysis have resulted in various EMC guidelines and Standards applicable to Space Systems. Even if the two dominant Standards, which are also explicitly described and discussed in this chapter, cover most of the EMC requirements, each mission or project is subject to tailoring regarding these requirements. That means equipment that has been deemed to comply with the EMC requirements of a mission is not necessarily compliant with the requirements of a different mission.

ELECTROMAGNETIC COMPATIBILITY TESTING: STANDARDS, METHODOLOGIES AND TRENDS

Conducted Emissions

Conducted emissions in general are defined as the noise currents generated by a device/equipment under test (DUT/EUT) and propagating through the cable wiring, either the power cord or the harness, to other system components within the spacecraft. The frequency range of interest is in most cases 30Hz-100MHz and is divided into two subranges based on the frequency usage and coupling mode of the measured noise. The lower subrange is 30Hz-100kHz and relates to audio and VLF frequencies up to 100kHz. The rationale behind the conducted emission requirements in such low frequencies is mainly based on the effort to ensure that the EUT does not contribute significantly to the voltage ripple on the power bus leading to potential corruption of the bus quality. The upper-frequency subrange relates to RF frequencies up to 100MHz and ranges from 100kHz to 100MHz. The scope of this requirement is the reduction of the crosstalk between cables and terminals, as well as the mitigation of RF pollution of the power bus or the signal lines.

30Hz-100KHz

Already in this first section of the emission measurements, there is a significant difference between the two Standards. MIL-STD-461G does not specify a requirement for emissions in the low frequency – audio range up to 10kHz (CE101) for Space Systems, as shown on the requirement matrix of Table 1.

However, ECSS-E-ST-20-07C includes an emission requirement for the 30Hz-100kHz range describing the measuring equipment and measurement setup in §5.4.2. It should be noted that in such low frequencies the emissions are measured as disturbance current (with a current probe) rather than disturbance voltage, because of the difficulty in controlling the power source impedance in this frequency range. Additionally, in this range, the emission source impedance inside the equipment under test is assumed to dominate the LISN (Line Impedance Stabilization Network) impedance used for the measurement and therefore the measured current levels are considered independent of the power source impedance.

The measurement procedure includes the following two steps:

Table 1. MIL-STD-461G Requirement Matrix (DoD: USA, 2015)

Installation conditions	Requirement Applicability																		
	CE101	CE102	CE106	CS101	CS103	CS104	CS105	CS109	CS114	CS115	CS116	CS117	CS118	RE101	RE102	RE103	RS101	RS103	RS105
Surface Ships	A	A	L	A	S	L	S	L	A	S	A	L	S	A	A	L	L	A	L
Submarines	A	A	L	A	S	L	S	L	A	S	L	S	S	A	A	L	L	A	L
Aircraft, Army, Including Flight Line	A	A	L	A	S	S	S		A	A	A	L	A	A	A	L	A	A	L
Aircraft, Navy	L	A	L	A	S	S	S		A	A	A	L	A	L	A	L	L	A	L
Aircraft, Air Force		A	L	A	S	S	S		A	A	A	L	A		A	L		A	
Space Systems, Including Launch Vehicles		A	L	A	S	S	S		A	A	A	L			A	L		A	
Ground, Army		A	L	A	S	S	S		A	A	A	S	A		A	L	L	A	
Ground, Navy		A	L	A	S	S	S		A	A	A	S	A		A	L	L	A	L
Ground, Air Force		A	L	A	S	S	S		A	A	A		A		A	L		A	

Legend:
A: Applicable L: Limited as specified in the individual sections of this standard. S: Procuring activity must specify in the procurement documentation.

a) A measurement system check where known calibrated signals of 1kHz and 100kHz are applied and measured. Additionally, a DC current, equivalent to the EUT supply current, is applied for the investigation of potential probe saturation due to this DC current.
b) The measurement of the EUT conducted emissions after applying any necessary correction factors calculated from the aforementioned system check.

The measurement setup, including the necessary measuring equipment, is presented in Figure 1 below.

Figure 1. Measurement system test setup according to ECSS-E-ST-20-07C §5.4.2 (ECSS, 2012)

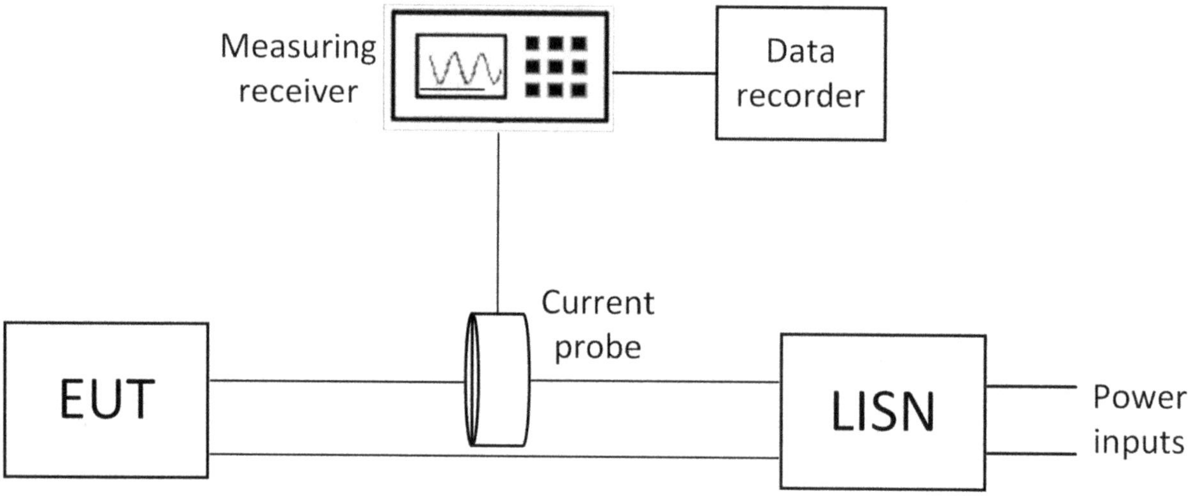

The bandwidth and measurement time parameters for the measuring receiver are presented in Table 2.

Regarding the operation of the EUT during the measurements, ECSS-E-ST-20-07C requires the EUT to be operated in the mode which produces the maximum emissions. In case of several available modes, including also modes controlled via software, the number of tested modes shall be such that every function and all circuitry are evaluated.

Table 2. Measuring receiver – Bandwidth, measurement time for different frequencies (ECSS, 2012)

Frequency Range	6 dB Resolution Bandwidth	Minimum Dwell Time		Minimum Measurement Time Analog-Tuned Measurement Receiver
		Stepped- Tuned Receiver (Seconds)	FFT Receiver (Seconds/ Measurement Bandwidth)	
30 Hz - 1 kHz	10 Hz	0.15	1	0.015 sec/Hz
1 kHz - 10 kHz	100 Hz	0.015	1	0.15 sec/kHz
10 kHz - 150 kHz	1 kHz	0.015	1	0.015 sec/kHz
150 kHz - 10 MHz	10 kHz	0.015	1	1.5 sec/MHz
10 MHz - 30 MHz	10 kHz	0.015	0.15	1.5 sec/MHz
30 MHz - 1 GHz	100 kHz	0.015	0.15	0.15 sec/MHz
Above 1 GHz	1 MHz	0.015	0.015	15 sec/GHz

10kHz-100MHz

In this upper-frequency range, both Standards include conducted emission requirements for the equipment under test. More specifically:

MIL-STD-461G, CE102 Conducted Emissions, Radio Frequency Potential, Power Leads

In this Standard, the frequency range for the measurements is 10kHz-10MHz and the requirement is applicable to all power ports/leads including returns, which are connected to power sources that are not part of the equipment under test. The measurement procedure is very similar to the procedure described above for the 30Hz-100kHz measurement, where a calibration prior to the final measurements is required to ensure the proper operation of the measuring equipment and the estimation of potentially required correction factors. The measurement setup is presented in Figure 2, while the bandwidth and measurement time values are selected according to Table 2.

ECSS-E-ST-20-07C Rev.1 §5.4.3. CE, Power and Signal Leads, 100kHz-100MHz

Similar to the procedure described above for the lower frequencies, this Standard requires conducted emission measurements in the frequency range 100kHz-100MHz according to the setup of Figure 1 (for differential mode measurements) and the receiver settings of Table 2. The varying point between these two procedures is the additional requirement for common-mode measurements in §5.4.3. For power lines and signal lines, the common-mode interference current is measured with current probes engulfing all power lines or all signal lines simultaneously, as also presented in Figure 3.

It should be noted here that, as it is also extensively explained in ECSS-E-ST-20-07C the "differential" mode measurements described in the Standard are not truly differential when measuring each wire separately. That is because differential and common-mode noises are combined on each wire and cannot be identified. Therefore, the ECSS-E-HB-20-07A handbook offers an alternative methodology in §7.4.2 utilizing two complementary current probe set-ups to measure the "true" differential and common mode conducted emissions on the power lines of the equipment under test.

Figure 2. MIL-STD-461G, CE102 Measurement setup (DoD:USA, 2015)

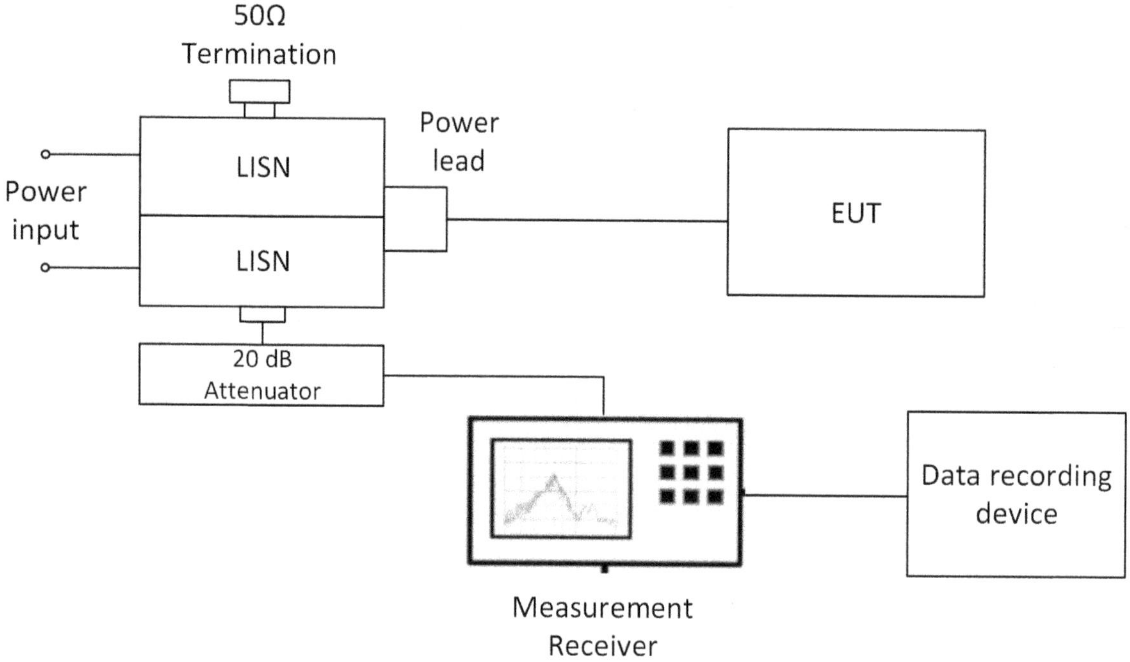

At this point, it is deemed important to highlight a crucial difference between the two aforementioned Standards regarding the proposed measurement methodology in this frequency range. More specifically, MIL-STD-461G specifies the conducted emission measurements in terms of voltage while ECSS-E-

Figure 3. ECSS-E-ST-20-07C Rev.1 §5.4.3 – Common mode disturbance measurement setup (ECSS, 2012)

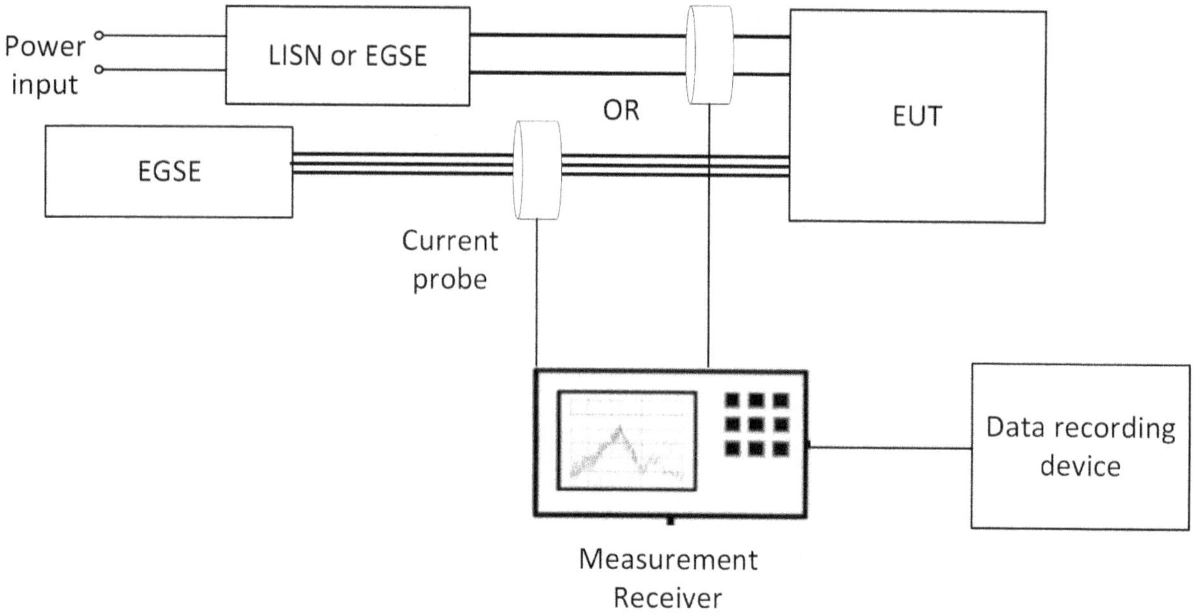

ST-20-07C specifies them in terms of current. The main argument for current measurements is based mainly on the rather high impedance requirement for LISNs (50μH) when measuring voltage values in this frequency range. While the requirement can be easily achieved it is considered nonrepresentative of the actual onboard impedance values, which are significantly smaller (~5μH for a harness on a small platform)

10kHz-40GHz

Conducted emission requirements above 100MHz are only specified in the MIL-STD-461G Standard and are applicable for the antenna ports of transmitters, receivers and amplifiers. Products designed to include antennas permanently mounted to the product are exempt from this requirement. Similar to radiated emission measurements in general, the upper-frequency limit for the measurements depends on the highest operating frequency (either generated or received by the equipment under test). For operating frequencies below 1GHz, the upper-frequency limit is either 20 times the highest frequency or 18GHz whichever is greater. For operating frequencies above 1GHz, the limit is 10 times the highest frequency or 40GHz whichever is greater. The specific measurement setups for low/high power transmitters and amplifiers, as well as for receiver and transmitters and amplifiers in standby mode are described in detail in §5.6 of MIL-STD-461G. It should be noted that a radiated emission measurement according to RE103 (described later in this chapter) can be used as an alternative methodology for measurements in this range.

Inrush Current

ECSS-E-ST-20-07C Rev.1 §5.4.4, CE, Power Leads, Inrush Current

All electrical/electronic units do not reach their steady-state instantaneously after getting powered on. During this switching process, current transients appear at the power input ports. These transients, which are most commonly referred to as inrush currents, need to be limited and controlled to ensure the compatibility of the equipment under test with the protection devices of the power conditioning/distribution unit, which can be either fuses or solid-state protection circuits and reduce the resulting transient voltage on the power bus. This inrush current includes the current peak due to the charging of the power input filter capacitors and the start-up current of various electronics. In a similar to the aforementioned conducted emission measurements manner, after initial calibration of the measurement setup this transient current is measured in the time domain with the use of an oscilloscope and the measured parameters are the amplitude of this current as well as its rise time. The measurement setup is presented in Figure 4.

Summarizing the Conducted Emissions section, an observation can be made, which is true for almost every certification or EMC evaluation process. The complete evaluation of the conducted emissions of a product for Space Applications can rarely be covered by only one Standard. This means that product manufacturers, certification consultants and program coordinators should thoroughly research potentially applicable Standards to best evaluate their product EMC performance. Additionally, it should be noted that not every procedure is suitable for every case. This can be easily observed by the impedance requirements, mentioned earlier, for the 10kHz-100MHz frequency range

Figure 4. ECSS-E-ST-20-07C Rev.1 §5.4.4 – Inrush current measurement setup (ECSS, 2012)

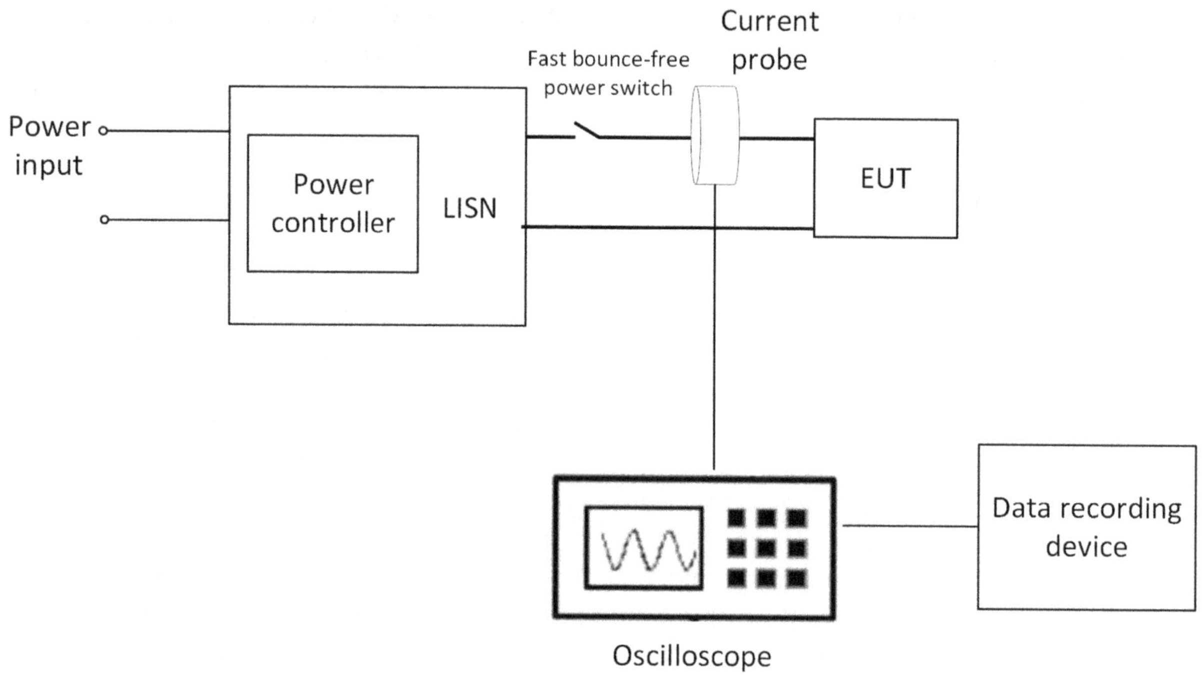

Radiated Emissions

Every equipment generates unintentional electromagnetic disturbances, measured as electromagnetic fields, due to the operation of its electronic components, especially when these utilize switching voltages and currents but also from the steady-state AC or DC voltages/currents. The measurement methodology for these emissions varies significantly depending on the nature of the measured field (electric or magnetic) and the applicable frequency range. In this section, the measurement methods for each type of field and frequency as described by Standards MIL-STD-461G and ECSS-E-ST-20-07C will be presented and discussed.

DC Magnetic Field Emissions, Magnetic Moment

The purpose of these measurements is to ensure the magnetic cleanliness of the spacecraft since magnetometers are part of the essential spacecraft equipment used for studying the structural composition of planets and the interaction of solar winds and planetary environments. They are also used in the Attitude Control System (ACS) of the spacecraft, where the angular error should not be affected by parasitic emissions beyond the acceptable limits and since these magnetometers are measuring very low magnetic field values the spacecraft needs to as magnetically clean as possible. Additionally, these requirements aim to protect other susceptible payloads inside the spacecraft, such as inertial sensors involving floating proof masses, ultra-stable oscillators and atomic clocks.

ECSS-E-ST-20-07C Rev.1 §5.4.5

The methodology proposed in this Standards aims to provide a rough estimation of the magnetic moment of the equipment under test at distances approximately three times its size. Initial magnetic field measurements are fed to numerical solvers for magnetic dipole and moment modeling. The larger the number of measurements and measurement positions around the equipment available the better the accuracy of the modeling procedure. ECSS-E-ST-20-07C suggests measurements at two different reference distances r_1 and r_2 from the equipment under test and on the 6 semi-axes of the orthogonal coordinate system XYZ, assigned to the equipment geometric center. The Standard also offers an alternative measurement configuration where the equipment is placed on a turn-table rotating in front of a fixed magnetometer. The test facility for the measurements should be set in an earth field compensated area providing zero-field conditions. Such a facility is called Magnetic Coil Facility (MCF) and two of them currently owned by the European Space Agency (ESA) are located at ESTEC (ESA's European Space Research and Technology Centre in Noordwijk, Netherlands) for the tests of payload units of LISA Pathfinder and Swarm and EADS Astrium Stevenage in the UK for testing units of the LISA Pathfinder mission.

The test sequence proposed in ECSS-E-ST-20-07C includes

a) the demagnetization of the unit under test (also known as deperming) by application, while the equipment is not operating, of a deperming field, as shown in Figure 5, and the measurement on all six semi-axes at both reference distances
b) while the equipment is not operating, the magnetization of the unit (perming) by application of a 300μT perm field and the measurement on all six semi-axes at both reference distances
c) measurement of the stray magnetic field while the equipment is operating on all six semi-axes at the reference distances.

The measurements are then processed with the Multiple Dipole Modeling method, modeling the DC magnetic field of the equipment under test as a set of magnetic dipoles defined by their individual location, orientation and amplitude.

Electric Field Radiated Emissions 10kHz-18GHz

The rationale behind setting requirements for electric field radiated emissions is mainly focused on the protection of antenna-connected receivers from interference coupled through their antennas. Measurements of radiated emissions from products are usually a very complex procedure since the emitted electromagnetic noise from the equipment under test does not display homogeneous spherical characteristics and spans in a very wide range of radio frequencies. Unintended (parasitic) emissions are mostly directional, therefore requiring a 360° observation/measurement in a variety of antenna heights. Moreover, the wide frequency range of these emissions results in the requirement of a variety of antennas suitable for each frequency region. The different types of linearly polarized antennas used are:

- 10 kHz to 30 MHz, rod antenna with impedance matching network
- 30 MHz to 200 MHz, biconical antenna
- 200 MHz to 1 GHz, double ridge horn, or log-periodic antenna
- 1 GHz to 18 GHz, double ridge horn antenna

Figure 5. ECSS-E-ST-20-07C Rev.1 §5.4.5 – Deperming field for DC magnetic field emission measurements (ECSS, 2012)

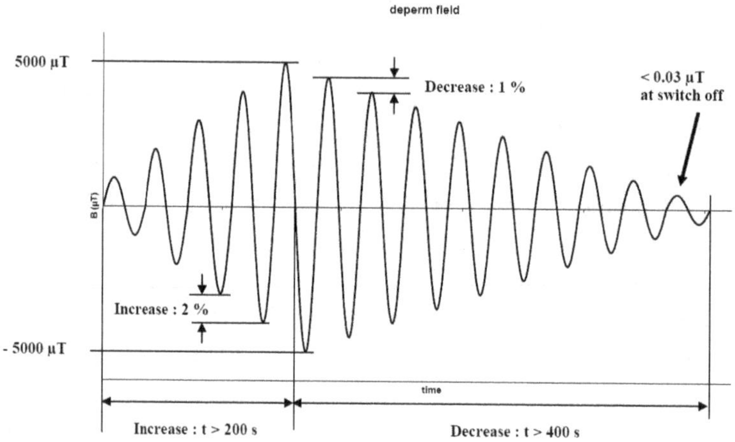

MIL-STD-461G, RE102, Radiated Emissions, Electric Field

MIL-STD-461G describes measurements in the whole frequency range of 10kHz p to 18GHz for Space Systems. As already discussed above the measurement procedures vary significantly based on the frequency band to be measured. So, apart from the different types of antenna required for each band, the measuring antenna positioning should also be adjusted accordingly. A simplified diagram of the test setup is presented in Figure 6.

The purpose of the signal generator in Figure 6 is for the verification of the measurement equipment and the associated cabling prior to the final measurements. More on this procedure is described in §5.18.3.4 c) and d) of MIL-STD-461G. The limits proposed for Space Systems in this Standard are shown in Figure 7.

The arrangement of the EUT and the associated cabling is presented in detail in Figures 1-5 of MIL-STD-461G. One of the main differentiating parameters of these setups is the use of a conducting or non-conducting table. In the latter case, the Standard specifies the requirement for the cables to be mounted on insulating support with a height of 5cm. The effect of the material composition of the insulating support, however, is often neglected and Standards do not always define specific requirements for this characteristic recommending the use of wood or foam to minimize the distortion of the electromagnetic fields. Research performed recently by Sen et al (Sen et al., 2016) has shown that during testing according to RE102 ofMIL-STD-461G foam type supports have displayed low measurement errors leading to lower uncertainty values. In their work, Sen et al., after measuring the dielectric constant (ε_r) values for four different materials (Wood, Molding Polyamide, Foam Rubber, Foam-Like Plastic, Styropor) at two different frequencies (1kHz and 2GHz), performed measurements in the frequency range of 200MHz-1GHz calculating the deviation between measured values (dBµV) with and without support as well as the associated uncertainty component. Supports made of foam-type materials displayed a maximum deviation of 1dB while supports made of wood or molding polyamide display a deviation of 15dB greatly reducing measurement accuracy and repeatability (Sen et al., 2016).

A final point regarding the radiated emission testing specified in MIL-STD-461G is the recommendation for utilizing the FFT-based time domain capabilities of EMI receivers. Contrary to conventional EMI

Figure 6. Electric Field measurement system test setup (DoD: USA, 2015)

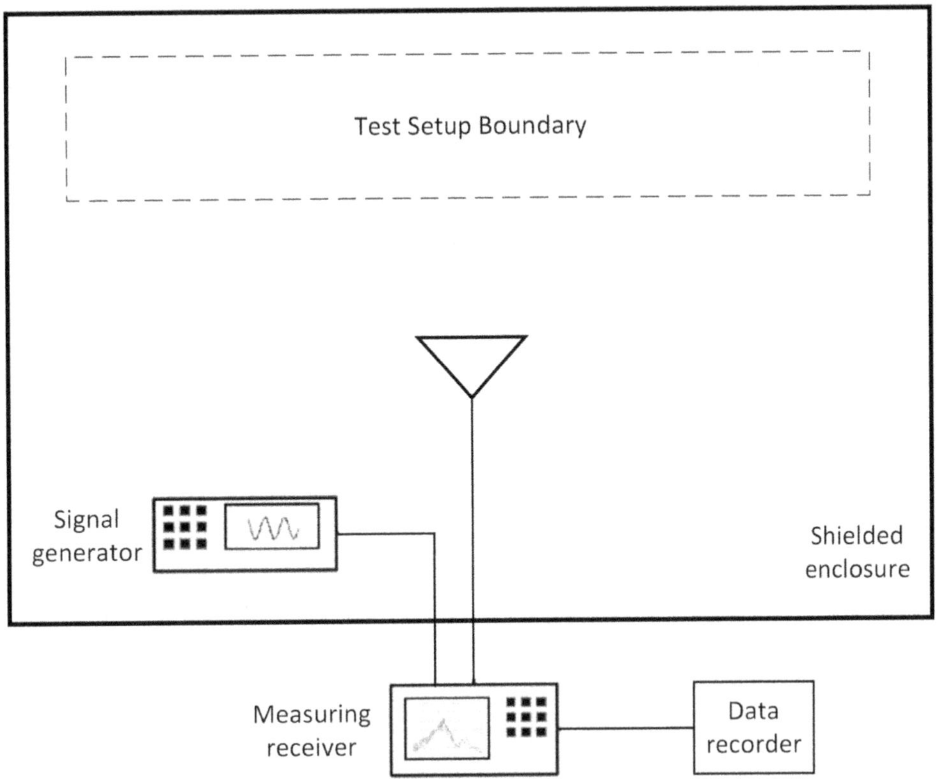

receivers which measure the signal within a pre-configured resolution bandwidth in a given measurement time window, FFT receivers utilize the parallel calculation capabilities of the receivers to measure spectral segments wider than the measurement bandwidth offering significantly lower scanning times, which allow an increase on measurement times, for intermittent signals to be identified and measured. NASA has already applied this type of radiated emissions measurement during the electromagnetic compatibility tests on the Integrated Science Instrument Module (ISIM), which is the science payload of the James Webb Space Telescope, at NASA's Goddard Space Flight Center (McCloskey, 2016). During radiated emissions testing, during an effort to capture the worst-case emissions of the equipment under test, mechanism movements needed to be performed. By nature, those movements are non-continuous operations of finite-time duration, therefore the conventional measurement technique (conventional EMI receivers) was not deemed appropriate for measuring these emissions and the FFT-based function of the EMI receiver was used (McCloskey, 2016).

ECSS-E-ST-20-07C §5.4.6, RE, Electric Field, 30MHz to 18GHz

The proposed methodology described in this Standard is almost identical to the methodology in MIL-STD-461G described above. The only significant difference is the absence of the requirement for measurements in the 10kHz-30MHz range. The rationale behind the discouragement of measuring in this frequency range with the use of rod antenna, unless equipment sensitive to low-frequency electric fields is expected to be installed in the specific spacecraft, is based on the reduced repeatability/reproducibility

Figure 7. Electric field emissions limit line according to MIL-STD-461G (DoD: USA, 2015)

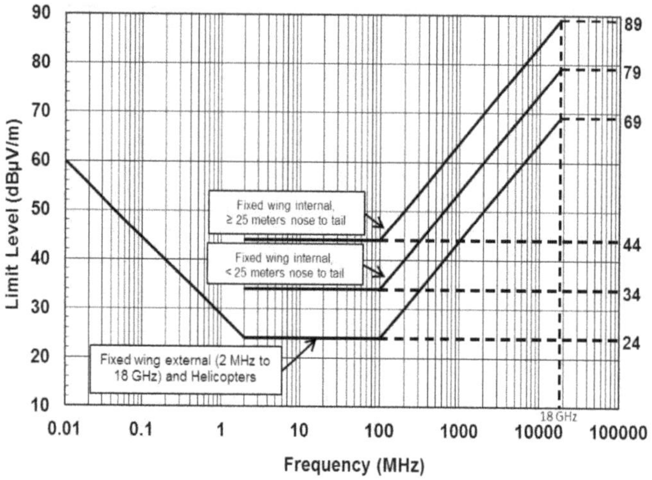

of this measurement procedure due to unwanted (parasitic) resonances of the test set-up. Recent work by Gandolfo et al. (Gandolfo et al., 2017) focused on the investigation of the contribution of certain elements to the reduced repeatability of the measuring procedure for measuring radiated electric fields in such low frequencies. Utilizing electromagnetic simulations and an innovative design for a rod antenna that includes an embedded receiver and the data transfer implementation via optical fiber, they studied different proposed setups. The results showed that the simplification of the test setup, by removing the coaxial cable as well as some components that were introduced to minimize its effects, contributed to the improvement of the test repeatability leading to more accurate measurements in relation to measurements under simulated undisturbed field conditions. In their work, Gandolfo et al. also investigated three different counterpoise configurations (isolated counterpoise, counterpoise bonded to floor configuration and counterpoise bonded to the ground plane). The results showed that the most consistent behavior can be achieved utilizing the isolated counterpoise configuration (Gandolfo et al., 2017).

ECSS-E-ST-20-07C §7.3.1.1 offers a specific precaution approach for limiting these resonances from ±(10-15) dB to ±1 dB, as presented in Figure 8.

Summarizing the radiated emissions section, the absence of magnetic field emissions requirement in the low-frequency range (30Hz-50kHz) needs to be addressed. Neither of the two main Standards includes such a requirement for Space Systems and that is because both Standards consider these measurements not applicable unless specified by the procuring activity. That means if equipment sensitive to low-frequency magnetic fields is to be installed in the spacecraft. Such equipment is AC magnetometers, inertial sensors, etc. The argument against the by default requirement of these measurements is based on the fact that magnetic field measurements in this frequency range are significantly affected by radiated emissions at the main frequency (50Hz) and its harmonics. Most common sources of magnetic emissions in this range are the solar array, which can be adequately tested in laboratory conditions and cryocoolers incorporating compressors as well as reaction wheels.

A great, recently emerged, challenge regarding the electromagnetic compatibility of space equipment is the rapid increase of scientific satellites, equipped with very sensitive measuring devices for the observation of deep space, planets and stars requiring measurements of very low electric and magnetic

Figure 8. Precautions for electric field measurements in the frequency range 10kHz-30MHz (ECSS, 2012)

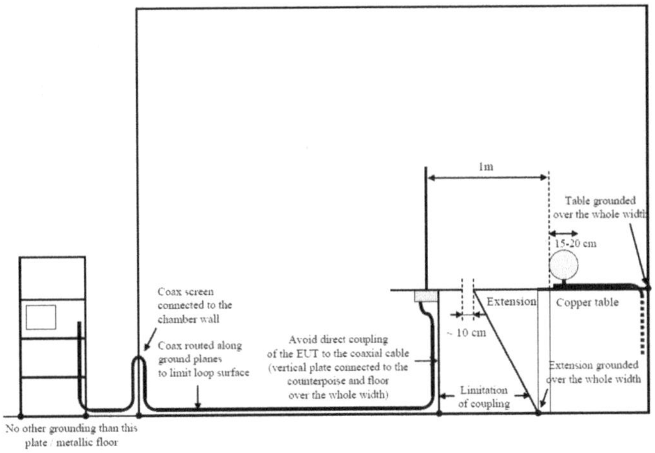

field values. In such cases, the requirements described earlier in this section are not strict enough to allow such measurement equipment to receive undistorted and noise-free measurements. For two of their most recent missions, Solar Orbiter and Juice, the European Space Agency (ESA) specified requirements for the electric and magnetic field emissions of space equipment, which are way lower (more strict) than those in their commonly used ECSS-E-ST-2007C Standard and cover a larger frequency range extending the lower frequency range limit to some Hz. While the E-Field Standard limits vary between 20dBμV/m up to 60dBμV/m, the limits for these missions reach in some cases values of -27dBμV/m. For equipment installed in the interior, the limits are a bit more relaxed, since they are "protected" by the shielding of the enclosure, however, they remain still way lower than those specified in the Standard. In their work, Hernandez et al. present these more stringent requirements in detail and describe some countermeasures and suggestions for the reduction of electric and magnetic field emission levels of equipment to be installed in scientific satellites. These suggestions are applicable not only during the design phase of the product but also during the system analysis of the mission. The designed equipment could potentially be completely "silent", regarding its unintended electromagnetic emissions, during isolated testing, but emit electromagnetically way over the limits when connected to the entire system via its interconnecting cables (Hernandez et al., 2018).

Conducted Susceptibility Testing

When performing conducted immunity testing, RF voltages or currents are injected into each of the cables connected to the equipment under test. Since it is technically difficult to generate uniform fields at low frequencies in typical test facilities, it is preferred to perform conducted testing since susceptibility problems at low RF frequencies are usually associated with the cable coupling. This type of testing is considered an alternative, although not entirely equivalent, to radiated immunity tests at lower RF frequencies, which also offer very good reproducibility.

The purpose of these tests is to simulate the behavior of the EUT and its attached cables to RF noise at low frequencies emanating from various sources within the spacecraft. It also aims to determine the level of susceptibility of the EUT to these interference signals applied to incoming and outgoing EUT cables to ensure that the functionality and performance of the EUT are not affected by interference noise on the power or signal lines.

Conducted susceptibility testing can be time-consuming depending on the number of cables connected to the EUT. Additionally, this test requires different types of testing equipment to cover the frequency range of interest from 30 Hz to 100 MHz or more for space applications. For equipment utilizing antenna ports, the upper-frequency limit can be extended up to 20GHz. The test is divided into multiple sub-ranges depending on the type of interference signal and the applicable port. The tests are applicable to all types of devices and require a screened room to be performed. The main purpose of this is to avoid ambient noise that could interfere with other nearby activities, but also to ensure that ambient signals do not interfere with the test.

Low Frequency Conducted Susceptibility Testing

This requirement is used to ensure that equipment performance is not degraded from ripple present on the power bus. The frequency range of the interference application is typically between 30 Hz to 100kHz or 150kHz. In this frequency range, an injection transformer is used to apply the sinusoidal interference signal to the EUT power leads as well as a current probe to measure the injected current. The tests defined in the two main Standards, mentioned earlier, are described and discussed below.

MIL-STD-461G, CS101, Conducted Susceptibility, Power Leads

The frequency of interest in this Standard is 30Hz-150kHz for AC equipment and systems with rated current £30A per phase and DC input leads. This requirement is also applicable to systems with rated current ^{3}30A, an operating frequency £150kHz and an operating sensitivity of £1μV.

The Standard offers two alternative methodologies for monitoring the applied signal during immunity testing and calibrating it before the test: a) using a power input isolation transformer connected to an oscilloscope or b) using a measuring receiver connected to a transducer, which is used for the isolation of the receiver from the EUT power and its protection. The first combination of instruments is used for the verification of the test signal in the time domain, while the second combination for the verification in the frequency domain. Recently, the benefits of utilizing the FFT function of an oscilloscope for the verification of the signal in the frequency domain have been investigated (Schnecker, 2016). Despite the useful voltage handling capability and the probing options provided by oscilloscopes, their limited resolution in the time-domain display leads to inaccurate measurements of the applied signal level limiting the measurement range and requiring a guard band. The results of this work show that utilizing the FFT function of oscilloscopes, measurements display more than adequate accuracy in accordance with the requirements of CS101 of the MIL-STD-461G Standard without the requirement for special coupling units or circuits which are necessary for measurements with a measuring receiver. (Schnecker, 2016).

The applied voltage used to simulate the interference signal is presented in Figure 9. The proposed setup for DC or single-phase AC systems is presented in Figure 10.

ECSS-E-ST-20-07C §5.4.7, CS, Power Leads, Differential Mode, 30Hz to 100kHz

In general, the requirements and test setup described in this Standard are very similar to the requirements of MIL-STD-461G (CS101). The European Standard differentiates itself from its American equivalent in some important points. The applied voltage is set to $1V_{rms}$ for the entire frequency range of 30Hz-100kHz

Figure 9. MIL-STD-461G, CS101 applied voltage (DoD: USA, 2015)

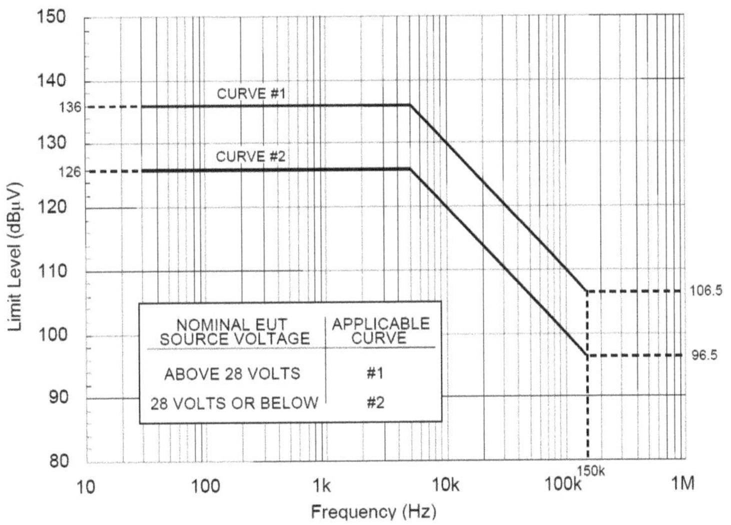

contrary to the voltage presented above in Figure 9 in MIL-STD-461G. Additionally, ECSS-E-ST-20-07C specifies a resistor connected in parallel with the primary winding of the coupling transformer to prevent the DC/DC converter from being damaged by experiencing oscillations as a consequence of inadvertent switch-off or disconnection of the amplifier. Another important difference between the two Standards is the absence of differential-mode conducted susceptibility requirements for frequencies above 100kHz. As already discussed in a previous section of this chapter, ECSS-E-ST-20-07C emphasizes a lot on the concept of "true" differential mode above 100kHz, therefore it introduces an alternative test setup for carrying out "true" differential mode conducted susceptibility testing for the frequency range of 100kHz-10MHz, which is described in the Standard not as a requirement but as a tailoring option for cases of Space Systems, where such testing is required. This alternative setup is presented in Figure 11.

MIL-STD-461G, CS109, Conducted Susceptibility, Structure Current, 60Hz to 100kHz

According to this requirement equipment and systems operating at frequencies below 100kHz and with a sensitivity threshold of £1μV, such as tuned receivers, shall be able to withstand structure currents. More specifically, this requirement is defined to ensure that the equipment under test does not respond to magnetic fields produced by currents in platform structure and through the enclosure materials of the EUT. The applied current values defined in the Standard and presented in Figure 12 are derived from recorded failure occurrences caused by conducted currents on equipment cabinets and laboratory measurements where the response of specific receivers is recorded.

The test setup described for this test is very similar to all conducted susceptibility tests described above with the addition of the test point selection procedure. An electrical connection on the external surface of the EUT should be made. Special attention should be paid to minimize the damage to the exterior finish. Screws or other protuberances with connection to ground at the outer corners of the EUT surface may be used as test points utilizing clip or clamp type leads. These leads need to remain

Figure 10. MIL-STD-461G, CS101 test setup for DC or single-phase AC systems (DoD: USA, 2015)

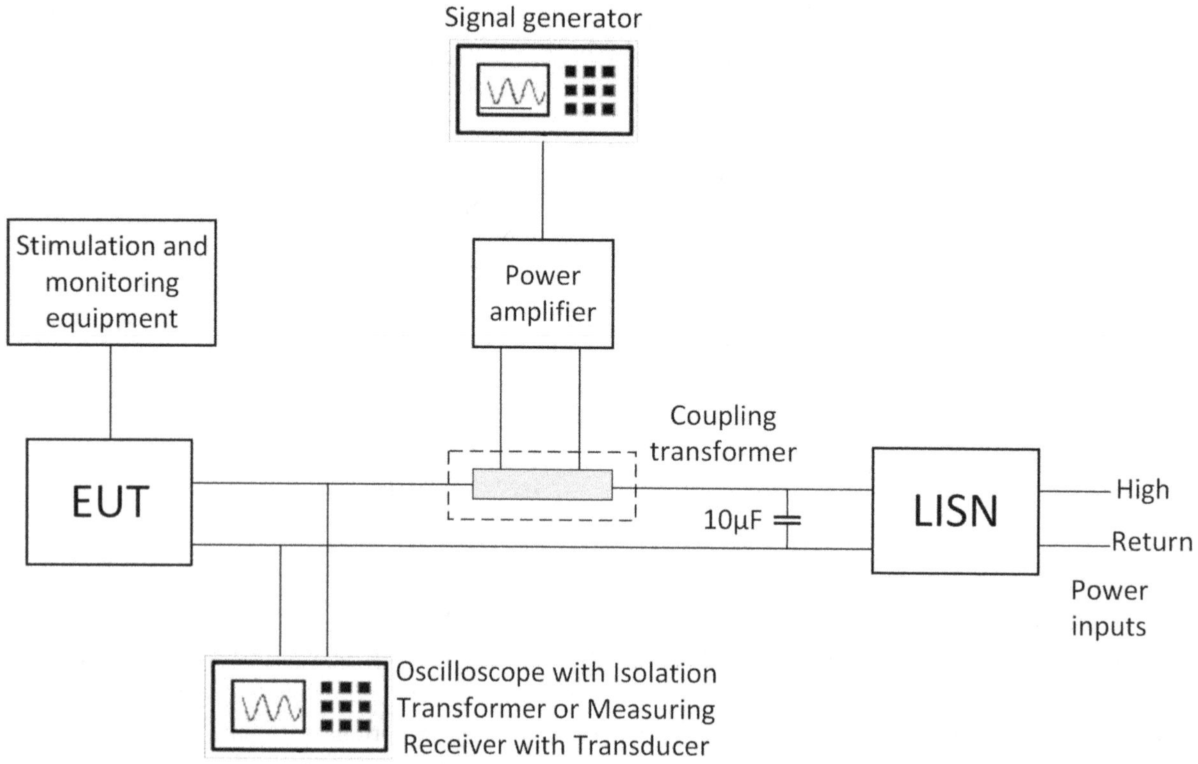

perpendicular to the surface for at least 50cm to minimize the effect of the lead currents. The proposed setup is presented in Figure 13.

Radio Frequency Conducted Susceptibility Testing

MIL-STD-461G, CS114, Conducted Susceptibility, Bulk Cable Injection, 10kHz-200MHz

The scope of this specific test is the simulation of currents induced on the cabling of the system under test from electromagnetic fields produced by antenna transmissions both on and off the platform. An important advantage of such a requirement is that its results can be directly associated with induced current values measured during evaluations at platform-level. During such evaluations, the platform is illuminated with a predefined low-level uniform field and the induced current levels are monitored. The results of these evaluations can lead to the revision of laboratory data or current injection with the measured currents scaled to the full threat level. The rationale for a conducted approach to susceptibility testing at such frequencies over the radiated approach that will be discussed later in this chapter is based on examples such as the HF (2-30MHz) radio transmissions, where due to size constraints and available field patterns, problems associated with cable coupling that was experienced at the platforms could not be reproduced in the laboratory environment during radiated susceptibility testing.

The proposed testing methodology includes the injection, via current injection probes, with predefined insertion loss characteristics, of a pre-calibrated modulated current. The different current limit curves for

Figure 11. ECSS-E-ST-20-07C Differential mode conducted susceptibility setup for frequencies 100kHz-10MHz (ECSS, 2012)

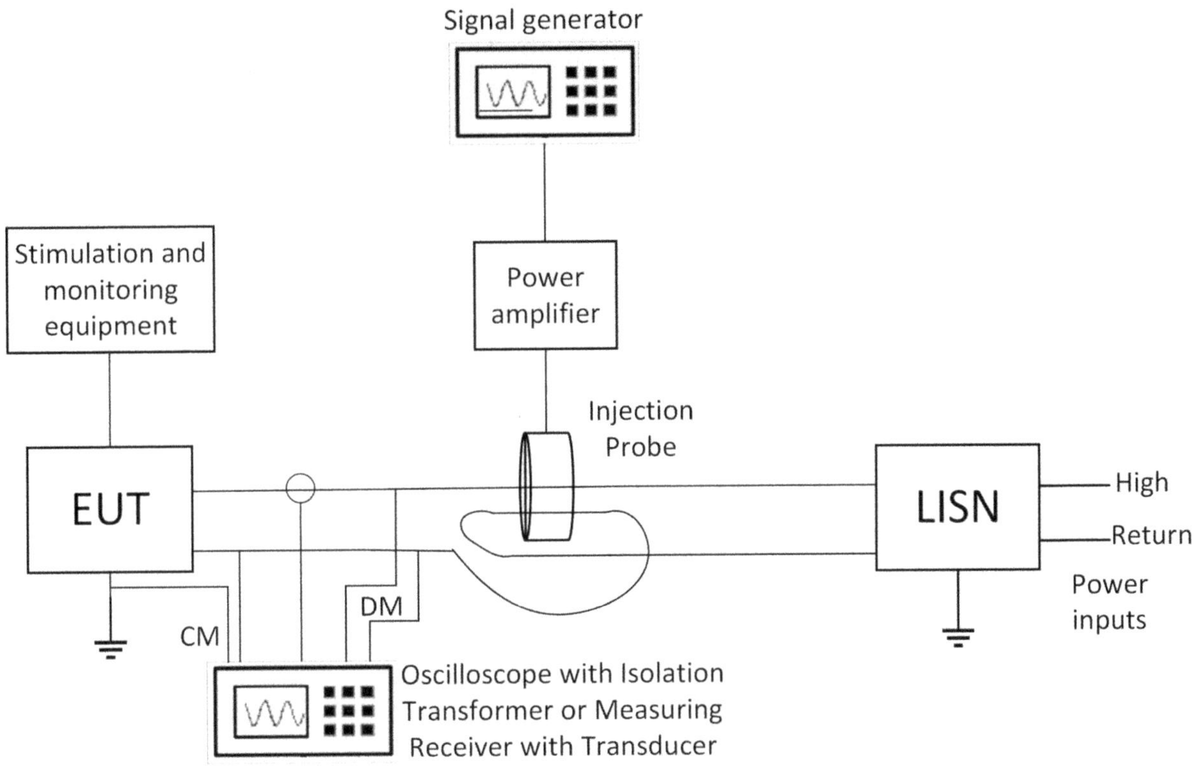

Figure 12. MIL-STD-461G, CS109, applied current limit (DoD: USA, 2015)

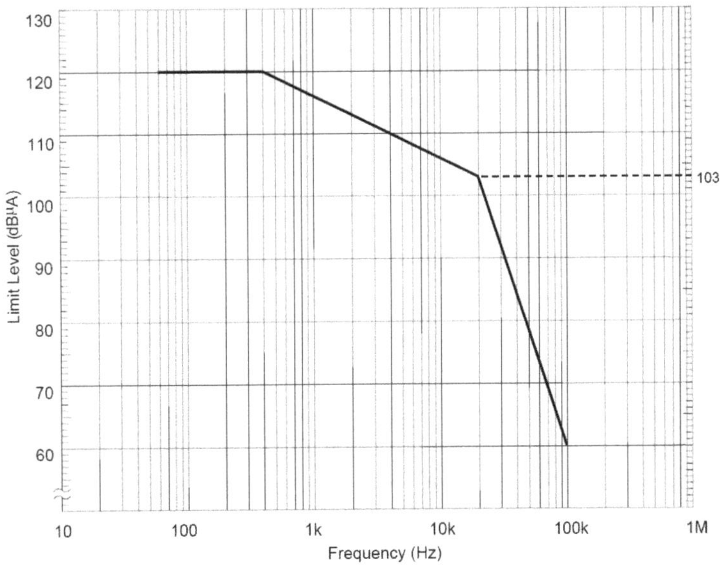

Figure 13. MIL-STD-461G, CS109, test setup (DoD: USA, 2015)

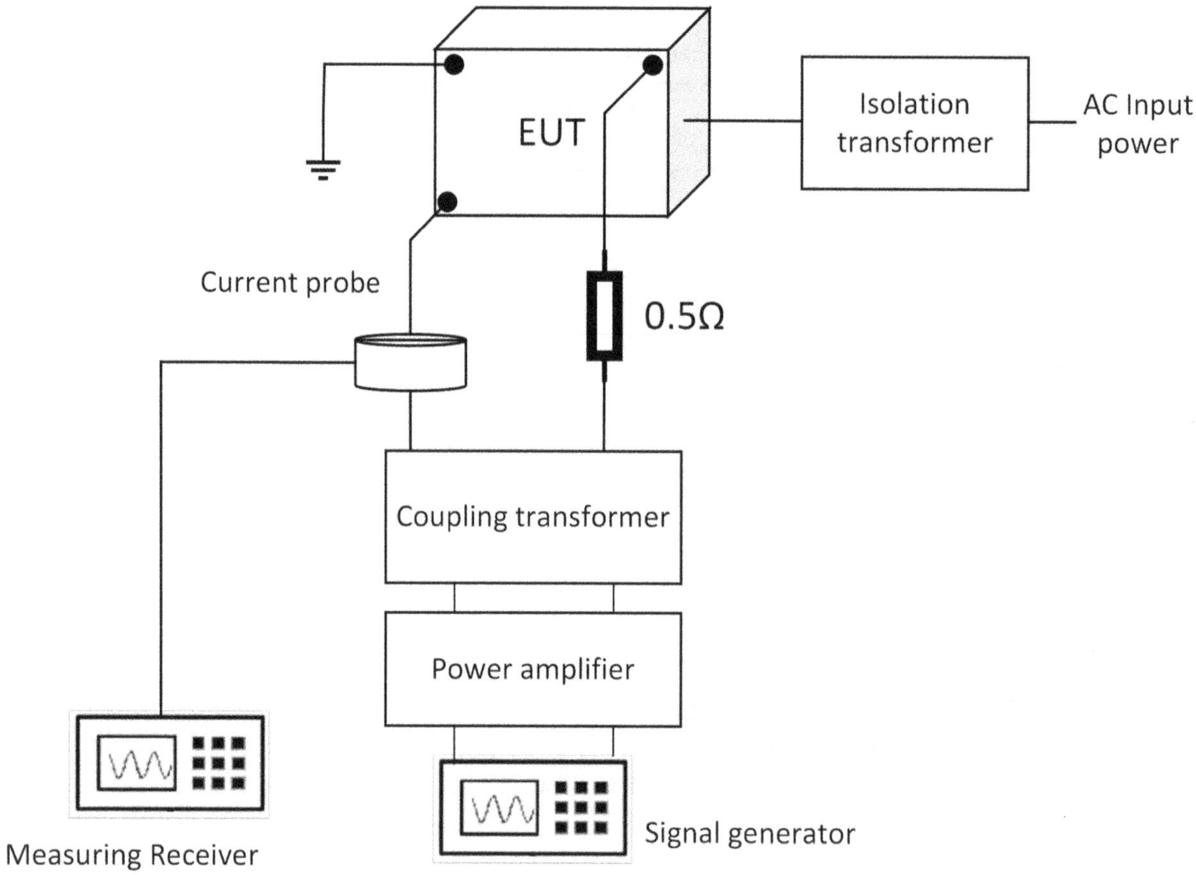

the different applicable environments are presented in Figure 14. Table VI of the Standard recommends using Curve #3 for Space Systems.

The test setup, presented in Figure 15, includes two current probes, one used for the injection of the required current and one for monitoring the EUT current. The injected current is provided by a signal generator through an amplifier and a directional coupler used for the connection of an additional measuring receiver for monitoring the injected current.

MIL-STD-461G, CS115, Conducted Susceptibility, Bulk Cable Injection, Impulse Excitation

Transients with fast rise and fall times are also included in the high frequency conducted susceptibility testing requirements of MIL-STD-461G due to the large frequency spectrum associated with such waveforms. The scope of this requirement is the investigation of the system's sensitivity to fast transient phenomena produced by platform switching operations and external environmental sources, such as lightning strikes and electromagnetic pulses. The proposed methodology replaces the old "chattering relay" approach, which has been criticized for its limited repeatability. The transient waveform, as also presented in Figure 16, has the following characteristics. A rise time of 2ns, simulating the rise times of the switching interruption of inductive devices, a 30ns pulse width which results in a predefined injected

Figure 14. MIL-STD-461G, CS114, injection current limit curves (DoD: USA, 2015)

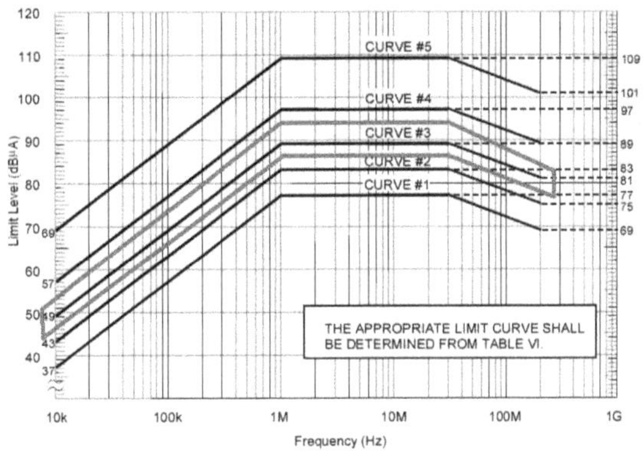

energy level and 30Hz pulse rate to increase the repeatability of the results and ensure the immunity of the system under test.

The proposed test setup is almost identical to the one presented in Figure 15 with two differences: a) The injection of the transient pulses is performed with a Pulse Generator and b) the monitoring is performed with a current monitor probe connected to the 50Ω input of an oscilloscope.

MIL-STD-461G, CS116, Conducted Susceptibility, Damped Sinusoid Transients, Cables and Power Leads

The frequency range of interest for this test is 10kHz-100MHz and it applies to all cables connected to the equipment under test as well as on each high side power input. The aim of this requirement is the susceptibility evaluation of the EUT to simulated current and voltage waveforms produced from the excitation of natural resonances in platforms.

Figure 15. MIL-STD-461G, CS114, test setup (DoD: USA, 2015)

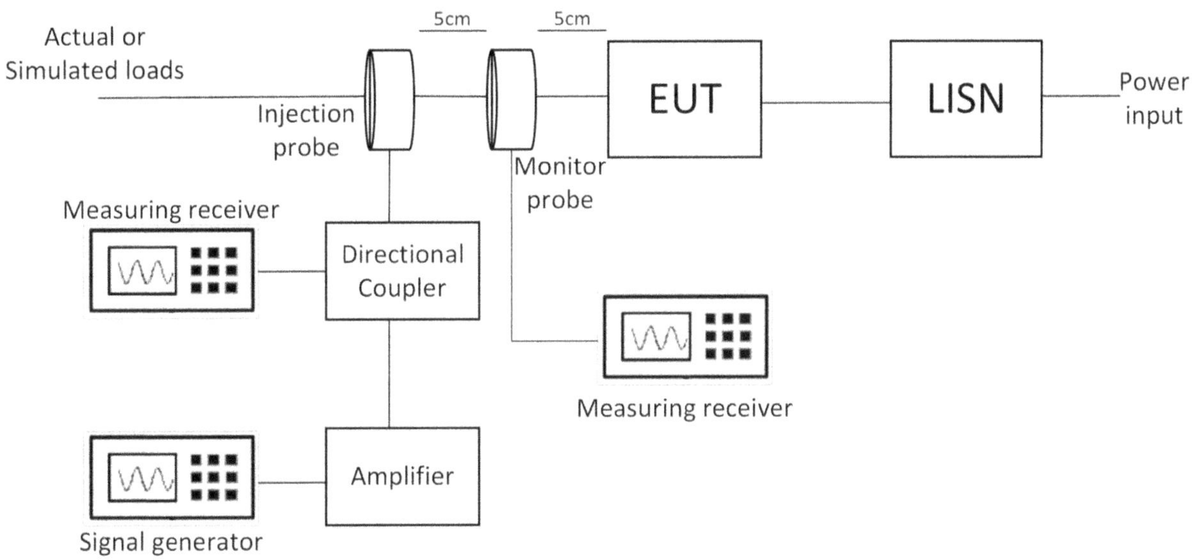

Figure 16. MIL-STD-461G, CS115, transient waveform characteristics (DoD: USA, 2015)

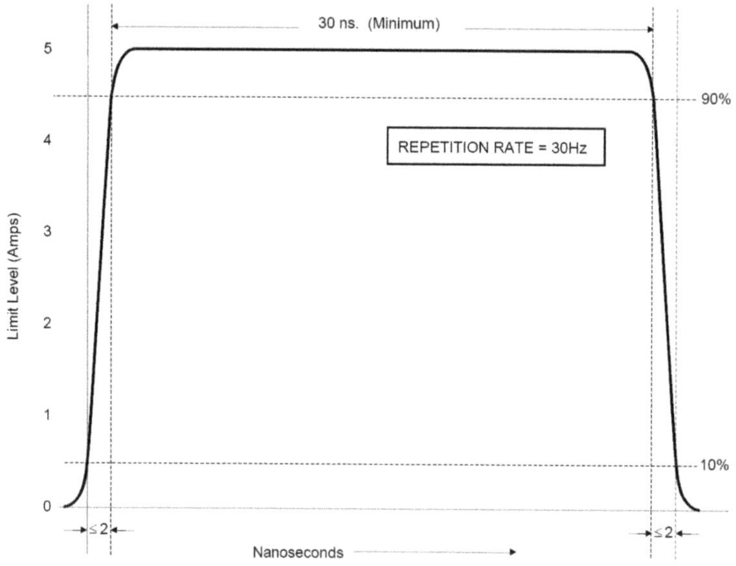

Similar to the CS115 requirement described above, but with a different approach, this conducted susceptibility test aims to control the waveform as a damped sine. Damped sinusoid transient, or oscillating transients, are the result of external stimuli such as lightning strikes, electromagnetic pulses or switching phenomena. Such phenomena induce noise on all types of cables attached to the equipment of the spacecraft, which take the damped sinusoid shape due to the various resonances along the coupling path. These resonances can be attributed either to the conducting cables or other resonating structures along the path. The proposed methodology subjects the equipment under test to the normalized waveform of Figure 17, where the peak current is derived from the limit line of the same Figure. The test setup is almost identical to the one presented previously for CS114 and CS115 with the only difference being (except for the use of a Damped Sinusoid Generator instead of the signal generator or the pulse generator) the usage of a Storage Oscilloscope for the connection of the monitor probe.

Figure 17. MIL-STD-461G, CS116, typical damped sinusoidal waveform and peak current limit line (DoD: USA, 2015)

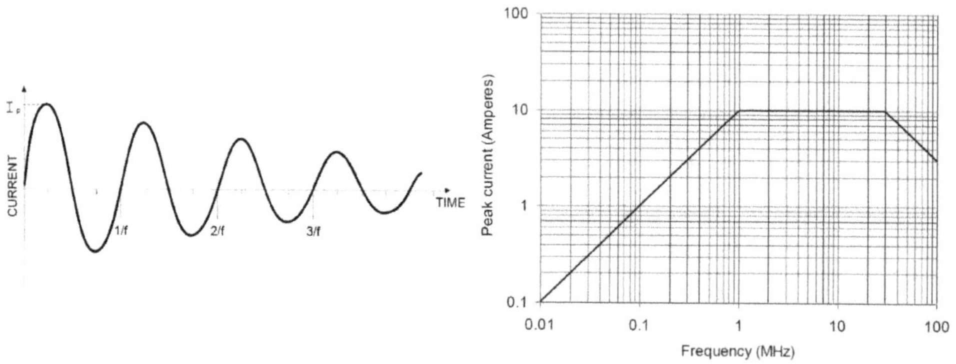

Regarding the applicable frequencies, the Standard requires that compliance should be demonstrated at the following frequencies:0.01, 0.1, 1, 10, 30 and 100MHz. If for any reason (e.g. investigation of specific resonating frequencies due to platform resonances) more resonating frequencies critical for the equipment performance have been identified, compliance shall be demonstrated at these frequencies also. The pulse duration shall be five minutes according to this requirement and the pulse repetition rate should be between 1pulse/second and 1pulse/2seconds.

ECSS-E-ST-20-07C §5.4.8, CS, Bulk Injection, 50kHz to 100MHz

The test described in this European Standard, focused entirely on Space Systems, exhibits many similarities with the tests described above in CS114, CS115 and CS116 of MIL-STD-461G. It has however been refined to the test requirements for Space Systems. The injection method is based on the bulk current injection (BCI) methods described for MIL-STD-461G with slightly modified signal shapes and test frequency characteristics. Figure 18 presents the proposed signal shape for the recommended level of 3V peak to peak calibrated on the BCI jig load with 50Ω, while Table 3 presents the signal test characteristics.

Figure 18. ECSS-E-ST-20-07C §5.4.8, signal shape (ECSS, 2012)

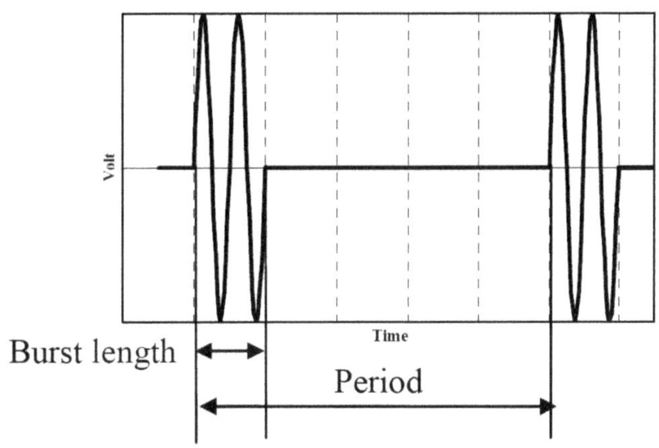

Regarding the differences between this requirement and the CS114-116 requirements of MIL-STD-461G, the ECSS-E-HB-20-07A Electromagnetic Compatibility Handbook attributes the difference of the covered frequency range of antenna transmissions from the CS114 requirement to the more limited transmit frequencies on a spacecraft, which are also almost always well-identified and above 1GHz. However, CS114 has been occasionally used on European Space projects as well but with a different purpose. The purpose was to evaluate the susceptibility of the equipment under test against signals coupled on its bundled or common mode noise produced in the spacecraft internal circuitry, such as crosstalk. According to ECSS-E-HB-20-07A, the testing methodology described in ECSS-E-ST-20-07C §5.4.8 is supposed to be more suitable for the Electromagnetic Compatibility evaluation of typical spacecraft conditions aiming to discover susceptible points due to design shortcoming without experiencing “false alarm” cases which can be time-consuming and not representative of actual and possible problems. Regarding the CS115 and CS116 requirements of MIL-STD-461G, ECSS-E-ST-07C does not specify an equivalent or alternative requirement for the evaluation of the equipment under test immunity to the

Table 3. ECSS-E-ST-20-07C §5.4.8, CS signal test characteristics (ECSS, 2012)

Frequency range	Pulse repetition frequency	Duty cycle
50 kHz- 1 MHz	1 kHz	50% (square wave)
1 MHz – 10 MHz	100 kHz	20%
10 MHz – 100 MHz	100 kHz	5%

effects of external transient environments. It however specifies a dedicated ESD test, simulating the coupling of electrostatic discharges to bundles, which will be discussed later in this chapter.

Transient Conducted Susceptibility Testing

Immunity testing to transient electromagnetic phenomena (except immunity to electrostatic discharges, which is usually investigated and tested on its own, that means with a dedicated test) can be covered by various Standard requirements, some of which have already been presented and discussed in the previous section, such as CS115 and CS116 of MIL-STD-461G. More emphasis on the susceptibility of the equipment under test to transient voltage or current pulses with precisely defined waveforms, that means transient waveforms with specified voltage or current peak levels, rise/fall time and width, is provided by MIL-STD-461G in the CS117 section, which however is not applicable to Space Systems. ECSS on the other hand includes in its ECSS-E-ST-20-07C Standard a dedicated section for the susceptibility testing of the spacecraft equipment to voltage transients that can occur either due to the switching on or off of other systems connected to the spacecraft power bus or due to a short circuit event on the power bus, before the reaction of the central protections.

Figure 19. ECSS-E-ST-20-07C §5.4.9 test setup (ECSS, 2012)

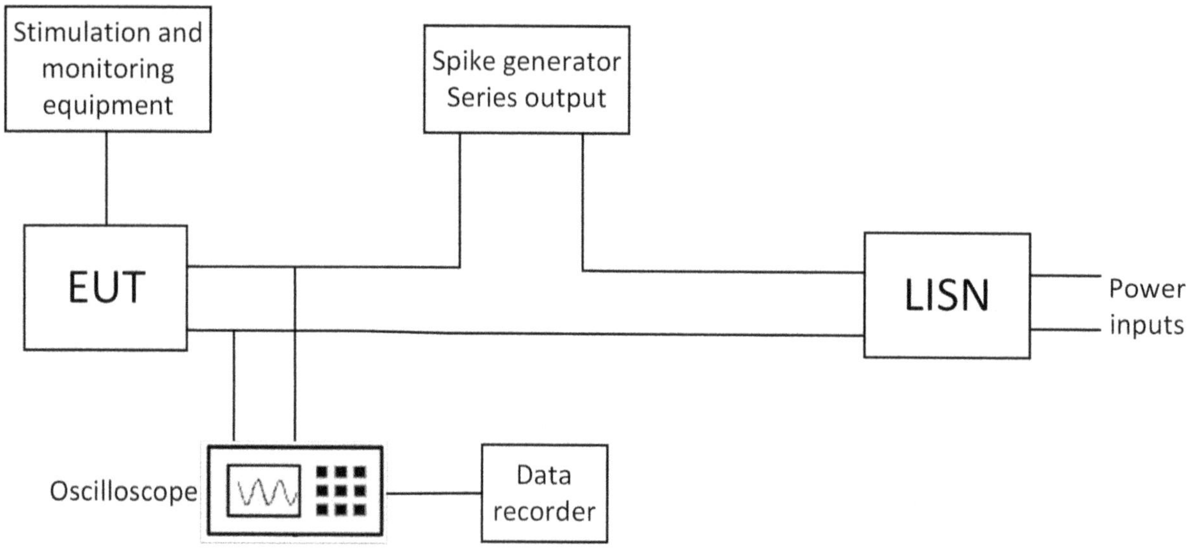

ECSS-E-ST-20-07C §5.4.9, CS, Power Leads, Transients

The proposed test setup (Figure 19) is very similar to the one presented in CS115 of MIL-STD-461G. Figure 20 presents the time normalized voltage spike waveform. The applied peak voltage values are specified as a percentage of the line voltage.

Figure 20. ECSS-E-ST-20-07C §5.4.9 voltage spike waveform (ECSS, 2012)

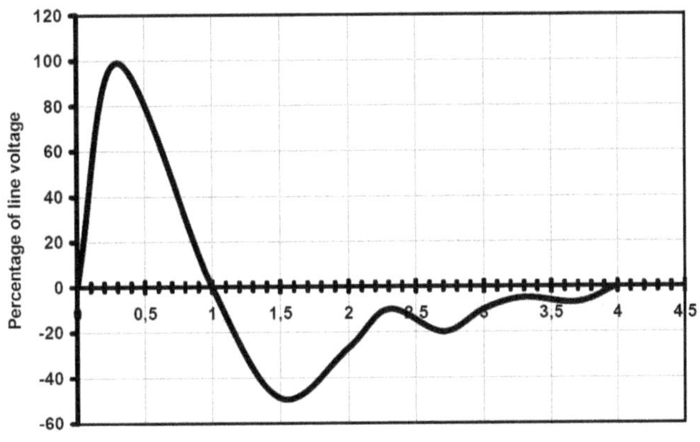

The proposed methodology specifies some of the necessary parameters for the conduction of the aforementioned transient immunity tests. First, a series of positive spikes superimposed on the input power voltage shall be applied and then a series of negative spikes. The spike amplitude for the positive polarity spike shall be +100% of the actual line voltage if the rated bus voltage is lower than 100V or 50% if the rated bus voltage is equal or greater than 100V. For the negative polarity spikes, the amplitude shall be -100% for every rated bus voltage value. Two sets of pulse application are specified in the Standard for two different spike duration values. The spike duration, which is defined as a first zero-crossing point, shall be 150ns and 10μs.

Radio Frequency Conducted Susceptibility Testing up to 20 GHz

Requirements for conducted susceptibility evaluation for frequencies up to 20 GHz in the MIL-STD-461G Standard are focused on the evaluation of the antenna port of specific RF equipment, such as communication receivers, radar and acoustic receivers, transceivers and RF amplifiers. The CS103, CS104 and CS105 sections of the Standard offer guidance on the necessary testing conditions and parameters. More specifically, CS103 is used to determine the presence of intermodulation products due to undesired signals at the antenna input ports of the equipment under test. CS104 investigates the presence of spurious responses caused by such undesired signals and CS105 is used to determine the presence of cross-modulation products. The applicability of these requirements and the corresponding testing limits depend on the subsystem design being developed, which can vary significantly between different subsystems. Therefore, testing parameters need to be determined for each procurement. Annex A (A.5.8, A.5.9 and A.5.10) of MIL-STD-461G offers guidance towards this effort. Since these requirements are very design-specific, this chapter will not focus on more details on this matter and will refer the reader to the aforementioned Standard Annex.

Radiated Susceptibility Testing

The evaluation of the performance of equipment, when subjected to radiated electromagnetic fields, has always been a core element in Electromagnetic Compatibility testing. Due to the nature of electromagnetic fields, there is a categorization of immunity tests based on the frequency range of each type of field. That means, that for low frequencies, i.e. up to couple hundred kilohertz, the equipment is subjected to magnetic fields, while for higher frequencies to electric fields. In the following section, the immunity requirements specified by the two main Standards, discussed in this chapter, will be presented for each frequency region.

Radiated Susceptibility Testing to Low-Frequency Magnetic Fields

Requirements in this frequency region are used to evaluate the performance of equipment considered potentially sensitive to such magnetic fields which are installed in the vicinity of a source of such fields. In the MIL-STD-461G Standard, the associated requirement (RS101), is not applicable to equipment included in Space Systems. However, its European equivalent (ECSS-E-ST-20-07C) involves such a requirement.

ECSS-E-ST-20-07C §5.4.10, RS, Magnetic Field, 30Hz to 100kHz

The test setup proposed in this Standard, as presented also in Figure 21, includes the injection of a pre-calibrated current from a signal generator via a power amplifier in a radiating loop of specific characteristics (Diameter: 12cm, Number of turns: 20, etc.) resulting in magnetic field values according to the limit line of Figure 22. The loop antenna should be positioned in a 5cm distance from the face of the equipment being tested.

Figure 21. ECSS-E-ST-20-07C Radiated susceptibility to magnetic fields setup (ECSS, 2012)

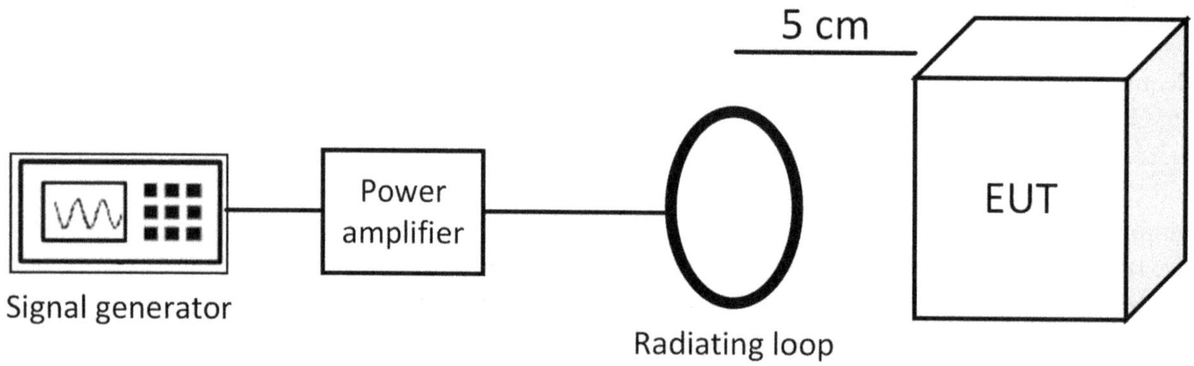

Figure 22. ECSS-E-ST-20-07C Radiated susceptibility to magnetic fields limits (ECSS, 2012)

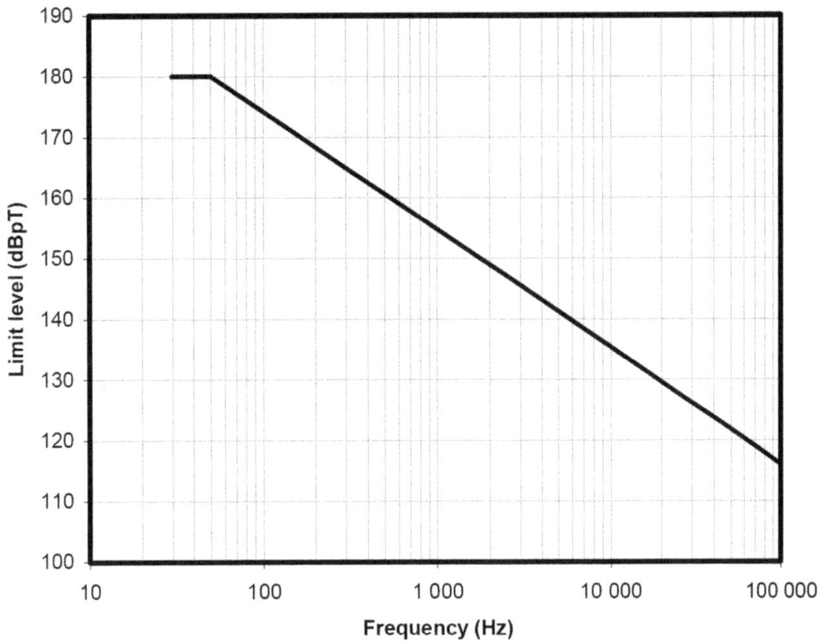

It should be noted that due to the large size of the used radiating loops, this methodology can not be effectively applied for the susceptibility testing of the cable harness to the coupling of low-frequency magnetic fields. When this type of testing is required, for example, if a sensitive chain harness can be potentially affected by magnetic fields produced by solar arrays, specialized tests need to be designed and implemented. Additionally, this requirement does not cover the frequency range below 30Hz, which may be of interest when specialized equipment, such as magnetotorquers, need to be evaluated. For these cases, specialized tests and planning are also required.

Radiated Susceptibility Testing to Electric Fields

Contrary to the limited frequency range of radiated susceptibility testing to magnetic fields, the electric field equivalent process is more complicated and time-consuming due to the large variety of operating frequencies of various equipment that could be potentially installed in the vicinity of the system under test. Taking also into consideration the intentional RF transmitters (wireless communication systems) the testing frequency range expands from the couple megahertz region up to 40 gigahertz (depending on the implemented design). This leads to the requirement of different transmitting antenna types, similar to the Radiated Emissions requirement described earlier in this chapter. Both reference Standards (MIL-STD-461G with the RS103 requirement and ECSS-E-ST-20-07C) specify requirements for the immunity of equipment used in the space environment. The test setup proposed by both Standards is almost identical. The concept requires the tests to be carried out inside a shielded room to establish a uniform radiated field for the window under test and to avoid the interactions with the equipment in the vicinity of the test site. The system under test is installed in its normal operating condition with all associated cables connected to it and inside the test setup boundary (Figure 23). Monitoring and auxiliary equipment are installed outside the shielded chamber.

A small but important differentiating detail regarding the test setup between the two Standards is the sensor position. While the ECSS-E-ST-20-07C Standard explicitly specifies that the E-Field sensor is positioned 1m from, and directly opposite the transmit antenna at a minimum height of 30cm above the ground plane and not directly near the edges or corners of the equipment under test, MIL-STD-461G does not strictly specify the position of the E-Field sensor. The effect of the E-field sensor positioning has been studied by Çakır et al. (Çakır et al, 2014). In this work, dummy equipment under test with dimensions 140cm x 65cm x 64cm and a metallic enclosure was tested in the frequency range 1GHz-18GHz, since resonance effects caused by the equipment under test are more commonly observed at frequencies above 1GHz. Utilizing directional couplers Çakır et al. measured the forward power values injected to the transmitting antennas for different E-Field sensor positions (10cm distance between each

Figure 23. Radiated Susceptibility testing to electric fields test setup (DoD: USA, 2015)

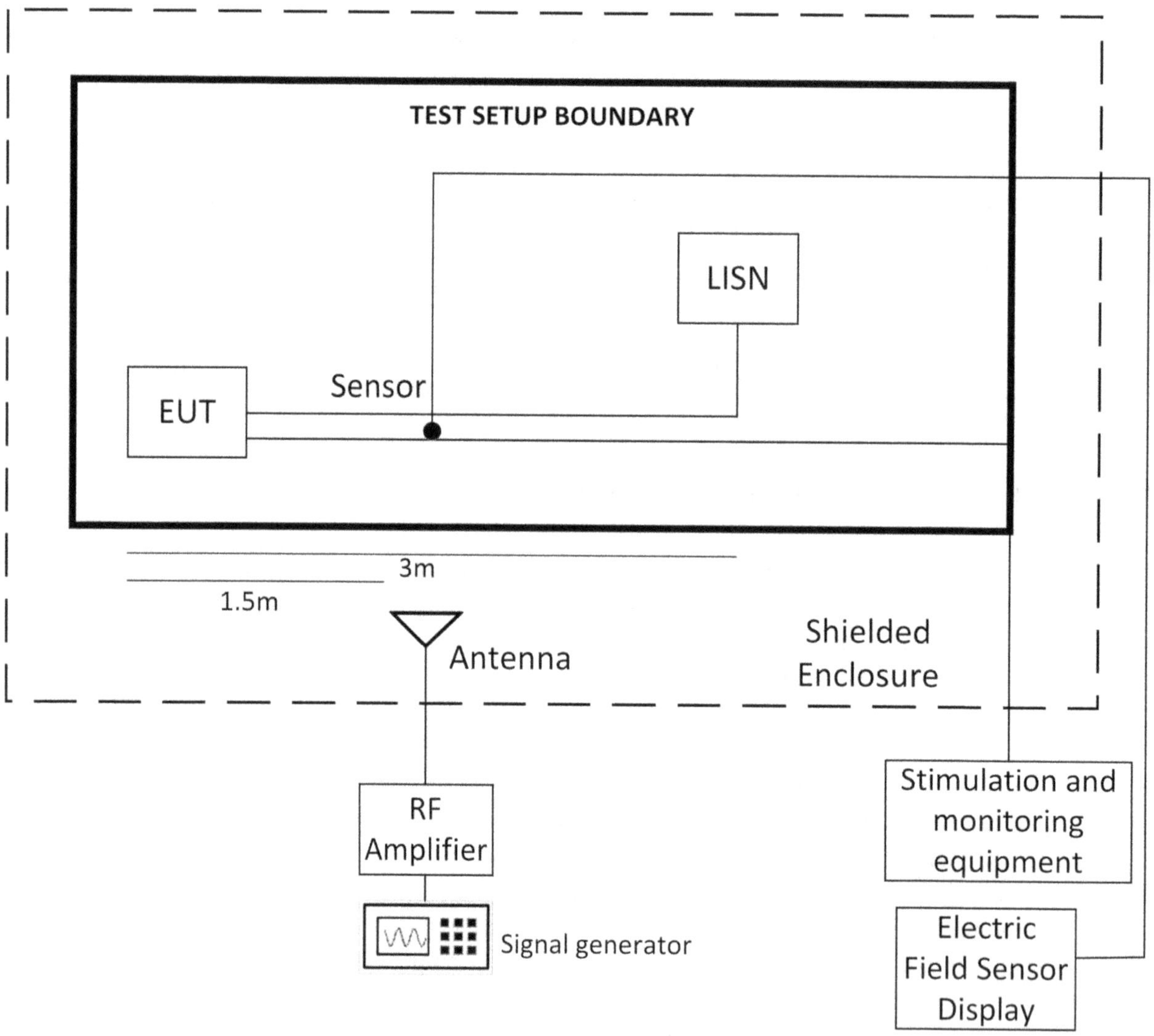

E-Field sensor position). This distance was considered by the authors as representative of the variation during the E-Field sensor positioning process performed by a test engineer during a routine compliance test. The results clearly show a significant variation of the measured forward power values for the different positions despite the small distance difference between the points and although all points are well within the 3dB beamwidth of the transmitting antennas. More specifically, the deviation as expected increases with the increase of frequency reaching values at specific resonance frequency above 10GHz of approximately 16dB.

The test levels for each frequency range according to MIL-STD-461G for Space equipment is 20V/m in the whole frequency range of 2MHz-40GHz. ECSS-E-ST-20-07C, however, takes a different approach and differentiates the test levels based on the vicinity of the equipment, in normal installation conditions, to main/secondary lobes of radiating sources and depending on whether it is to be installed inside or outside the main frame, considered as a Faraday Cage. Table 4 presents these tests and the relative conditions.

Table 4. ECSS-E-ST-20-07C Radiated Susceptibility testing to electric fields test levels

Frequency range	Electric field test level	Equipment installation condition
30 MHz-18 GHz	10 V/m	Outside the main frame and near beams
	>10 V/m	If RF analysis estimates field levels greater than 10 V/m
	1 V/m	Outside the main frame and far from beams
	1 V/m	Inside the main frame
	X V/m	Special test level for specific RF transmit/receive frequencies

Conducted and Radiated Susceptibility Testing Overlap

Substituting the radiated susceptibility testing procedure with a conducted approach has always been very appealing to EMC and test engineers. Radiated immunity testing can be time-consuming, quite expensive since apart from the expensive equipment necessary for the conduction of such tests (RF amplifiers, able to cover the forward power requirements in a very large frequency range, as well as a large variety of antennas are needed to accurately establish a uniform field of predefined field level inside an anechoic chamber) a very expensive infrastructure is required, i.e. an anechoic chamber and last but not least not repeatable since the coupled interference levels are very setup dependent, that means for complex systems under test different setup implementation can provide different test results.

In general, the dominant coupling mechanisms for frequencies up to a couple megahertz is the coupling of the field via the interconnecting cables of the equipment under test with external sources, while the equipment apertures effects become dominant in higher frequencies. Therefore, the upper-frequency limit of a potential radiated susceptibility testing substitution is between 500MHz and 1GHz (Badini et al., 2016). In their work, Badini et al. present a statistical approach for the Radiated-Conducted susceptibility testing equivalence, proposing an equivalence scheme, in accordance with the practice and requirements of aerospace testing, which does not require information regarding the impedances and other uncontrolled factors utilizing only a Bulk Current Injection probe, which feeding parameters are specified by the standard calibration procedure through a 50Ω calibration fixture. Their work utilizes statistics to compensate for the lack of knowledge about uncontrolled parameters and to quantify the over-testing probability of the alternative conducted susceptibility test. According to the authors of this work, the proposed methodology offers flexibility regarding the tailoring of the test levels to different setup requirements (Badini et al., 2016).

Susceptibility to Electrostatic Discharges

Immunity testing against electrostatic discharges, either personnel-borne in the case of manned space missions or indirect due to environmental effects or interior discharging mechanisms, is investigated extensively in both reference Standards for EMC testing. The authors would like to refer the reader to another chapter of this book, where this specific requirement is investigated and discuss in more detail.

FUTURE WORK

As already discussed earlier in this chapter, Electromagnetic Compatibility testing is progressing and evolving rapidly in recent years and will continue to do so in a probably faster pace. At the same time, Space Systems are becoming more complex and demanding with scientific missions quickly increasing in numbers. Space organizations, institutions, international and national committees should therefore be able to react immediately on the topic of EMC testing tailoring and improving the relative Standards according to future needs, trying however at the same time to improve the proposed methodologies regarding their accuracy, repeatability and reproducibility. Utilizing the progress in scientific knowledge methodologies complementing or replacing those described in this chapter need to be implemented to cope with the limitations of current methodologies. Finally, progress in the electromagnetic simulation of space environmental conditions could attribute significantly to the better approximation and visualization of electromagnetic phenomena in space.

CONCLUSION

Space missions and projects present great challenges regarding the electromagnetic compatibility analysis and testing of the associated systems and equipment. In this chapter, the EMC requirements and proposed methodologies of the two main Standards, MIL-STD-461G and ECSS-E-ST-20-07C, are presented in detail and recent trends associated with these requirements are discussed. The scope of this chapter was to provide the reader with a detailed overview of the requirements and highlight the points where specific tailoring of the Standard methodologies is required. It can be used as a compact guide to EMC testing of Space Systems and applications according to the guidelines of two of the most active organizations in the field.

REFERENCES

Badini, L., Grassi, F., Pignari, S. A., Spadacini, G., Bisognin, P., Pelissou, P., & Marra, S. (2016). Conducted-susceptibility testing as an alternative approach to unit-level radiated-susceptibility verifications. *2016 ESA Workshop on Aerospace EMC (Aerospace EMC), Valencia, 2016*, 1-5, 10.1109/AeroEMC.2016.7504582

Çakır, S., Şen, O., Çınar, M., Ayaydın, A., Çetintaş, M., & Üstüner, F. (2014). Effects of sensor positions on military radiated susceptibility tests. *29th Conference on Precision Electromagnetic Measurements (CPEM 2014)*, 584-585, 10.1109/CPEM.2014.6898520

Department of Defense: United States of America. (2015). *MIL-STD-461G: Requirements for the Control of Electromagnetic Interference Characteristics of Subsystems and Equipment*. Author.

European Cooperation for Space Standardization (ECSS). (2012). *ECSS-E-HB-20-07A: Space engineering, Electromagnetic compatibility handbook*. Author.

European Cooperation for Space Standardization (ECSS). (2012). *ECSS-E-ST-20-07C Rev. 1: Space engineering, Electromagnetic compatibility*. Author.

Gandolfo, A., Azaro, R., & Festa, D. (2017). Improving the accuracy of radiated emission measurements for frequency below 30 MHz by using a fiber optic isolated rod antenna. *2017 IEEE International Symposium on Electromagnetic Compatibility & Signal/Power Integrity (EMCSI)*, 63-68, 10.1109/ISEMC.2017.8077842

Hernandez, M. A., Lorenzo, J. M., Escolar, G. J., Sanz, L. D., Trillol, A. A., & Cancela, A. M. (2018). EMC Challenges for ESA Scientific Missions. *2018 International Symposium on Electromagnetic Compatibility (EMC EUROPE), Amsterdam, 2018*, 34-39. 10.1109/EMCEurope.2018.8485137

McCloskey, J. (2016). EMC testing on the Integrated Science Instrument Module (ISIM): A summary of the EMC test campaign for the science payload of the James Webb Space Telescope (JWST). *2016 IEEE International Symposium on Electromagnetic Compatibility (EMC)*, 138-143. 10.1109/ISEMC.2016.7571632

Schnecker, M. (2016). Verifying MIL-STD-461G CS101 using frequency domain measurements on an oscilloscope. *2016 IEEE International Symposium on Electromagnetic Compatibility (EMC)*, 847-851. 10.1109/ISEMC.2016.7571760

Sen, O., Cakir, S., Celep, M., Cinar, M., Hamid, R., & Cetintas, M. (2016). Influence of dielectric support on military radiated emission tests above 30 MHz. *2016 Asia-Pacific International Symposium on Electromagnetic Compatibility (APEMC)*, 709-711. 10.1109/APEMC.2016.7522843

KEY TERMS AND DEFINITIONS

Conducted Emissions: Emissions conducted through the cable of the equipment under test and measured either via specific probes or Line Impedance Stabilization Networks and a measuring receiver.

Electromagnetic Disturbance: Unintended electromagnetic emission, radiated or conducted from a source.

Electromagnetic Susceptibility (Immunity): The ability of a system or device to operate without performance degradation under the influence of electromagnetic disturbances.

Electrostatic Discharge (ESD): A transient current pulse as a result of the breakdown of a dielectric due to charge accumulation and different potential levels between conductors or non-conductive surfaces.

Inrush Current: The maximum instantaneous input current of an electrical device during its initial powering on.

Magnetic Cleanliness: An environment where the ambient magnetic field levels are adequately low for magnetic field measurements of ultra-low values.

Radiated Emissions: Emissions radiated from the system under test as a whole, measured with a receiving antenna and a measuring receiver.

Transient Electromagnetic Phenomena: Phenomena occurred as a result of switching operations of capacitive or inductive loads, direct or indirect lightning strikes or in general abrupt and rapid changes of currents or voltage levels.

Chapter 5
Aspects of Extremely Low Frequency Electric and Magnetic Cleanliness on Space Platforms

Alexandra P. Mavropoulou
National Technical University of Athens, Greece

Alexandros D. Bechrakis Triantafyllos
National Technical University of Athens, Greece

Christos D. Nikolopoulos
https://orcid.org/0000-0003-1344-4666
Department of Electronic Engineering, School of Engineering, Hellenic Mediterranean University, Greece

ABSTRACT

Nowadays, a wide range of space missions accommodate ever-stricter electromagnetic cleanliness requirements arising either from the need for more precise measurements or from the implementation of highly sensitive equipment. Therefore, the establishment of a methodology that ensures the minimization of the electric and/or magnetic field in specific areas inside or outside the spacecraft structure is crucial. Towards this goal, the current chapter proposes that utilizing the results of a process completed during the early design stages of a mission, that is, the measurement and characterization of each implemented device, the desired elimination of the field can be achieved. In particular, the emerged electromagnetic signatures of the units are proven essential for the proposed methodology, which, using a heuristic approach, defines the optimal ordinance of the equipment that leads to system-level electromagnetic field minimization in the volume of interest. The dimensions of the devices and the effect of the conductive surfaces of the spacecraft's hull are also taken into account.

DOI: 10.4018/978-1-7998-4879-0.ch005

INTRODUCTION

The necessity as well as the difficulties of achieving electromagnetic cleanliness in space missions are well known issues that constantly concern EMC engineers ((Junge & Marliani, 2011), (Polirpo & Cucca, 2012), (Weikert, Mehlem, & Wiegand, 2012) (Boschetti, Gervasio, & Marziali, 2012) (Lassakeur & Underwood, 2019) (Michelena, Rivero, Frutos, Ordóñez-Cencerrado, & Mesa, 2019) and other). In fact, every electronic device situated inside a spacecraft emits electromagnetic radiation that in conjunction with the properties of the materials selected for the design, can pose a serious threat to the whole system's performance. Therefore, at the early phases of the design, a lot of effort is made in order to achieve acceptable levels of EM radiation at certain regions and prevent system failures, such as permanent damage of power supplies, false commanding, noisy measurements etc. A typical example of equipment pieces that require strict electromagnetic cleanliness constraints are field and particle sensors. Those, aiming to widen humanity's understanding regarding the laws governing the universe, from planet formation to particle interactions, require precise and accurate measurements that cannot be obtained if the measuring environment is not free from spacecraft-induced interference.

The techniques that can be utilized in order to achieve this challenging objective of electromagnetic cleanliness vary from the implementation of adequate shielding on the onboard equipment, to the use of low-emissions interconnecting wiring methods between the mission's units (ECSS Secretariat, 2019). But except for the unit-level work, certain design choices can also contribute to the attainment of this goal. In particular, based on the electromagnetic theory and the modeling approximation of the problem, the authors of (Kountantos, Nikolopoulos, Baklezos, & Capsalis, 2019), (Nikolopoulos, Baklezos, & Capsalis, 2020) and (Baklezos A., Nikolopoulos, Vardiambasis, Kapetanakis, & Capsalis, in press) have developed a methodology that aims to the formation of an environment inside or outside the boundaries of the spacecraft, where the system's interference is minimized, if not eradicated. This methodology deploys an optimization algorithm, namely the Differential Evolution (DE) algorithm (Storm & Price, 1995), which calculates the proper ordinance of the equipment in order to achieve EM cleanliness either at a single observation point or on a set of points (boundaries of the area of interest). More specifically, the optimal spatial arrangement (positions) as well as the proper orientation angles of the devices (DUTs) are calculated. In order for this process to be realized, the electric and magnetic signatures of the units comprising the spacecraft's equipment should be known. Usually, measuring and modeling campaigns are performed at the early stages of the design, therefore such information is available.

The algorithmic process also accounts for the dimensions of the equipment, thus avoiding the overlap of the units in the proposed solution, as well as for the Spacecraft Hull effect, i.e. the contribution of the conductive surfaces of the spacecraft to the distribution of the system's field. In addition, if the strict cleanliness requirements cannot be achieved with the proper placement of the onboard equipment, additional auxiliary sources can be included in the design, the ordinances of which are calculated similarly to that of the DUTs. It should be noted that in a wide range of missions, with the arrangement of the units, a trade-off aims to be achieved, between not only EM cleanliness restrictions but also thermal, mechanical, accessibility etc. Hence, those may take precedence over the currently discussed cleanliness requirements, such that the spatial arrangement of the equipment cannot be addressed as proposed. But even in those cases, this methodology can be implemented for determining the position and orientation of the potentially used compensation sources.

Overall, this stochastic methodology, given a set of previously characterized devices, can calculate the optimal arrangement of the equipment in a predefined search space (spacecraft boundaries), aiming at the minimization of the total field in a given area. In the framework of this chapter, DC and extremely low frequency (ELF) sources are examined, operating in the frequency range up to 250 KHz. Simulation results show that, in these cases, due to the strong dependence of the emissions on the relative distance between the source and the observation point(s), even small rearrangements can have noteworthy impact to the distribution of the system's field.

BACKGROUND

In the current analysis, the frequency range examined does not surpass the upper limit of 250 kHz. Therefore, the corresponding wavelengths are in the order of kilometers ($\lambda \geq 1.2$ km), while the potential observation points are in the vicinity of the spacecraft. In fact, the sensitive equipment can be placed either inside the spacecraft or in a boom extending up to 11 meters (Lassakeur & Underwood, 2019) away from it (e.g. Solar Orbiter, Figure 1) and thus the area of EM cleanliness interest, is limited in the district of the spacecraft. Consequently, in order to compute the electromagnetic fields produced by the system's devices, in one of the aforementioned positions, the near field approximation should be taken ((Kountantos, Nikolopoulos, Baklezos, & Capsalis, 2017) and (Baklezos A., et al., 2019)).

Figure 1. Payload accommodation onboard Solar Orbiter. Credit: ESA – Areas with Magnetic Cleanliness Requirements (Nikolopoulos, Baklezos, & Capsalis, 2020)

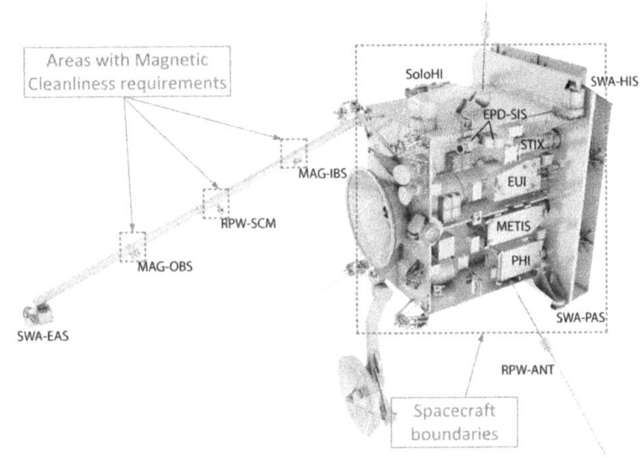

The methodology deployed for the modeling of the electric field emissions of the equipment is the Multi-Frequency Electric Dipole Model (Kountantos, Nikolopoulos, Baklezos, & Capsalis, 2017) and (Baklezos A., et al., 2019), according to which, each DUT can be replaced by one or more electric dipoles oscillating at each of the frequencies that contribute to the emissions. The same technique can also produce accurate results for the modeling of the magnetic radiation of the devices, if the electric dipoles are replaced by magnetic ones. Hence, if a single dipole is considered for each frequency of interest, the near field approximation is expressed as:

$$\begin{cases} E\left(\overrightarrow{n_e},t\right)=\dfrac{e^{-i\omega t}}{4\pi\varepsilon_0}\left\{3\overrightarrow{n_e}\left(\overrightarrow{n_e}\bullet\vec{p}\right)-\vec{p}\right\}\dfrac{1}{r_e^3} \\ B\left(\overrightarrow{n_m},t\right)=\dfrac{\mu_0 e^{-i\omega t}}{4\pi}\left\{3\overrightarrow{n_m}\left(\overrightarrow{n_m}\bullet\vec{m}\right)-\vec{m}\right\}\dfrac{1}{r_m^3} \end{cases}, \quad (1)$$

$$\overrightarrow{r_e}=\overrightarrow{R}_K-\overrightarrow{R}_e, r_e=\left|\overrightarrow{r_e}\right|, \overrightarrow{n_e}=\frac{\overrightarrow{r_e}}{r_e}$$

$$\overrightarrow{r_m}=\overrightarrow{R}_K-\overrightarrow{R}_m, r_m=\left|\overrightarrow{r_m}\right|, \overrightarrow{n_m}=\frac{\overrightarrow{r_m}}{r_m}$$

Where $\overrightarrow{R_e}\left(x_e,y_e,z_e\right),\overrightarrow{R_m}\left(x_m,y_m,z_m\right),\vec{p}\left(p_x,p_y,p_z\right)$ and $\vec{m}\left(m_x,m_y,m_z\right)$ are the cartesian coordinates of the position and moment vectors of the sources (electric and magnetic dipoles respectively) and $R_K(x_K, y_K, z_K)$ the position vector of an observation point K.

In the case of a system of N_e electric dipoles or N_m magnetic dipoles, for one specific frequency ωi (quasi-static approximation), the components of the electric and magnetic fields at the observation point K are:

$$\begin{cases} E_{s,Total}=\displaystyle\sum_{i=1}^{N_e}\frac{1}{4\pi\varepsilon_o}\left[\frac{3\left(s_m-s_{e,i}\right)\bullet C_e}{r_e^5}-\frac{p_{s,i}}{r_e^3}\right] \\ B_{s,Total}=\displaystyle\sum_{i=1}^{N_m}\frac{\mu_0}{4\pi}\left[\frac{3\left(s_m-s_{m,i}\right)\bullet C_m}{r_m^5}-\frac{m_{s,i}}{r_m^3}\right] \end{cases}, \quad (2)$$

Where s: x,y,z,

$$C_e=p_x\bullet\left(x_K-x_e\right)+p_y\bullet\left(y_K-y_e\right)+p_z\bullet\left(z_K-z_e\right),$$

$$C_m=m_x\bullet\left(x_K-x_m\right)+m_y\bullet\left(y_K-y_m\right)+m_z\bullet\left(z_K-z_m\right),$$

While the total electric and magnetic field amplitudes are:

$$\begin{cases} \left|E_{total}\right|_{N_e}=\sqrt{E_{x,Total}^{\;2}+E_{y,Total}^{\;2}+E_{z,Total}^{\;2}} \\ \left|B_{total}\right|_{N_m}=\sqrt{B_{x,Total}^{\;2}+B_{y,Total}^{\;2}+B_{z,Total}^{\;2}} \end{cases}, \quad (3)$$

In order for the equations (2) and (3) to be used for the calculation of the electric or magnetic field of a system, the moments and coordinates of each dipole should be expressed in accordance with the spacecraft's coordinate system. Therefore, it is considered essential to define the transformation method from the DUT's coordinate system (DCS) to the spacecraft's coordinate system (SCS). For simplicity purposes, let's examine the example of the j-th DUT, whose electric or magnetic emissions are modeled using the k-th (electric or magnetic) dipole, placed at (x_{Ojk}, y_{Ojk}, z_{Ojk}) and whose DCS is centered at (x_{Oj}, y_{Oj}, z_{Oj}), as illustrated in Figure 2.

In order to account for the orientation of the moment vector, when a rotation of the DUT is applied, the transformation matrices R_x,R_y,R_z (4) should be employed.

$$\begin{cases} R_x = \begin{bmatrix} 1 & 0 & 0 \\ 0 & cos\theta & -sin\theta \\ 0 & sin\theta & cos\theta \end{bmatrix} \\ R_y = \begin{bmatrix} cos\omega & 0 & sin\omega \\ 0 & 1 & 0 \\ -sin\omega & 0 & cos\omega \end{bmatrix}, \\ R_z = \begin{bmatrix} cos\Phi & -sin\Phi & 0 \\ sin\Phi & cos\Phi & 0 \\ 0 & 0 & 1 \end{bmatrix} \end{cases} \tag{4}$$

In particular, if the rotation is applied firstly on x-axis, then on y-axis and finally on z-axis (φ, ω and θ the respective angles, Figure 3), the moments (electric or magnetic) and the coordinates of the model are computed as shown in (5)-(7).

$$\begin{pmatrix} p_{xkj} \\ p_{ykj} \\ p_{zkj} \end{pmatrix}_{(rotated)} = R_z R_y R_x \begin{pmatrix} p_{xkj} \\ p_{ykj} \\ p_{zkj} \end{pmatrix}_{(initial)}, \tag{5}$$

$$\begin{pmatrix} m_{xkj} \\ m_{ykj} \\ m_{zkj} \end{pmatrix}_{(rotated)} = R_z R_y R_x \begin{pmatrix} m_{xkj} \\ m_{ykj} \\ m_{zkj} \end{pmatrix}_{(initial)}, \tag{6}$$

Figure 2. Translation form DUT's coordinate system to the Spacecraft coordinate system (Nikolopoulos, Baklezos, & Capsalis, 2020)

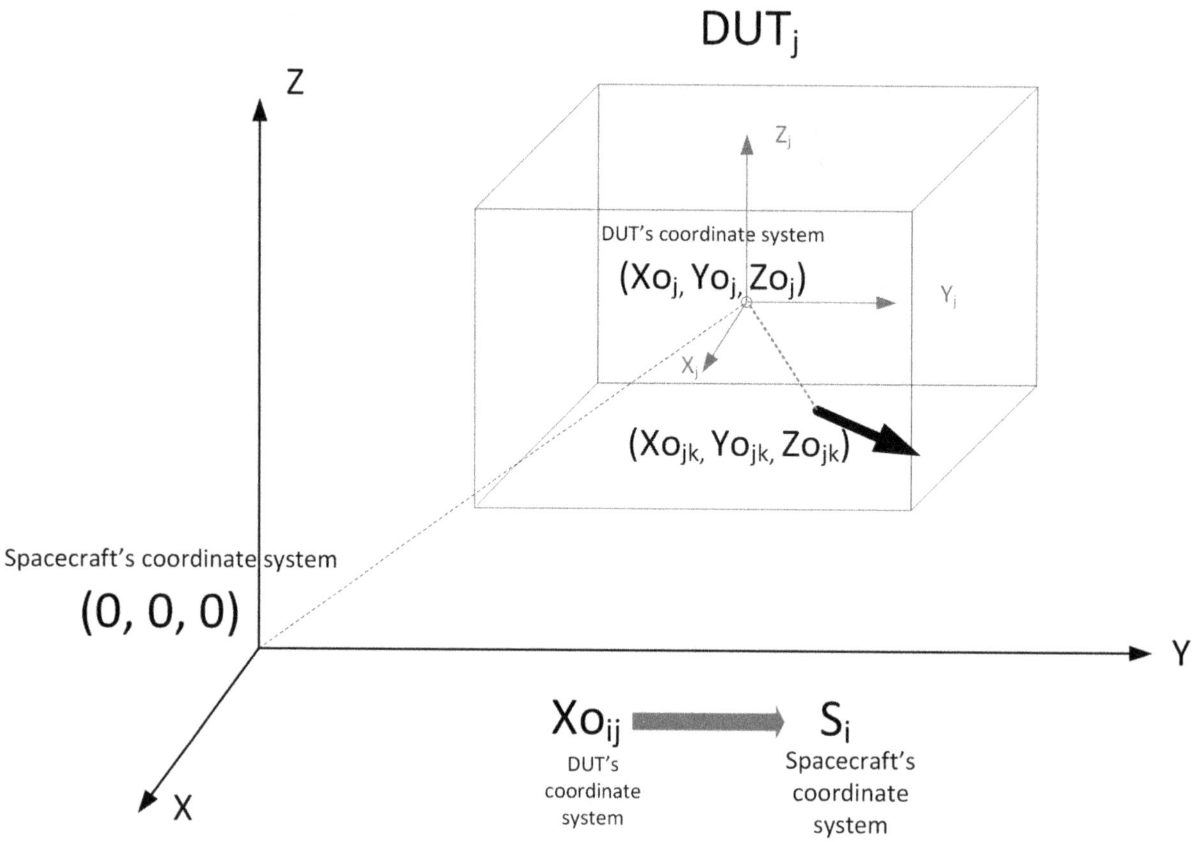

$$\begin{pmatrix} x_{okj} \\ y_{okj} \\ z_{okj} \end{pmatrix}_{(rotated)} = R_z R_y R_x \begin{pmatrix} x_{okj} \\ y_{okj} \\ z_{okj} \end{pmatrix}_{(initial)}, \tag{7}$$

The final step towards the transformation from the DCS to the SCS is to calculate the coordinates of the dipole in the SCS (8).

$$\begin{pmatrix} x_{oij} \\ y_{oij} \\ z_{oij} \end{pmatrix} = \begin{pmatrix} x_{okj} \\ y_{okj} \\ z_{okj} \end{pmatrix} + \begin{pmatrix} x_{oj} \\ y_{oj} \\ z_{oj} \end{pmatrix}, \tag{8}$$

It is noted that the electric and magnetic cleanliness problems differ with each other and are not related in terms of this methodology. They are presented simultaneously due to the similar processes followed in the attainment of the two problems. Also, different frequencies produce different ordinances of the

Figure 3. DUT's ordinance representation. Example for $(x_{Ojk}, y_{Ojk}, z_{Ojk})=(0,0,0)$ *and magnetic cleanliness testing (Nikolopoulos, Baklezos, & Capsalis, 2020)*

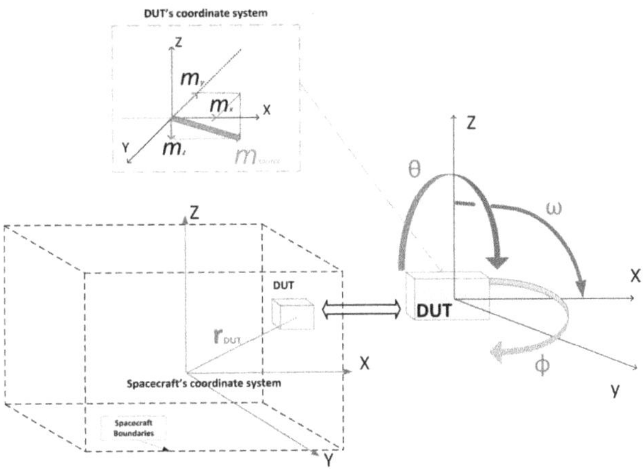

same system, therefore, with the exception of harmonics, the solution for a single frequency may not be apposite for other contributing frequencies.

Problem Formulation

Before the description and discussion of the heuristic algorithm that calculates the optimal placement of the equipment inside the spacecraft, some definitions should be made regarding the approach of the problem.

Important Definitions

- **Number and Type of DUTs taken into consideration**

In actual mission designs, a variety of devices produce electromagnetic emissions and may interrupt the cleanliness of the spacecraft, but only three or four of them are the most contributing ones that mostly define the overall behavior of the system. For the purposes of this work, actual DTUs have been measured and modeled in (Baklezos A., et al., 2019) and (Nikolopoulos, Baklezos, & Capsalis, 2018), and the four most contributing of them are used in the simulations. Each DUT is modeled with a single dipole at one frequency.

- **Spacecraft Boundaries**

The spacecraft is considered as a rectangular container (cuboid) with dimensions $L \times W \times H$ defined individually in each of the tests. The spacecraft's boundaries define the search space inside which the DUTs can be placed.

- **Observation Points**

In actual missions, the sensitive equipment pieces are placed in a volume inside or outside the spacecraft's boundaries, therefore, the area that requires electric or magnetic cleanliness is in the vicinity of the spacecraft. This area is called the "observation area", it is referred by its center (observation point - OP) and is considered a cuboid with its eight peaks referred as OP1 to OP8.

- **Orientation of the DUTs**

The DUTs are allowed to rotate only in z-axis ($0° \leq \varphi \leq 360°$), in order to remain in the horizontal plane regarding the spacecraft's azimuth (ω, θ: 0°, 90°, 180°, 270°).

- **Effect of Spacecraft Conductive Hull on the Field Distribution**

The conductive surface of the spacecraft significantly alters the distribution of the field inside its volume. Therefore, in the case of the observation point being situated inside the spacecraft, this effect should be considered, by including a number of dipole images in the analysis (image theory). Then, for a single source, the total field is calculated as the sum of the field produced by the source and the ones produced by six image sources (only first order images), as depicted in Figure 4.

Figure 4. A single dipole source and the image dipoles from the spacecraft's conductive surfaces: the case of an electric dipole (Baklezos A., Nikolopoulos, Vardiambasis, Kapetanakis, & Capsalis, in press)

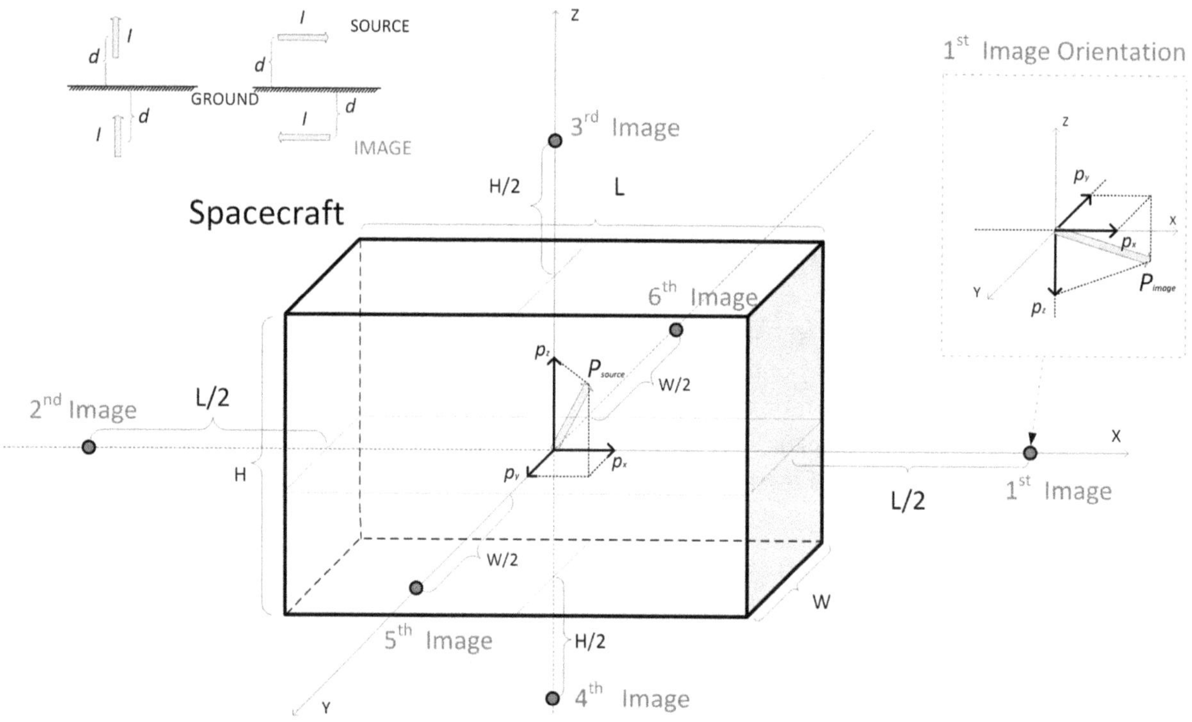

Even evident from the strong dependance of the field on the relative distance between observation point and source (2), the impact of the rearrangement of the equipment can also be showcased as follows. If a large number of random placements of the dipoles (DUTs) are tested, wide fluctuations of the fields produced at the OP are observed. In fact, the calculation results of (Kountantos, Nikolopoulos, Baklezos, & Capsalis, 2019) showed that electric field emissions vary from $10^{-4}\frac{V}{m}$ to nearly $1\frac{V}{m}$, while the ones in (Nikolopoulos, Baklezos, & Capsalis, 2020) that magnetic field emissions can take values from $10^{-11}T$ to $10^{-3}T$. Through this simple test, the improvement that can be achieved with the employment of this methodology is highlighted.

Algorithm Description

The algorithm described in this chapter employs a well-known and proven heuristic algorithm, Differential Evolution (DE), which, having an optimization character, aims to minimize or maximize a given criterion (objective function). In the current analysis, DE extracts a solution that consists of the cartesian coordinates (SCS) as well as the orientation angles of the DUTs' moments ($4DUTs \times 6coordinates = 24variables$) and aims to minimize the field across an observation volume. Thus, the objective function used is the sum of the (electric or magnetic) fields' amplitudes at the peaks of the observation cuboid (OP1, OP2, …, OP8), (9). This way, it is ensured that the whole cuboid accommodates a suitable environment.

$$\begin{cases} o_e = \sum_{i=1}^{8} \left| E_{total} \right|_{N_i} \\ o_m = \sum_{i=1}^{8} \left| B_{total} \right|_{N_i} \end{cases}, \tag{9}$$

The known parameters of the problem are the dimensions and moments of the contributing sources, the boundaries of the spacecraft (search space), the volume inside which cleanliness needs to be achieved as well as the cleanliness conditions.

Let's consider some interesting points about some of these parameters.

- **Dimensions of the DUTs**

The pieces of equipment that need to be placed inside the spacecraft are represented by the position and orientation of their moments. Therefore, if not prevented, two or more devices may overlap at the given optimal solution. For this reason, a condition that prevents such a collision by taking into consideration the dimensions of the devices, is included in the process. In fact, if a $a_i \times b_i \times c_i$ cuboid DUT is bounded by a sphere, the radius of this sphere will be:

$$R_i = \sqrt{\left(\frac{a_i}{2}\right)^2 + \left(\frac{b_i}{2}\right)^2 + \left(\frac{c_i}{2}\right)^2}, \tag{10}$$

And the condition that ensures that two neighbor spheres (Figure 5) and thus two neighbor devices, will not overlap is:

$$d \geq R_1 + R_2, \tag{11}$$

Where d is the distance between the centers of the spheres, as illustrated in Figure 5.

Similarly, the potential placement of part of the device outside the spacecraft's boundaries is also avoided.

- **Cleanliness Requirements**

Regarding the maximum accepted interference, each mission has each own victim-equipment related cleanliness requirements; however, due to the stochastic nature of the solutions, an additional criterion that assesses their quality is implemented. This criterion correlates the level of the total emissions with the emissions produced individually by the DUTs. This is achieved via the quantity $\frac{|E_{total}|}{\min E_i}$ or $\frac{|B_{total}|}{\min B_i}$. In (Kountantos, Nikolopoulos, Baklezos, & Capsalis, 2019) and (Nikolopoulos, Baklezos, & Capsalis, 2020), the criteria are defined as: $\frac{|E_{total}|}{\min E_i} < 1\%$ and $\frac{|B_{total}|}{\min B_i} < 5\%$ respectively.

Overall, the algorithm is initialized with randomly generated solutions, drawn from the valid 24-Dimensional solution space according to the uniform distribution. Then, it attempts to iteratively improve the candidate solution by minimizing the objective function. If the generated solution is a valid placement, in terms of device-to-device collision (11) and device-to-spacecraft boundaries collision, it is evaluated based on the aforementioned cleanliness criteria. If either the collision or the cleanliness conditions are not satisfied, the solution is discarded and a new one is generated. The whole algorithmic process is depicted in flowchart form in Figure 6.

A modification of this algorithmic process can be implemented to serve two additional scenarios. In particular, in cases where the rearrangement of the spacecraft's equipment in inadequate to fulfill the strict cleanliness requirements of the mission, an (unknown) auxiliary source can be employed to further minimize the field (active field cancellation). Also, if in a predefined design, an additional device is to be inserted, its optimal ordinance needs to be defined. In the two aforementioned scenarios, the Differential Evolution algorithm can be deployed to calculate the final configuration. In fact, taking as inputs the 24 variables extracted from the initial problem, the algorithm is modified to calculate the 6 new variables inserted. In the case of the auxiliary source, these variables represent the location and moment vectors of the 5th dipole, while in the case of the additional DUT, its location and orientation vectors need to be computed. The rest of the procedure remains as discussed previously.

Figure 5. DUT's placement condition to prevent overlap (Nikolopoulos, Baklezos, & Capsalis, 2020)

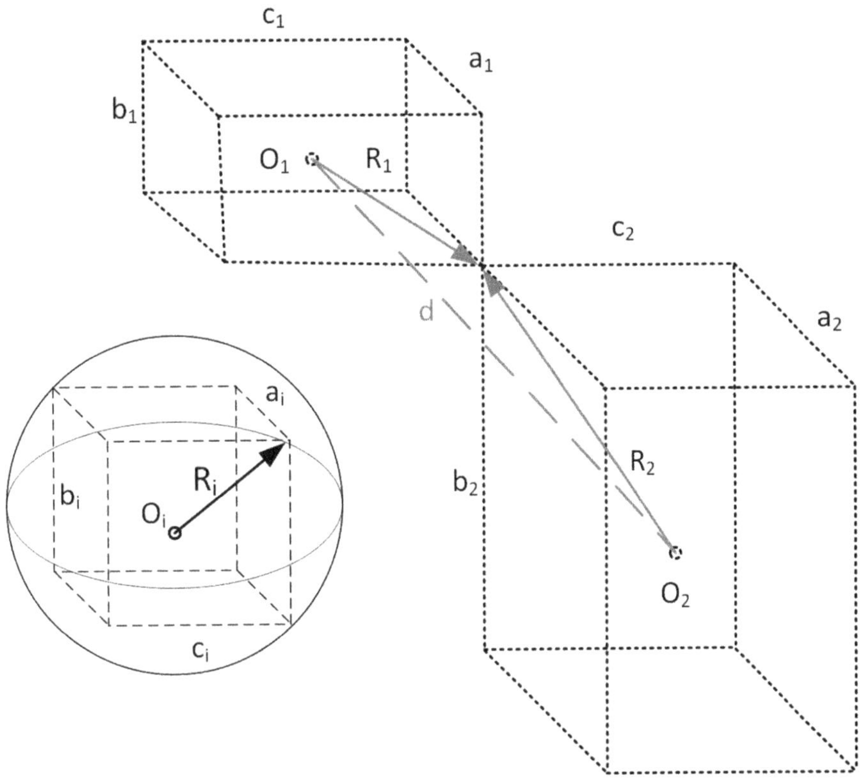

SIMULATION RESULTS AND DISCUSSION

In the framework of the development of the discussed methodology, a variety of simulations have been realized. Through these simulations, not only the success of the practice to create a suitable environment is emphasized, but also the contribution of some key parameters to the solution derived by the algorithm, is examined. These parameters are: the actual position of the observation point, the dimensions of the observation volume, the effect of the spacecraft's conductive enclosure and the impact of adding an auxiliary source to the system.

The electric and magnetic moments used for the simulations are products of measurement and modeling of actual space equipment performed in (Nikolopoulos, Baklezos, & Capsalis, 2018) and (Baklezos A., et al., 2019) and are presented in Table 1 and Table 2 respectively.

The Effect of the Observation Point's Actual Position

In order to examine the applicability and effectiveness of this methodology in any region insight or outside the spacecraft, a variety of simulations have been performed, utilizing different setups in terms of observation point location (Figure 7 and Figure 8). The results showed that the algorithm performs correctly in any case, as illustrated in the contour plots in Figure 9 and Figure 10. These plots showcase the field distribution in the YX plane, with the boundaries of the spacecraft and the observation cuboid being marked in green and red correspondingly. Therefore, the fact that the red rectangle falls into the

Figure 6. Algorithm description for proper equipment ordinance

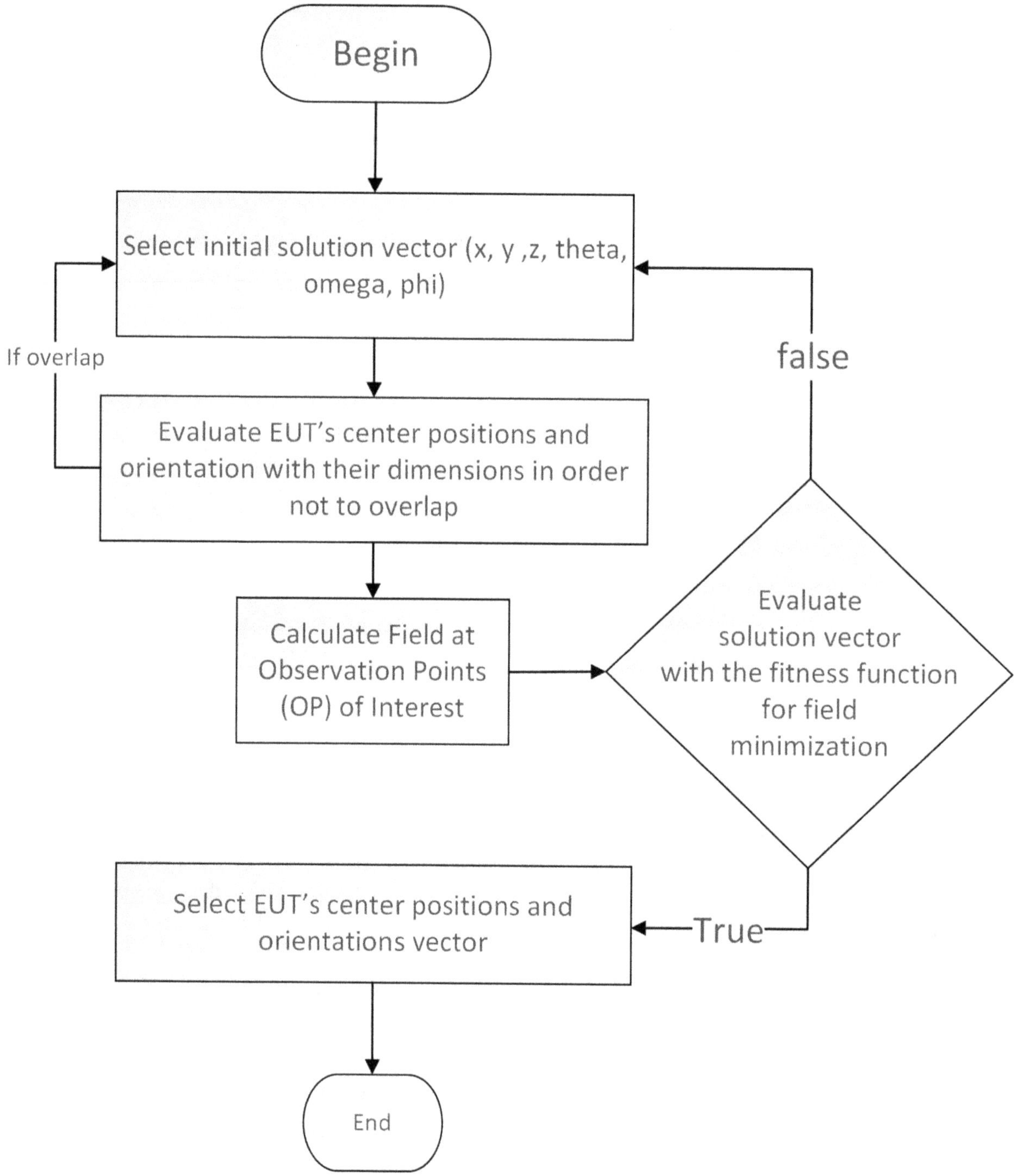

minimum-field area renders the process successful. The same results can be extracted from Table 3, in which the field's and criterion's values are presented for all cases.

Table 1. Dipoles'/DUTs' electric moments

Dipole/DUT	$p_x(fC \bullet m)$	$p_y(fC \bullet m)$	$p_z(fC \bullet m)$
1	3	12	4
2	8	-9	25
3	7	-5	13
4	-23	-11	6

Table 2. Dipoles'/DUTs' magnetic moments

Dipole/DUT	$m_x(mA \bullet m^2)$	$m_y(mA \bullet m^2)$	$m_z(mA \bullet m^2)$
1	-0.9	-27	-41
2	4.1	-87	9.4
3	-3.6	-53	1.6
4	-23	9.3	11

Table 3. Minimization field criterion

Observation point location		$B_{Total}(T)$	Criterion Value
Case 1	Center of the spacecraft (Figure 7a)	8.7368×10^{-11}	0.0307
Case 2	Random position inside the spacecraft (Figure 7b)	7.873×10^{-11}	0.0431
Case 3	Random position outside of the spacecraft, but in close proximity to it (Figure 8)	2.821×10^{-11}	0.0526

Figure 7. Location of two observation points situated insight the spacecraft (case 1-a and case 2-b), magnetic cleanliness testing (Nikolopoulos, Baklezos, & Capsalis, 2020)

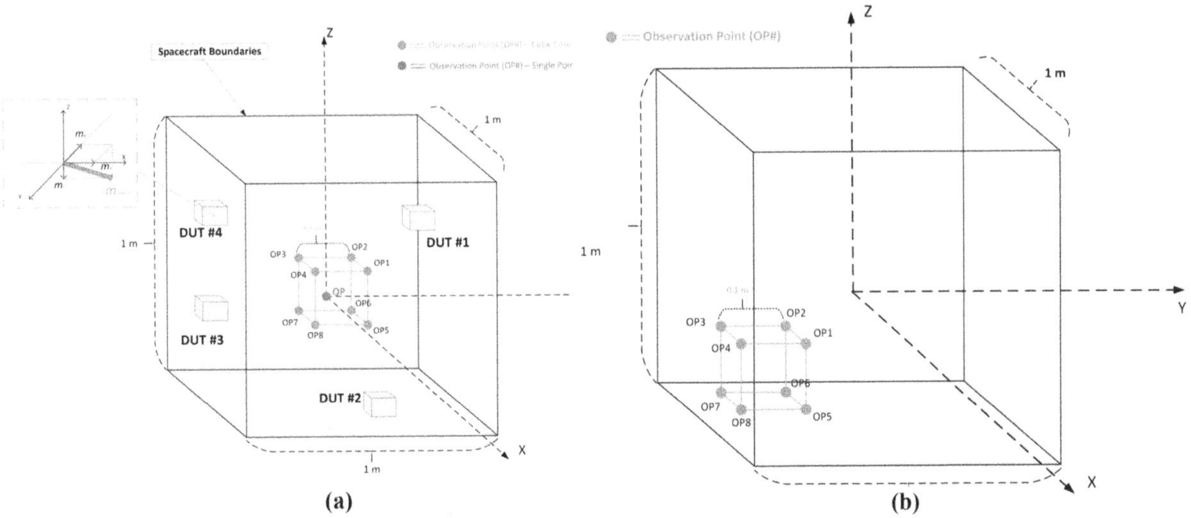

(a) (b)

Figure 8. Location of observation point situated outsight the spacecraft (case 3), magnetic cleanliness testing (Nikolopoulos, Baklezos, & Capsalis, 2020)

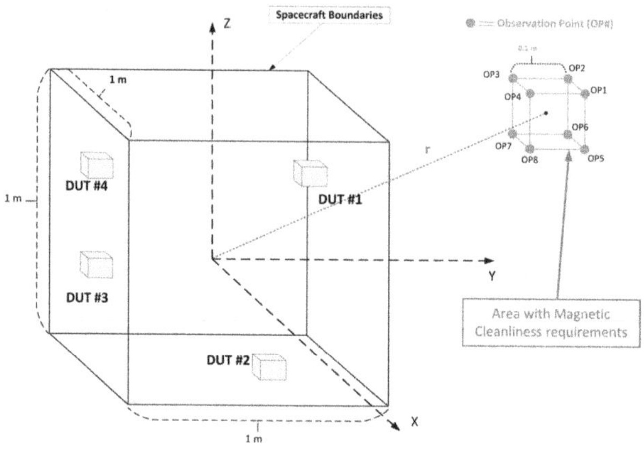

It is worth noting that the cleanliness examination outside the conductive enclosure of the spacecraft is only performed for the DC and ELF magnetic cleanliness problem. This is justified due to the fact that the spacecraft's hull acts as a Faraday Cage for the DC and ELF electric fields, therefore the electric cleanliness is ensured in any region outside it.

Figure 9. Distribution of the magnetic field at the YX plane (case 1-a, case 2-b) (Nikolopoulos, Baklezos, & Capsalis, 2020)

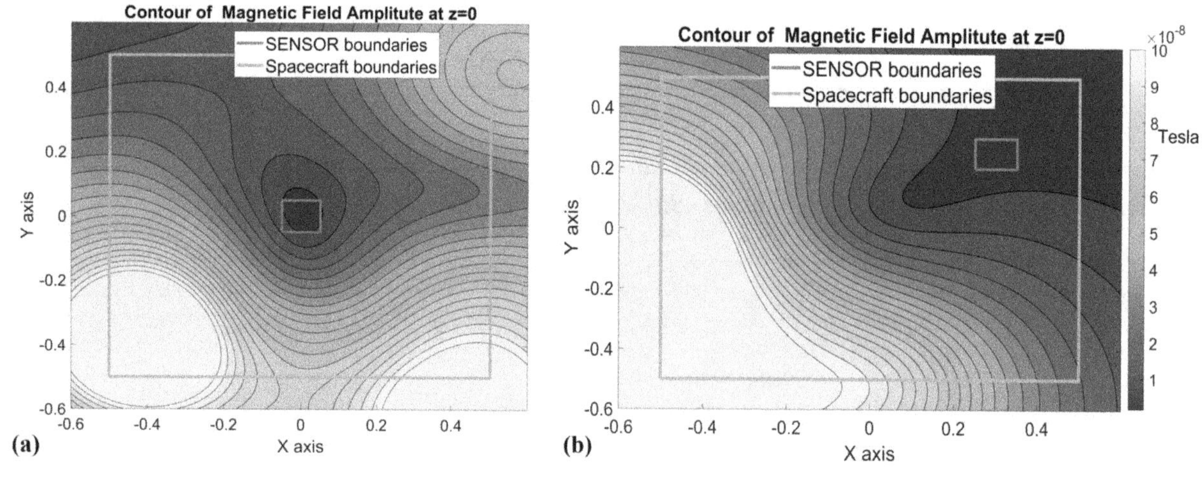

The following figures (Figure 11 and Figure 12) depict the ordinance of the DUTs computed by the algorithm.

Similar results are presented for the electric cleanliness problem in (Kountantos, Nikolopoulos, Baklezos, & Capsalis, 2019), where it is also verified that small spatial displacements of the DUTs do not affect the fulfillment of the criterion.

Figure 10. Distribution of the magnetic field at the YX plane (case 3) (Nikolopoulos, Baklezos, & Capsalis, 2020)

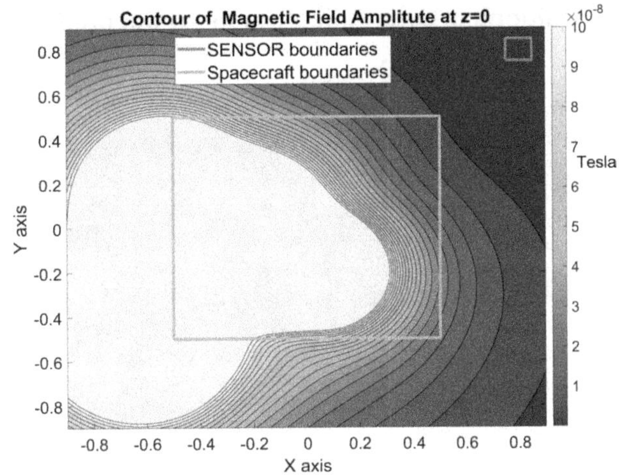

The Effect of the Addition of an Auxiliary Source

In the case where further decrease of the field is required, the method of including an auxiliary source is implemented, as described earlier in this chapter. The effectiveness of this method is validated via simulations in the setup of case 1 (Table 3), from where it is derived that the compensation magnet achieves an improvement of the total field by 87% (Table 4) (Nikolopoulos, Baklezos, & Capsalis, 2020)

Figure 11. DUTs ordinance for minimum magnetic field in two different observation points insight the spacecraft (Nikolopoulos, Baklezos, & Capsalis, 2020)

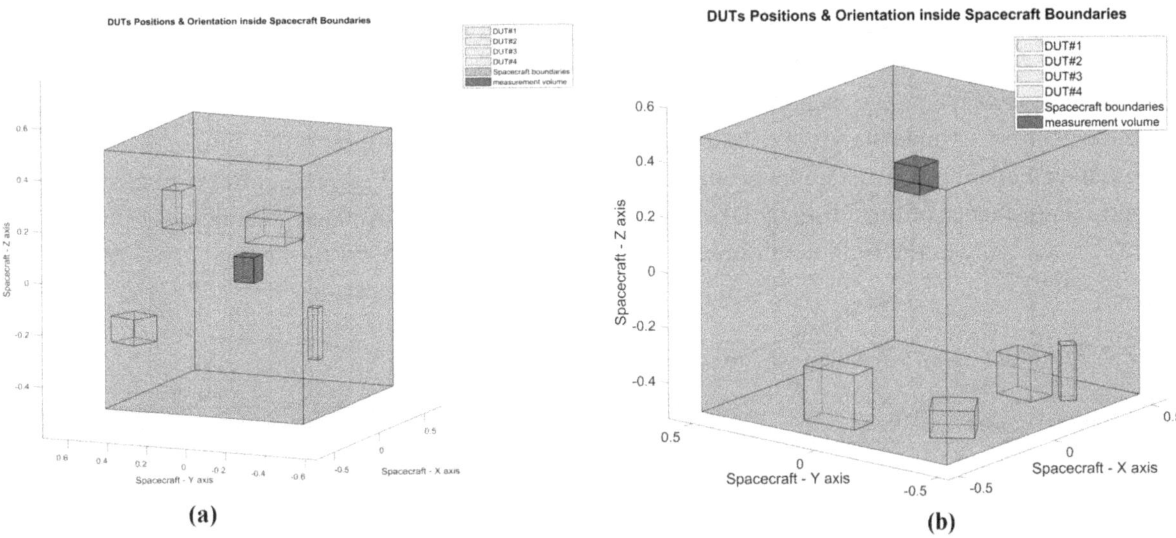

Figure 12. DUTs ordinance for minimum magnetic field in observation point outsight the spacecraft (Nikolopoulos, Baklezos, & Capsalis, 2020)

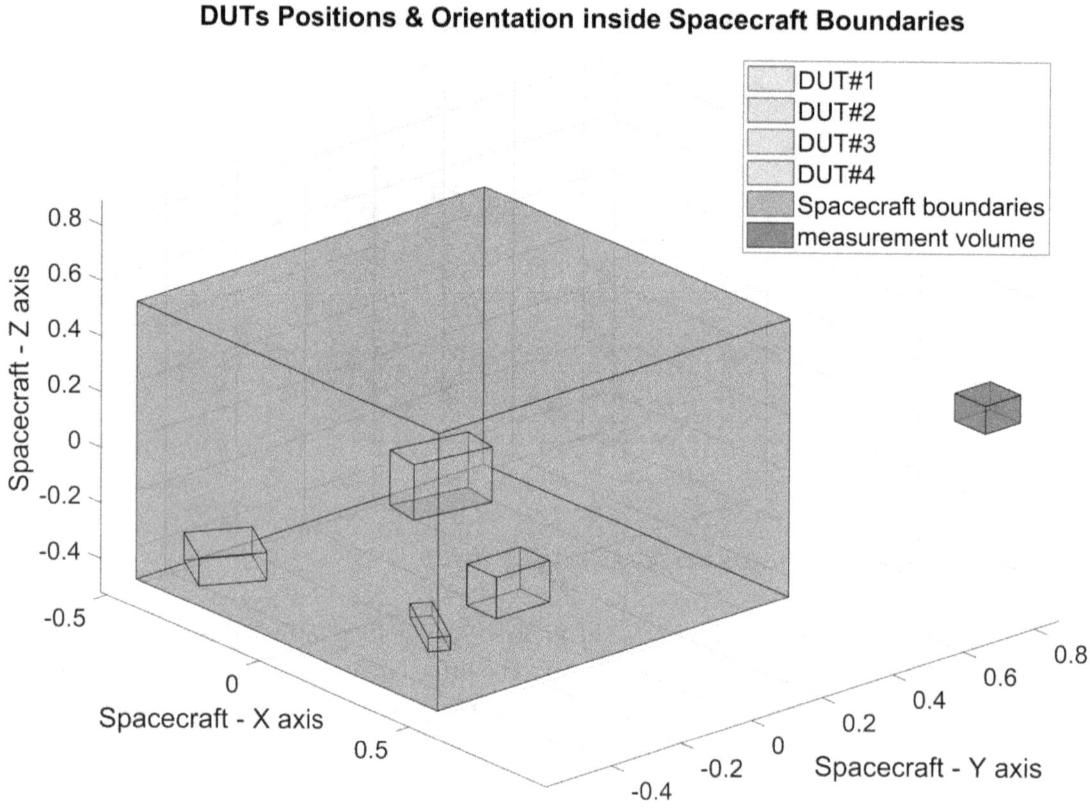

The Effect of the Spacecraft's Hull Conductive Surfaces

In order for the effect of the conductive surfaces of the spacecraft to be taken into consideration while examining the electric cleanliness inside it, a number of image dipoles should be included in the analysis (Figure 4). To evaluate the contribution of these conductive surfaces, simulations were performed in (Baklezos A., Nikolopoulos, Vardiambasis, Kapetanakis, & Capsalis, in press) with and without the implementation of the image theory and the results were compared.

In fact, for better understanding the alteration of the field's distribution if the conductive surface is taken into consideration, an example is employed. Given the setup of Figure 13, after the calculation of the optimal ordinance of the DUTs by DE, the field's distribution in YX plane is depicted at Figure 14a, where it is shown that the field is minimized in the vicinity of the observation cuboid (red rectangle),

Table 4. Minimization field criterion (including auxiliary source)

Observation point location	$B_{Total}(T)$	Cleanliness Criterion Value
Case 1, including auxiliary source	2.45×10^{-11}	0.0291

Figure 13. Location of observation point, electric cleanliness testing (Baklezos A., Nikolopoulos, Vardiambasis, Kapetanakis, & Capsalis, in press)

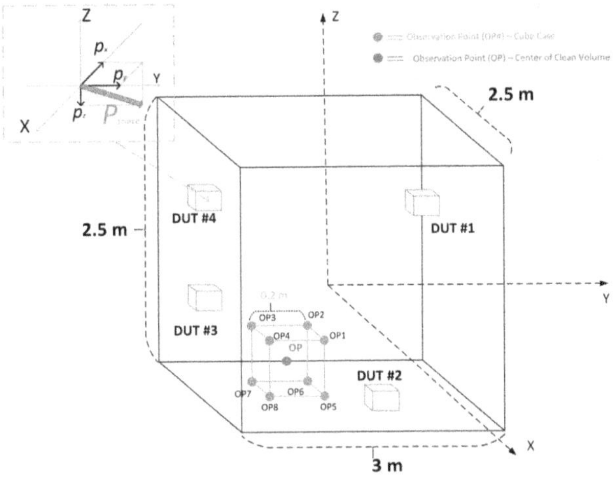

as required. If the field is now computed, also accounting for the image sources, the red rectangle falls outside of the minimum field area (Figure 14b), showcasing that if this effect is not included in the process, its results may be insufficient.

Figure 14. Distribution of the electric field at the YX plane, without (a) and with (b) considering the image sources (Baklezos A., Nikolopoulos, Vardiambasis, Kapetanakis, & Capsalis, in press)

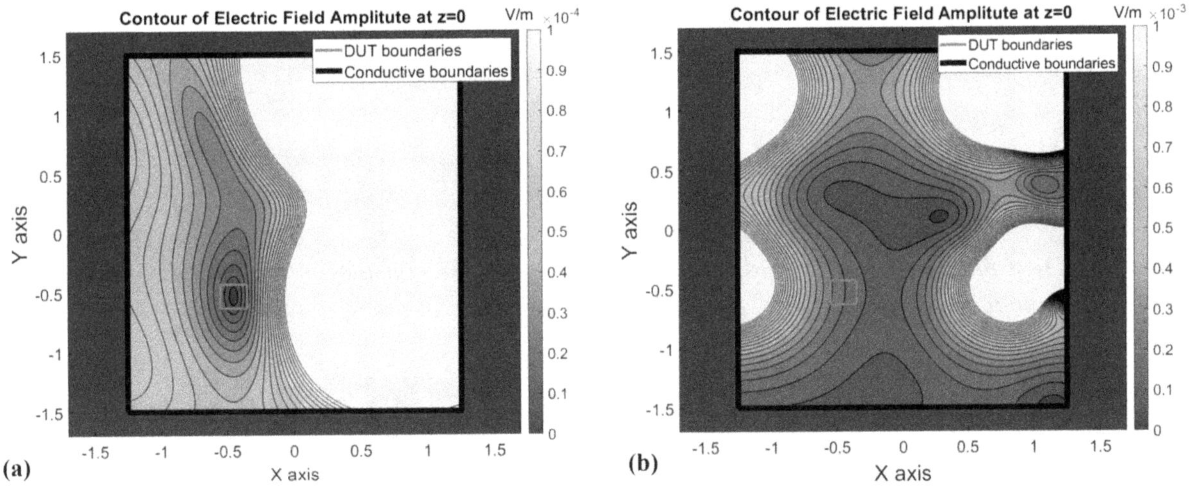

The results of a new simulation in which the objective function is reconsidered in order to include the spacecraft hull effect are shown in Figure 15. This arrangement of the equipment produces electric field distributed as illustrated in Figure 16 (YX plane). It is obvious that this distribution of the field meets the cleanliness requirements, in contrast with the case shown in Figure 14b.

Future Research Directions

The methodology discussed in this chapter has still to be implemented in an actual mission design, where measurements can verify the simulations' results. Also, future work may be directed towards the expansion of the algorithm in order to examine multiple conditions that need to be fulfilled in the framework of a mission's goals. One such case is the need for simultaneous electric and magnetic cleanliness in an area inside the spacecraft. Then, a modification of this algorithm could optimize the DUTs' placements in order to achieve a trade-off between low electric and low magnetic field interference in the area of

Figure 15. DUTs ordinance for minimum electric field in an observation point, taking into consideration the spacecraft's hull effect (Baklezos A., Nikolopoulos, Vardiambasis, Kapetanakis, & Capsalis, in press)

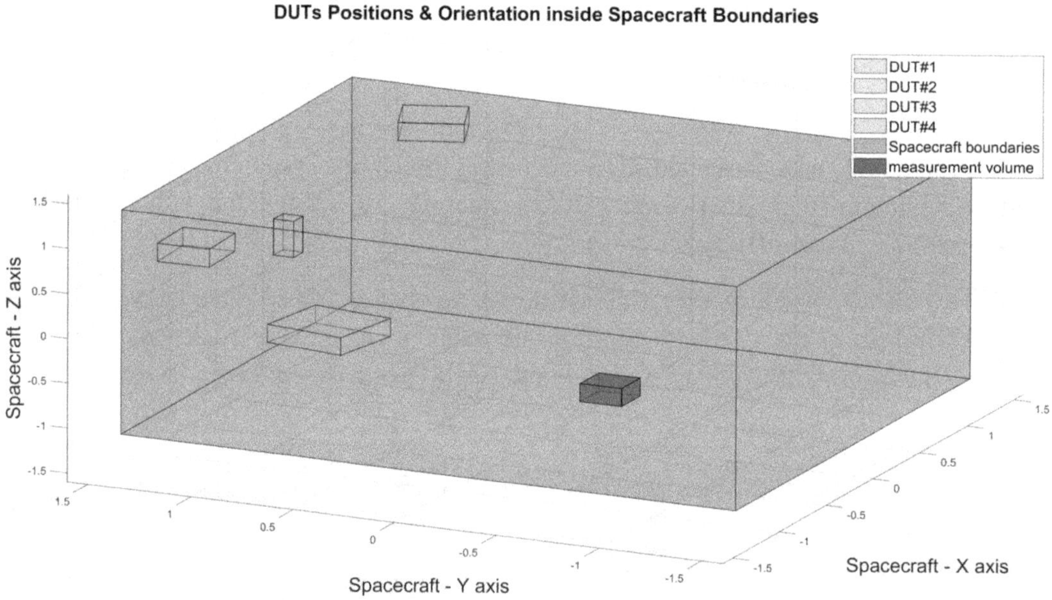

Figure 16. Distribution of electric field at the YX plane, optimized considering the image sources (Baklezos A., Nikolopoulos, Vardiambasis, Kapetanakis, & Capsalis, in press)

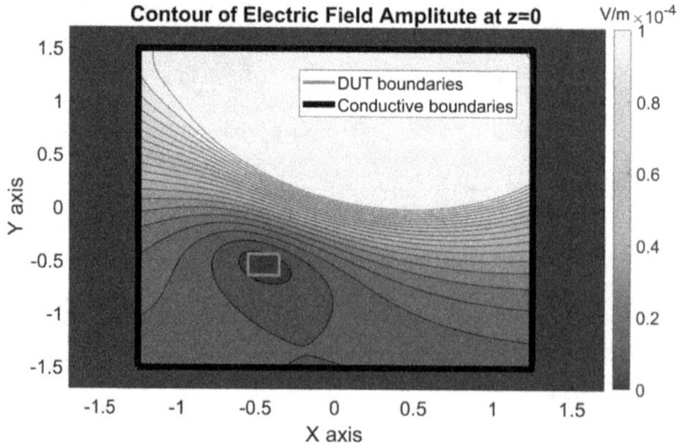

interest. Furthermore, the perspective of a multi-frequency methodology that proposes an arrangement which achieves the objective of low interference in a specific frequency range, instead of a single frequency, would be very useful for the EMC testing of space missions.

CONCLUSION

Overall, the current chapter deals with a very challenging aspect of current and future space missions, namely their electric and magnetic cleanliness. The approach presented proposes that certain simple design choices, such as the proper placement of the payload of a mission, can prove determining for the EM environment created, not only in the interior of the spacecraft, but also in regions outside it (in the case of DC and ELF magnetic fields). The methodology that addresses this problem implements a well-established heuristic algorithm, Differential Evolution, which, using stochastic methods, calculates the optimal configuration of the spacecraft's equipment in order to minimize the field in a given region. Although in a variety of missions the cleanliness requirements are less prioritized than operational-related requirements and thus the rearrangement of the units is not possible, this methodology can also be implemented to actively achieve cleanliness after the initial design has been defined. In fact, the moment and position of an auxiliary source that compensates for the desired field's distribution, can be calculated using a modified version of the discussed algorithm. Another application of the developed process is the proper placement of a device inside the spacecraft, after the initial stage of the design, with the aim of disturbing the balance of the system as less as possible. Some key parameters taken into consideration during the process, are the Spacecraft Hull Effect (using image theory) as well as the dimensions of the equipment.

REFERENCES

Baklezos, A., Nikolopoulos, C., Vardiambasis, I., Kapetanakis, T., & Capsalis, C. (in press). Spacecraft Hull Effect on Radiated Emissions and Optimal Onboard Payload Allocation. *2020 International Symposium on Electromagnetic Compatibility (EMC EUROPE).*

Baklezos, A., Nikolopoulos, C. D., Spantideas, S., Chatzineofytou, E., Nicoletto, M., Marziali, I., & Capsalis, C. N. (2019). *Steady State Emissions Modeling of Low Frequency Magnetic and Electric Fields Generated by GOCE CDMU. In 2019 ESA Workshon on Aerospace EMC.* Aerospace EMC.

Boschetti, D., Gervasio, G., & Marziali, I. (2012). Montecarlo Approach for Magnetic Cleanliness Evaluation on Spacecraft. *2012 ESA Workshop on Aerospace EMC.*

Junge, A., & Marliani, F. (2011). Prediction of DC Magnetic Fields for Magnetic Cleanliness on Spacecraft. *2011 IEEE International Symposium on Electromagnetic CompatibilityWorkshop on Aerospace EMC.* 10.1109/ISEMC.2011.6038424

Kountantos, G., Nikolopoulos, C., Baklezos, A., & Capsalis, C. (2017). On the Modeling of ELF Electric Fields for Space Mission Equipment. *IEEE Transactions on Electromagnetic Compatibility, 59*(5), 1457–1464. doi:10.1109/TEMC.2017.2685199

Kountantos, G., Nikolopoulos, C., Baklezos, A., & Capsalis, C. N. (2019). Proper Equipment Ordinance for Achieving EM Cleanliness in Space Missions: The case of ELF Electric Sources. *IEEE Transactions on Electromagnetic Compatibility*, *62*(5), 1686–1692. doi:10.1109/TEMC.2019.2937379

Lassakeur, A., & Underwood, G. (2019). Magnetic Cleanliness Program on Cubesats For Improved Attitude Stability. *2019 9th International Conference on Recent Advances in Space Technologies (RAST).*

Michelena, M. D., Rivero, M. A., Frutos, J. d., Ordóñez-Cencerrado, A., & Mesa, J. (2019). Adaption of Magnetic Cleanliness Facilities and Procedures to Overcome the New Challenges of the Scientific Missions. *2019 ESA Workshop on Aerospace EMC (Aerospace EMC).* 10.23919/AeroEMC.2019.8788959

Nikolopoulos, C., Baklezos, A., & Capsalis, C. (2018). Measuring Transient and Stedy State Electric Field Emissions of Space Equipment for EMC and Cleanliness Purposes. *IEEE International Conference on High Voltage Engineering and Application (ICHVE)*, 1-4.

Nikolopoulos, C., Baklezos, A., & Capsalis, C. (2020). On Achieving Spacecraft Level Magnetic Cleanliness with Proper Equipment Ordinance of DC and ELF Magnetic Sources. *IEEE Transactions on Electromagnetic Compatibility*, 1–11. doi:10.1109/TEMC.2020.2992682

Polirpo, A., & Cucca, M. (2012). New Facility For S/C Cleanliness Program. *2012 ESA Workshop on Aerospace EMC.*

Secretariat, E. C. S. S. (2019). *ECSS-E-ST-50-12C Rev. 1, Space Engineering – SpaceWire- Links, nodes, routers and networks.* ESA-ESTEC Requirements & Standards Division.

Storm, R., & Price, K. (1995). Differential Evolution - a simple and efficient adaptive scheme for global optimization over continuous spaces. *Journal of Global Optimization*, 23.

Weikert, S., Mehlem, K., & Wiegand, A. (2012). Spacecraft Magnetic Cleanliness Prediction and Control. *2012 ESA Workshop on Aerospace EMC.*

Chapter 6
Magnetic Sensors for Space Applications and Magnetic Cleanliness Considerations

Anargyros T. Baklezos
https://orcid.org/0000-0002-0532-6216
National Technical University of Athens, Greece

Neoclis G. Hadjigeorgiou
National Technical University of Athens, Greece

ABSTRACT

This chapter is composed by three parts. The first is an introductory part, providing general information about magnetism and related phenomena. Magnetic materials are also discussed and presented. Afterwards, the magnetic field and various measurement techniques are discussed. In the second part, different magnetic sensors used in a laboratory or space are presented. Magnetic sensors that are discussed include anisotropic magneto-resistance (AMR), giant magneto-resistance (GMR), giant magneto-impedance (GMI), flux-gate, and superconducting quantum interference device (SQUID). Although some of them may be outdated and well known, they are widespread, and they still pose an excellent choice for certain applications. Advances in magnetometers also presented in order to provide the reader with the recent trends in the field. Magnetic cleanliness is an important factor both in calibration and in normal operation of a system; in the third part, current techniques to isolate a system from the external magnetic field providing cleanliness are discussed.

DOI: 10.4018/978-1-7998-4879-0.ch006

INTRODUCTION

During the last few decades, the continuously increasing demand for accurate and reliable AC/DC magnetic field measurements has paved the way for the development of various types of magnetic sensing systems as well as different measurement techniques. The sensor's sensitivity and linearity, signal-to-noise ratio, measurement range, cross-talk between sensors in multi-sensor applications are only some of the aspects that have been examined in the past.

Magnetic sensors are categorized primarily based on their sensing principle, which impacts directly the performance characteristics like measurement range, resolution, frequency response, working temperature, and manufacturing cost. For example, Hall Effect and "pick up coils" based sensors have been proven to be practical and cost-effective solutions in many common applications. Nevertheless, "pick up coils" are not appropriate for measuring low-frequency magnetic fields, whilst Hall-effect devices cannot provide the resolution levels needed in demanding applications. On the other hand, Anisotropic Magnetoresistance (AMR), Giant Magnetoresistance (GMR), Giant Magneto-Impedance (GMI), Magnetostrictive, and Flux-Gate sensors, although available at an elevated cost, are frequently employed in cases where higher resolution is required. Moreover, Superconducting Quantum Interference Devices (SQUID) offer the highest resolution and lower noise levels but requires cryogenic refrigeration.

Many industrial applications nowadays use magnetic sensors as measuring systems. These sensors have several advantages such as the fact that they can easily operate under harsh environmental conditions and elevated temperatures, they are very reliable due to the absence of moving parts and they can be embedded inside building materials. Magnetic Sensors are commonly used for laboratory measurements and satellite magnetic measurements. Understanding the various types of magnetometers and their limitations is crucial for the interpretation of the satellite magnetometer results.

For space use, magnetic sensors must have some unique characteristics. For example, SQUID sensors may not be suitable for this type of application due to the need for continuous cooling in cryogenic temperatures. In space, magnetometers can observe the fluctuations of a planet's magnetic field, from the planet's core or the solar wind. On the other hand, in some cases, there is a need to protect certain instruments from electromagnetic radiation. That is achieved by using a magnetic sensor to measure this disturbance and a closed-loop technique to cancel it. The most common types of magnetic sensors for space, are the fluxgate and search coils magnetometers.

In the laboratory, usually, there is the flexibility of using more complex and sensitive equipment. For accurate laboratory measurements, there is a need for a magnetically clean environment. The magnetically clean room or facility can be used for numerous applications such as:

- calibration of magnetic sensors
- noise measurements for the magnetic sensor
- identification of the magnetic signature of the Equipment Under Test (EUT)

There are passive and active techniques to create a magnetically clean room and they depend on the shielding specifications/requirements.

This chapter will explore the different types of magnetic sensors and their applications, as well as some techniques for electromagnetic shielding at low frequencies.

BACKGROUND

Good knowledge of magnetic fields and magnetization is paramount to understand the operating principle of the magnetometers. This information is also important because it will help identify the best sensor type for each specific application.

MAGNETIC FIELD

The definition of the magnetic field is most probably the most fundamental concept in magnetism. The magnetic field, which is generated by a magnetic source, covers a certain (infinite but reciprocal dampened) area around the source. In the vicinity of the magnetic field, there is an energy gradient, which can lead to the generation of a force in the presence of another magnetic field. For example, the needle of a compass is a magnetic dipole, which has its own magnetic field, which in turn interacts with the earth's magnetic field. Therefore, a torque is applied to the needle in order to come to rest, at the point with the least energy. (Daniel, 1988; Fraden, 2010; Jiles, 2015).

Magnetization and Magnetic Moment

The response of magnetic materials in a magnetic field is expressed by the magnetic moment and the magnetization. According to Ampere's theorem, the current in an electrical circuit generates a magnetic field. The simplest circuit that can generate a magnetic field, is a loop of a conductor carrying an electric current. If the cross-sectional area of the loop is $A(m^2)$ and the loop carries a current $I(A)$, the magnetic dipole moment m (Am) is $m=IA$. In a magnetic field B (*Tesla*), the torque τ (N*m)* applied on a magnetic dipole moment m (A*m)* is:

$$\tau = m \times \mathrm{B} \tag{1}$$

and in free space (vacuum):

$$\tau = \mu 0(_{\mathrm{m}} \times \mathrm{B}) \tag{2}$$

The magnetization M (A/*m)* is the vector field that expresses the density of permanent or induced magnetic dipole moments in a magnetic material. The magnetization is generated by the magnetic moments in the material that can be either microscopic electric currents resulting from the motion of electrons in atoms, or the spin of the electrons or the nuclei. The material responds to an external magnetic field by rotating its unbalanced magnetic dipole to reduce the energy of the magnetic field (e.g. in ferromagnets). Magnetization is not uniform within a solid due to imperfections such as dislocations and solute atoms or even vacancies. Magnetization also describes how a material responds to an applied magnetic field as well as the way the material changes that magnetic field (Jiles, 2015; Ripka 2001).

Magnetization can be defined as the magnetic moment per unit volume of a solid:

$$M = \frac{m}{V} \tag{3}$$

where V is the volume of the solid (m^3). Magnetization defines the auxiliary magnetic field H (A/m) as:

$$B = \mu_0(H + M) \tag{4}$$

According to the Jiles-Atherton model for the magnetic hysteresis that appears in a material, the magnetization is composed of two components: one hysteretic and one anhysteretic. The hysteretic magnetization is the cause of hysteresis in a B-H loop. The anhysteretic magnetization is composed of isotropic magnetization and anisotropic magnetization. (Jiles, 1984).

Measuring Magnetism

The magnetic field is measured with a variety of methodologies. Different methods present different drawbacks regarding the measurement parameters such as the frequency range, the measurement range, the resolution, the sensitivity, and the noise. Magnetometers or magnetic sensors can be classified into two major categories. The first category consists of the sensors that depend on magnetic induction using coils. The second category includes those which depend on measuring changes in various properties of materials due to the presence of the magnetic field. (Tumanski, 2011; Ripka, 2001).

Induction Methods for Measuring the Magnetic Field

All induction-based methods operate based on the same principle; they are measuring the magnetic flux based on electromagnetic induction - Faraday's Law. The electromagnetic force (E.M.F) or voltage V (V) that is induced in a circuit, is equal to the time derivative of the magnetic flux Φ (Wb) in the magnetic circuit.

$$V = -\frac{d\Phi}{dt} \tag{5}$$

If the coil has N turns and a cross-sectional area $A(m^2)$, the magnetic induction is $B=\Phi/A$. Therefore, equation 5 becomes:

$$V = -NA\frac{dB}{dt} \tag{6}$$

Method with a Stationary-Coil

A stationary-coil method can only measure time-variant magnetic induction by measuring the induced voltage. This type of sensor measures the magnetic induction according to the mathematical expression:

$$B = -\frac{1}{NA}\int V dt \tag{7}$$

The main purpose of this type of sensor is to measure the magnetic properties (e.g. hysteresis graph) for soft magnetic materials. Nevertheless, this type of sensor needs to be adjusted to avoid integration errors, which may mislead to a linear dependency of the magnetic induction. This problem can be ignored in cases where the magnetic flux level is smaller than 10^{-10} Wb (0.1 Maxwell).

Method with a Moving-Coil

In this case, instead of a time-varying magnetic flux passing through a coil, a coil moves inside a magnetic field. The integral of the voltage provides the difference of the magnetic fields as described in the following equation, where B_i is the initial magnetic field and B_f is the final magnetic field.

$$B_f - B_i = -\frac{1}{NA}\int V dt \tag{8}$$

An application of this moving-coil is the diversion of a galvanometer in which, the deflection is proportional to the change of the magnetic field.

Method with a Rotating-Coil

In this method, the calculation of the magnetic induction becomes possible from the use of various moving-coil sensors. The simplest type of this type of sensor is the rotating coil. A coil is rotating at a fixed angular velocity $\omega(rads/s)$. The magnetic field is calculated by:

$$B(t) = B_o(t)\cos(\omega t) \tag{9}$$

where t (s) is the time. The generated voltage is then:

$$V(t) = -NA\sin(\omega t)\frac{dB_o(t)}{dt} \tag{10}$$

Typically, this type of sensor can measure magnetic fields in the range of 10 down to 10^{-7} Tesla. The precision of these sensors is 100 ppm. The major source of error in dealing with small voltages is the electrical connection to the rotating coil, such as slip rings.

Method of Vibrating-Coil

The same principle of electromagnetic induction (Faraday's Law) is applied also in vibration-coils. This type of coils is used for determining magnetization and consequently the primary and minor B-H loops of a material. When a coil is around a magnetic material, the magnetic field is:

$$B_m = \mu 0(H + M) \tag{11}$$

Whenever the coil moves away from the sample, then the magnetic induction is:

$$B_0 = \mu 0_{H} \tag{12}$$

The change in the magnetic flux is then:

$$\Delta B = \mu 0_{M} \tag{13}$$

This technique to sense magnetization is sensitive to noise caused by the variation of the auxiliary magnetic field. It is preferable to vibrate the sample instead of the coil if this is possible.

Method of Vibrating the Sample

A sensor that employs the method of vibrating the sample, is known as a Vibration Sample Magnetometer (VSM). It follows the same principle as the vibration-coil magnetometer with the exception that the moving component is the sample instead of the coil. The VSM was first described by Foner (Foner, 1959) and it is widely used for determining the saturation magnetization M_s of the sample.

The VSM has an accuracy better than 2% and has the capability to sense a magnetic moment as small as $50kAm^2$. However, it should be noted that in cases of small samples, this method is not suitable for determining the magnetization curve or the hysteresis loop.

Fluxgate-Magnetometers

The last sensing technique that utilizes the electromagnetic induction effect is the Saturable-core magnetometry or fluxgate magnetometers. They were first developed in the 1930s for measurements of the Earth's magnetic field. They were used in magnetic mines for ships and later in space applications to examine the planets' magnetic field.

Fluxgates can only measure the component of the magnetic field vector that is parallel to the coils. They are capable of measuring magnetic induction in the range of 10^{-11} up to 10^{-2} Tesla. Their accuracy is better than 4‰. This type of sensor will be thoroughly analyzed later in this chapter.

Measuring the Magnetic Field Using Material Properties

The magnetic field does not affect only ferromagnetic materials or coils. It can also affect the trajectory of an electron. As was presented earlier, a magnetic field is produced by the movement of an electric current, but a magnetic field can also affect or produce an electric current (Jiles, 2015; Fraden 2010; Ripka 2007; Prudenziati 1999).

Magneto-Optic Magnetometers

Magneto-optic magnetometers are based on three physical effects, the Faraday effect, the Kerr effect, and the Cotton-Mouton effect.

Faraday Effect

Faraday effect or Faraday rotation refers to the rotation of the plane of polarization of polarized light that is transmitted through a medium and is subjected to a magnetic field along the direction of propagation. It was discovered in 1845 by Faraday, and Maxwell completed the theoretical analysis in 1870. The implementation of the Faraday effect is straightforward in measuring the magnetic field strength since the polarized light rotates as it passes through a transparent paramagnetic material, such as $MgCe(PO_4)_2$ and this can give a measurement of the magnetic field. The angle of rotation of the polarization plane θ is given by:

$$\theta = VHl \tag{14}$$

where V is the Verdet constant (0.001 to 0.1 *minute/A*), H is the magnetic field strength (*A/m*) and l is the length of the path that the light travels in the material (*m*). In ferromagnetic or ferrimagnetic materials, the angle of rotation θ can also be related to magnetization by:

$$\theta = \mathrm{K}Ml \tag{15}$$

where K is the Kundt's constant (typically up to $350 \bullet 10^{6\circ}/Tm$). Both, the Faraday effect and the Cotton-Mouton effect can be used for domain observation.

Kerr Effect

The Kerr effect, also called the Surface Magneto-Optic Kerr Effect (SMOKE), is the change of rotation of a linearly polarized light on the surface of the sample. The light beam should not be normal to the surface of the ferrimagnet or ferromagnet, since there must be a vector component of the magnetization on the surface of the sample parallel to the direction of the light. It should be noted that the surface domains are oriented with their magnetization to the plane of the surface. When a light beam is reflected in a magnetic domain, its polarization plane rotates by an angle θ, which is related to the magnetization of the domain by:

$$\theta = \mathrm{K}r_M \tag{16}$$

where K_r is the Kerr constant (mV^{-2}).

A typical rotation is 54° for saturated nickel and 120° for saturated cobalt and iron. However, this technique can be used if and only if, the magnetization can be directly related to the magnetic field.

Method of Magnetic Resonance

Sensors of magnetic resonance refer to all resonance techniques used to measure the magnetic field, including electron spin resonance, nuclear magnetic resonance, and proton precession. They are extremely precise and sensitive and have a resolution of up to 10^{-14} Tesla. An important note about this type of magnetic sensors is that they are measuring the total magnetic field and not the vector components.

The electron spin resonance (ESR) was first discovered by Zeeman (Ripka et al., 2007). When a magnetic field is applied to electrons, they can be exited to one higher state, and depending on the spin polarity and quantum number, the rotation frequency can be increased or decreased. The same theory is also applied to nuclear magnetic resonance (NMR) but instead of electron spin resonance, the magnetic field resonates the nucleus.

Hall Magnetometers

The Hall effect was discovered by Dr. Edwin Hall in 1879. The Hall effect refers to the phenomenon that a voltage is been generated perpendicular to the current direction when the conductive medium is being placed in a magnetic field.

Today Hall effect sensors are the most versatile and widely used magnetic sensors. They are being used for measuring the linear or angular velocity in closed-loop systems. (Honewell, 2011).

Thin-Film Magnetometers

Thin-film magnetic sensors are constructed by a thin layer of 200-5000 ångström in thickness. The material of the layer must come from a non-magnetostrictive alloy such as $Ni_{20\%}$-Fe. During manufacturing, a magnetic field is applied parallel to the easy magnetization axis. The sensing direction is the hard magnetization axis. This type of sensors has uniaxial anisotropy and can be used as a variable inductor in circuits. The sensing range is from 10^{-7} to 10^{-3} Tesla.

Magneto-Resistor Magnetometers

Magnetoresistance is the effect of change in the electrical resistivity of the material when subjected to an external magnetic field. The change in resistivity is about 2% at saturation in nickel and 0.3% in iron. Other materials can perform better than 100% change in resistivity but they cannot sense low magnetic fields. This type of sensors, like Hall sensors, can be easily fabricated in a chip. They can easily achieve a resolution of 10^{-5} A/m.

Magnetostrictive Magnetometers

Magnetostrictive sensors change their shape when an external magnetic field is applied. The disadvantage of magnetostriction is that it is nonlinear and has hysteresis. The phenomenon of magnetostriction is known as the fraction of length change over the initial length ($\Delta l/l$) of the sample.

Both magnetostriction and magnetoresistance are interconnected. They are being generated by the spin-orbit coupling so that the electron distribution at each ionic site is rotated. Magnetoresistance is a phenomenon in which the conductive electrons scatter by the change in electron distribution. Magnetostriction phenomenon is the change in ionic spacing when the moments of the electrons are rotated (Hristoforou et al., 1997; Hristoforou et al., 1991).

SQUID Magnetometers

The SQUID sensor (Superconducting Quantum Interference Device) is the ultimate magnetometer. This type of magnetometer is extremely sensitive and it is used to measure extremely low magnetic fields, as low as 4 fTesla (Drung, 2007). The SQUID magnetometers are based on the Josephson effect and

Josephson junction. The Josephson effect is the phenomenon of a supercurrent that flows indefinitely without any voltage applied in a superconductive ring. The Josephson junction is where two superconductors are coupled by a very thin layer of a non-superconductive metal (Josephson, 1962).

The SQUID is a split superconductive ring and is connected with a weak join. When there is no magnetic field, the current *I* splits into the two branches equally. If an external magnetic field is applied, a small current will appear in the ring to counteract the external magnetic field. If the magnetic field is more than half of the magnetic flux quantum, then the current changes direction and tries to increase the magnetic flux to be an integer number of the magnetic flux quanta.

This increase of the current and subsequently of the magnetic flux is more energetically preferred than trying to reduce it. The current-voltage characteristic of the SQUID has a hysterical behavior and several techniques are used to reduce this effect.

NOISE IN SENSORS AND CIRCUITS

Noise is defined as the unwanted electric signal that is arising within the circuit (inherent noise) or is picked up from outside the circuit (interference noise). The noise is present everywhere in electronics, it appears in the sensing element, in the transition lines, and in the electric circuit itself. The main source of noise occurred when the electron travels in the conductor and is scattered by the movements of the phono lattice. The noise can only be prevented from accumulating in a circuit and cannot be eliminated afterward.

Even the best sensor design is susceptible to noise, which means it is unable to make an identical representation of the impute physical stimulus. The noise is probably the most important factor determining the accuracy of a sensor. Many parameters introduce errors in a sensor reading, which have to do with the drift of the calibration parameters of the sensors. This part will only emphasize the noise sources, classifications, and some techniques to prevent the noise from entering the system. As mentioned earlier, there are two major noise classifications, the inherent noise, and the interference noise.

Inherent Noise

The sensing element, passive or active, rarely produces a voltage or current signal large enough to be picked up by an analog to digital converter or a meter. This small signal needs to be amplified by a series of operational amplifiers (OPA) however, each new electronic component introduced in a circuit, generates its own noise. The noise of the sensor elements and electronics can be classified according to the source and the frequency range. The main inherent noise sources are thermal noise, shot noise, flicker noise, and burst noise.

Thermal Noise

The thermal noise or Johnson-Nyquist noise is generated by the random thermal motion of the electric charge carriers (e.g. electrons), inside an electrical conductor due to scattering from the vibration of the photon lattice.

Thermal noise is unavoidable and is approximately white noise, which means that is nearly equal and independent of the frequency. The noise amplitude has a probability density function close to Gaussian.

The amplitude of the noise can be modeled by a voltage source (or current source) representing the noise of the non-ideal resistor in series (or parallel) with an ideal noise-free resistor. The one-sided voltage power spectral density, or voltage variance (mean square value) per hertz of bandwidth $\overline{\upsilon_n^2}$ (*V/Hz*), is given by (Fraden, 2010):

$$\overline{\upsilon_n^2} = 4k_B TR \tag{17}$$

where k_B is Boltzmann constant (*J/K*), T is the absolute temperature (*K*) and R is the total resistance (Ω). The current power spectral density, or current variance (mean square value) per hertz of bandwidth $\overline{\iota_n^2}$ (*A/Hz*), is given by:

$$\overline{\iota_n^2} = \frac{4k_B T}{R} \tag{18}$$

Some sensitive to noise electronics are kept in cryogenic conditions to suppress the effect of the thermal noise. This type of electronics can be found in very sensitive laboratory pre-amplifiers or radio telescopes.

Flicker Noise

Flicker noise; also referred to as *1/f* noise or pink noise, is a type of electronic noise that has 1/f behavior in its power spectral density (Fraden, 2010). The origin of this noise is unknown and there are only hypothetical mathematic models to explain it. Flicker noise is very important in electronic devices, especially at low frequencies. It is also a factor that produces phase noise in oscillators since it can be mixed up to frequencies close to the carrier. In most cases, the flicker noise has a Gaussian distribution. It has a linear behavior in resistors of FETs and a non-linear behavior in BJTs and diodes. The flicker noise has a stronger influence on carbon-based and thick-film resistors as opposed to the wire-wound resistors. It is characterized by the corner frequency (f_c), which is the threshold value when it becomes smaller than the thermal noise.

In some cases, the flicker noise can be minimized by converting the signal at a higher frequency (chopping) where the electronic circuit is less influenced by this type of noise. After the signal is conditioned, it can be down-converted to the baseband.

Shot Noise

Shot noise originates from the discrete nature of electric charges. Shot noise is less significant compared to the flicker and thermal noise while it is temperature and frequency independent. This noise is introduced mainly in the p-n junction (diodes, transistors). This noise can be modeled as a Poisson process and calculated as the mean square value of current fluctuation σi (A) (Fraden, 2010):

$$\sigma_i = \sqrt{2qI\Delta f} \tag{19}$$

Where q is the charge of an electron (C), Δf is the bandwidth in (Hz) over which is considered, and I is the DC current flowing through the conductor. The voltage created from the shot noise at the end of a resistor is

$$\sigma\upsilon_{=}\sigma \mathrm{iR} \quad (20)$$

Burst Noise

Burst noise or popcorn noise is a sudden step in the base current of a BJT or the threshold voltage of a FET. This name was given due to the sound produced when played over a speaker; it sounds like popcorn popping. It usually occurs at low frequencies of less than 1kHz. It has a non-Gaussian distribution. The repeatability of a burst is varying and it may happen several times in a second or it may take several minutes to occur.

The most probable causes of the popcorn noise are assumed to be the charge traps or microscopic defects in the semiconductor material. Heavy metal atoms or heavy ion implantation or even the unintentional side-effects such as surface contamination can produce such defects.

Interference Noise

The interference noise or transmitted noise is an external signal that it is picked up by the system. It can be produced by numerous sources such as power supplies, magnetic fields, electrostatic fields, electromagnetic emissions, thermal variations, mechanical vibrations, humidity, ionizing radiation, chemical agents, parasitic capacitances or inductive coupling, bad wire coupling, and non-ideal metal connections (Fraden, 2010).

These types of noise can be either additive or multiplicative. An additive noise e_n is added to the useful signal V_s and mixed up with an independent voltage (or current).

$$V_{out} = V_s + e_n \quad (21)$$

A multiplicative noise affects the transfer function of the nonlinear component of the circuit in a way that the signal V_s is altered or modulated by the noise source.

$$V_{out} = [1 + e_n]V_s \quad (22)$$

Although the interference noise can cause severe problems to the circuitry and the stability of a system, it can be eliminated or suppressed. Twisted pair or coaxial cables, the used of shielding, and the avoidance of the use of metals with different Fermi energies (thermoelectric effect) are some of the ways to suppress this type of noise.

TYPES OF MAGNETIC SENSORS

Magnetic sensors (magnetometers) are a key element for space applications. Geostationary satellites can be affected by the earth's magnetic field. However, sometimes the earth's magnetic field is destroyed by the solar wind (flare) and magnetic sensors can detect this phenomenon. Solar wind can destroy electronic circuits because it transfers electronic charges and therefore, circuits should be shut down in these circumstances. However, magnetic sensors are also used to measure the magnetic fingerprint of rock in other planets. Sometimes, there is a need to determine the magnetic field that is produced by the electronics of the spacecraft.

Through the years, a variety of methodologies have been developed in order to measure magnetism. In this section, only the following sensor types will be explored:

- Anisotropic Magnetoresistance (AMR)
- Giant Magnetoresistance (GMR)
- Giant Magneto-Impedance (GMI)
- Flux-Gate sensors
- Superconducting Quantum Interference Device (SQUID)

Each sensor type has certain properties and each application favors the use of a specific sensor type. Sometimes, there is also the need to use more than one sensor type due to the application requirements. Some of the application requirements to be identified to evaluate the suitability of the sensor types are:

- Noise levels and resolution
- Offset and loge-term stability
- Sensitivity, nonlinearity, and range of measured values
- Frequency bandwidth
- Temperature stability
- Immunity to orthogonal excitations
- Hysteresis (mostly due to magnetic material)
- Cost, availability of materials and production equipment
- Available installation space
- Sensor maintenance
- Power consumption

It is well known that the highest resolution possible is achieved by the SQUID magnetometer. However, the SQUID magnetometers need cryogenic temperatures, and the cost to maintain that state is extremely high. For unmanned space missions, there is the need for the components to be working perfectly for decades and while the sensor element of a SQUID magnetometer can last, the cooling system may not be able to work seamlessly.

Nevertheless, one of the most important parameters is the noise level that leads to the resolution of the sensor. The noise is measured through the power spectral density (PSD) which is the power of the noise concentrated around the frequency bandwidth. It is customary to refer to it as the power spectrum even when there is no physical power involved. This PSD is the sum of all noise sources, but the most important noise source is the internal noise of the sensing element. PSD, $S(f)$, is defined as:

Table 1. List of magnetic sensors PSD noise (Tumanski, 2011).

Sensor Type	S(f)
SQUID	$5fT/\sqrt{Hz}$
Hight T_C SQUID	$100fT/\sqrt{Hz}$
AMR	$15nT/\sqrt{Hz}$
GMR	$100pT/\sqrt{Hz}$
GMI	$3pT/\sqrt{Hz}$
Flux-Gate	$2pT/\sqrt{Hz}$

$$S(f)=\frac{V_n^2}{\Delta f}=\left(\frac{V_n^2}{\sqrt{\Delta f}}\right)^2 \tag{23}$$

where $V_n(V)$ is the voltage integral and $f(Hz)$ is the frequency range. In Table 1, the average PSD values that are usually found in each sensor type, are presented.

Anisotropic Magnetoresistance (AMR)

Anisotropic Magnetoresistance (AMR) was discovered in 1897 by William Thompson (Lord Kelvin). The evolution continued through the years and in 1960, it was fully modeled. Materials that exhibit AMR were used in the first magnetic random-access memory (MRAM) that was developed by IBM. Nowadays, AMR magnetometers are used in high current detection, position sensing, tachometry, low-magnetic-field anomaly detection, and multi-axis compasses. Philips Electronics and Honeywell are the main manufacturers of this type of sensors. In order to test this type of magnetometers and calibrate them, a 3D Helmholtz coil is used. This is necessary since AMR sensors have a non-linear behavior and there is no mathematical expression to calculate the AMR behavior in advance. The most common material used in AMR magnetometers is NiFe permalloy.

Physical Model

Anisotropic Magnetoresistance can be divided into two types, ordinary and anisotropic. Usually, non-magnetic metals and semiconductors appear to have ordinary magnetoresistance. This effect is due to the scattering of the electric carriers from the magnetic field (Hall effect) which increases the path length of the carriers that subsequently increases the resistance. The equation for the ordinary magnetoresistance is:

$$\frac{\Delta\rho}{\rho_0}=C\mu B^2 \tag{24}$$

Figure 1. AMR element behavior in an applied magnetic field

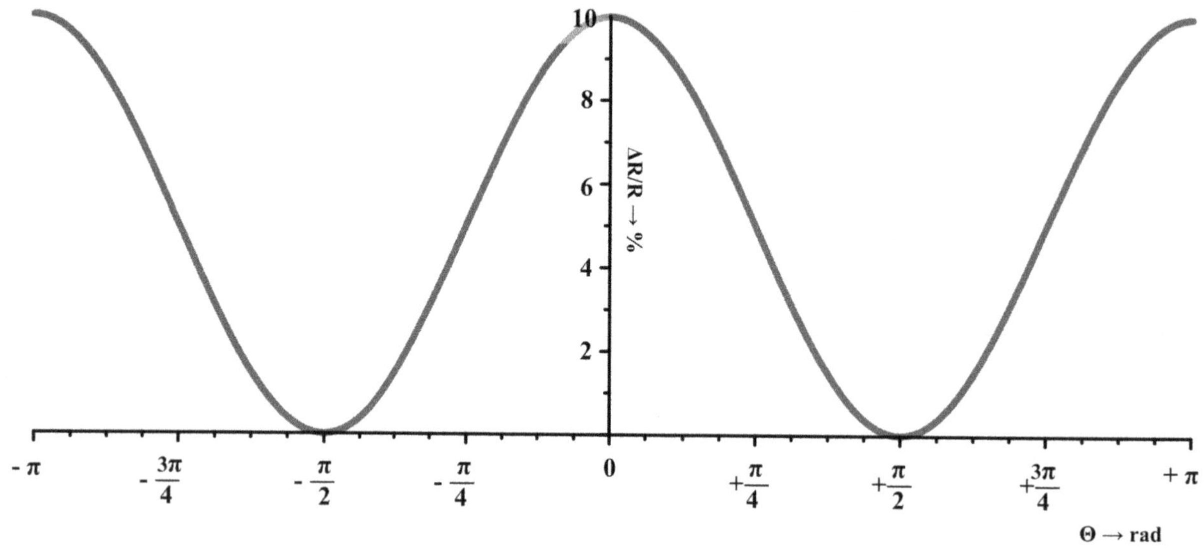

where C is a constant ($sT^2m^{-2}V^{-1}$), μ is the carriers' mobility (m2^{V}s-$^{1)}$, ρ is the electric resistivity of the material (Ωm) and B *is* the magnetic field (Te*sla*). An AMR magnetometer reaches a saturation level when a 25-30 Oersted (Oe) magnetization field strength or 25-30 Gauss (Gs) for magnetic field density is applied. For an AMR sensor, this behavior can be simplified according to the following equation:

$$\frac{\Delta R}{R_0} = \Delta R_{max} cos^2\left(\theta\right) \tag{25}$$

Figure 1 depicts this behavior.

The Construction of an AMR Barber-Pole

As stated previously, an AMR element has a highly non-linear behavior. Moreover, the AMR element does not sense the difference in the direction of the magnetic field due to the square law in Equation (25). To overcome this problem, the sensing element should be biased and the simplest way to do that is to make the current travel at a shallower angle into the magnetic field instead of 90°. The best angle to have the element biased is in the angle of 45°. In order to achieve this bias, as depicted in Figure 2, a series of alternating straps of aluminum and the magnetic material is placed as a thin film in a non-contractive substrate like a barber-pole (Ripka et al., 2007).

The linear region is moved to the 45°. The magnetic domain of the magnetic material is not necessarily polarized to the direction of the current. The linear region is moved to the 45° due to the orientation of the aluminum stripes as depicted in Figure 2. The 45° angle is very convenient because it turns the linear region in the middle of the rotating angle. All the magnetic domains must be polarized in the same direction to achieve maximum resolution. To achieve this, a series of set/reset pulses must be applied to a coil around the barber-pole in the easy magnetization axis (long axis). A set pulse polarizes

Figure 2. AMR (Barber-Pole) sensor element, aluminum shorting straps for redirecting the current (Ripka et al., 2007).

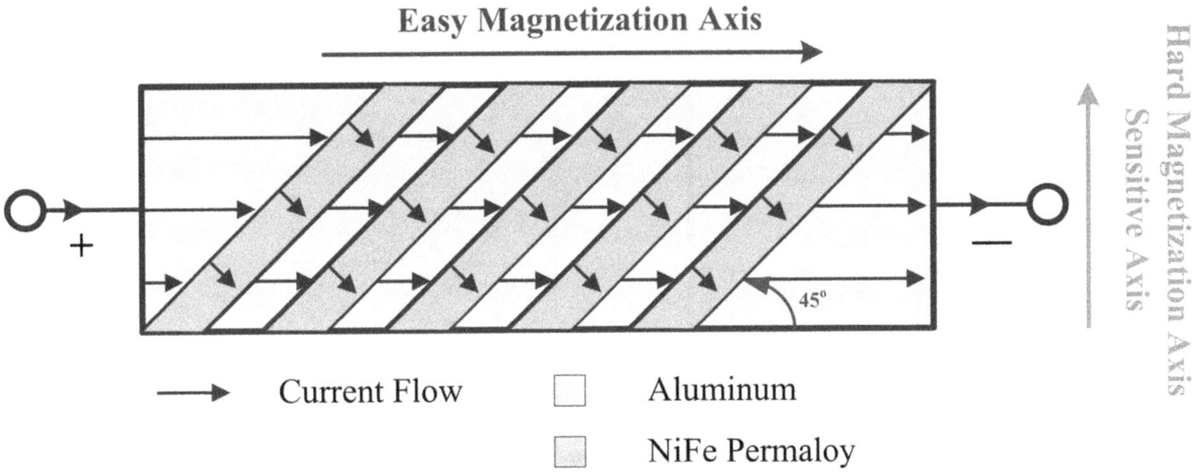

the magnetic domain in the direction of the current flow and a reset pulse polarizes it in the opposite direction. Figure 3 presents the change in the magnetic domain and the two operating regions of the magnetic material (Grosz et al., 2017).

Construction of an AMR Sensor Bridge

An AMR sensor is composed of the four AMR barber-pole sensing element, which is placed in a Wheatstone bridge configuration (Figure 4). For a complied magnetometer, there is a need for two extra coils in different orientations. The set/reset coil ($L_{S/R}$) is oriented in such a way that the magnetic field is in the direction of the easy magnetization axis of the elements. The offset strap (L_{offset}) is a coil that produces a magnetic field parallel to the sensitive axis. The use of this coil is to bias the bridge or to cancel the external magnetic field for closed-loop applications.

The AMR Wheatstone bridge is now a fundamental part of all AMR sensors. The construction of the sensing element is very important, especially for low noise applications. Operational amplifiers (OPA) can be noisier than the sensor element itself (Hadjigeorgiou, 2014). The use of the set/reset coil in a closed-loop architecture can further improve the resolution and suppress the 1/f noise. The AMR sensing element has a power spectral density of $15nT/\sqrt{Hz}$ (Tumanski, 2011) as tabulated in Table 1. However, the techniques discussed earlier show that the power spectral density can be as low as $4.5nT/\sqrt{Hz}$ (Hadjigeorgiou et. al, 2017). The AMR sensors have high bandwidth starting from DC and going up to several MHz, but when a closed-loop architecture is used, the frequency is reduced. This bandwidth limitation occurs due to several factors, including the response of the magnetic material in the fast set/reset pulse (Jiles, 2015; Tumanski, 2011) and the set/reset strap that can overheat and permanently damage the sensor. These materials need high magnetic fields to rotate and consequently, high current pulses are needed. The method of closed-loop architecture (Hadjigeorgiou et. al, 2017) is another possible factor. The AMR is a versatile, very widely used magnetic sensor that offers a good compromise between resolution and cost.

Figure 3. Resistivity behavior of a barber-pole with two directions of polarization.

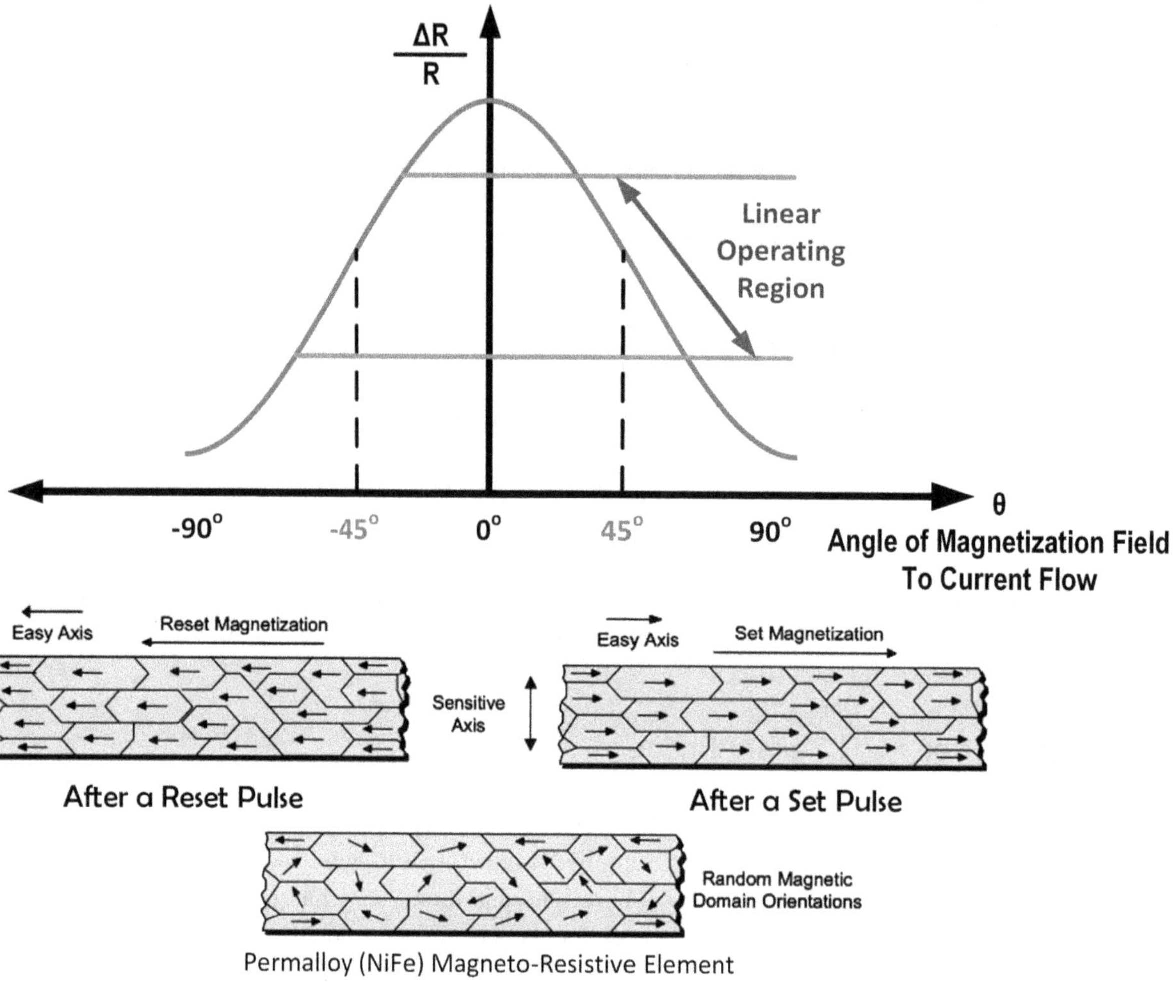

High-Frequency Behavior and System Model

AMR sensors are popular in sensitive applications, especially in the planar on-chip integrated form which implies the presence of several on-chip capacitive and inductive parasitic elements. However, there is a lack of equivalent circuit models capturing these parasitics and allowing the modeling of sensors' behavior at higher frequencies or in stability-sensitive closed-loop operation. The work of Hadjigeorgiou & Sotiriadis, (2020) attempts to fill this gap in the literature by introducing a high-frequency equivalent circuit model (Figure 5), for planar on-chip AMR sensors with integrated offset and Set/Reset coils. The model employes the main capacitive and inductive parasitic elements, namely:

- the capacitances between the offset coil and the bridge,
- the capacitances between the offset and the S/R coils, and,
- the magnetic coupling between the offset and the S/R coils.

Figure 4. The AMR Wheatstone bridge with the Set/Reset and Offset coils (Hadjigeorgiou et al., 2017).

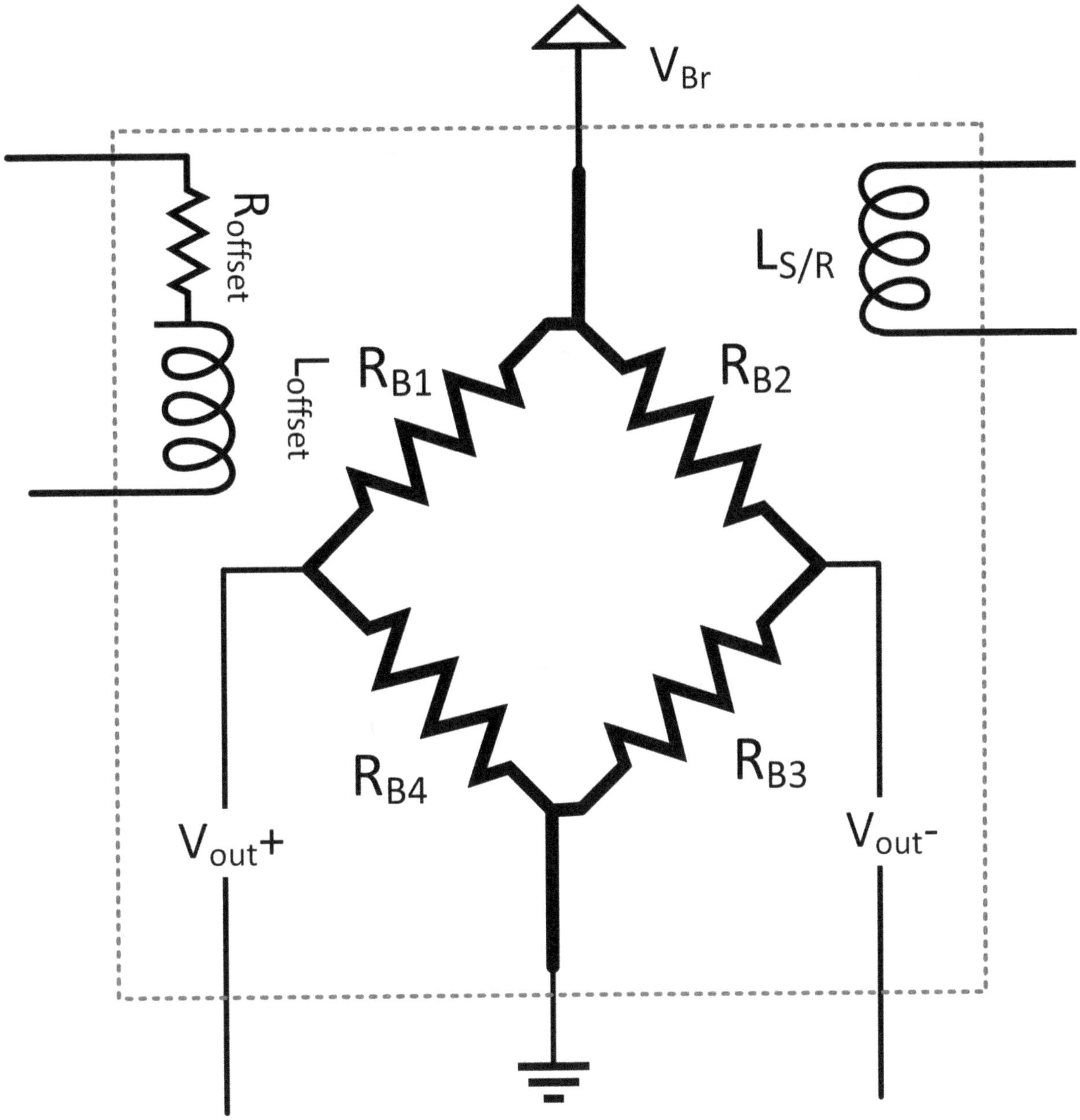

The parasitic capacitances between the S/R coil and the bridge are considered negligible and are ignored. This is because the bridge is placed at the top, the S/R coil is placed at the bottom and the offset coil is between them providing electric field isolation. This covers all pair-wise parasitic elements between the bridge and the offset and S/R coils. The parasitic capacitances between them and the substrate or the package are considered less important. Furthermore, it is assumed that the sensor operates in the linear region. This is typically the case when the sensor operates in a closed-loop configuration where the current of the offset coil is the feedback signal that ideally zeros the total magnetic field on the sensing elements. Also, it is assumed that the sensor's magnetic state has been set to maximum sensitivity before the measurements. To this end, a sequence of positive (Set) current pulses have passed through the S/R coil.

A methodology for estimating the values of the parasitic elements via measurements is presented and applied to a particular commercial AMR sensor (HMC100X). Using the introduced circuit model, the transfer function from the offset coil voltage to the output voltage of the sensor, via the parasitic capacitive coupling path, is derived. The transfer function from the S/R coil voltage to the offset coil voltage is also derived. Both are used to create a system-level model (Figure 6) of the AMR sensor which can be used in a system-level analysis to help the designer predict and optimize the open or the closed-loop sensor circuit system behavior. The introduced high-frequency equivalent circuit model and the two related transfer functions have been verified experimentally using the component values derived from the examined commercial AMR sensor. Finally, a system diagram of the AMR sensor was introduced which can be used in high-level system analysis. The introduced model can be used as a design and analysis tool in both open- and closed-loop architectures. It can help to derive the sensor system's transfer function and verifying stable operation in the closed-loop cases. Finally, it can be used as a guideline in deriving high-frequency equivalent models for other AMR sensors with offset and S/R coils. (Hadjigeorgiou & Sotiriadis, 2020)

Giant Magnetoresistance (GMR)

Giant Magneto Resistance magnetometers are much more sensitive than AMR magnetometers but at higher magnetic fields. This type of sensor is easily fabricated on a chip. They are widely used in hard disk drive heads to read the data from the disk plates. The simplest structure of a GMR is constructed by two layers of a magnetic material, which are separated by a conductor. GMR may consist of more magnetic and conductive layers. In each case, a conductor is sandwiched between two magnetic layers. Magnetic layers are constructed by Fe-Co-Ni alloy or NiFe permalloy. Conductive layers are made of Cu or Al. The Giant magnetoresistance effect was discovered by P. Grunberg and A. Fert. Figure 7 provides a simplified cross-section of a GMR.

Physical Model

The flow of an electric current in a magnetic multilayer structure, as shown in Figure 7, is influenced by the relative magnetization orientation of the magnetic layers. The resistivity of the magnetic layers is low when the magnetization is parallel to the field and it is higher when the magnetization of the neighboring magnetic layers is antiparallel. This is happening due to the appearance of the effect of the spin-dependent scattering. To observe this phenomenon, the thickness of the separating conductive material must be in the order of nanometers, and whilst its thickness is a major factor in the sensor's design. When an external magnetic field is applied, a change from antiferromagnetic to ferromagnetic (and vice-versa) coupling can be achieved, which leads to a change in the resistance value. The magnetoresistance (MR) ratio is defined as:

$$MR = \frac{\Delta R}{R} = \frac{R^{\uparrow\downarrow} - R^{\uparrow\uparrow}}{R^{\uparrow\uparrow}} \quad (26)$$

Figure 5. The frequency equivalent circuit model of the AMR sensor (Hadjigeorgiou & Sotiriadis, 2020).

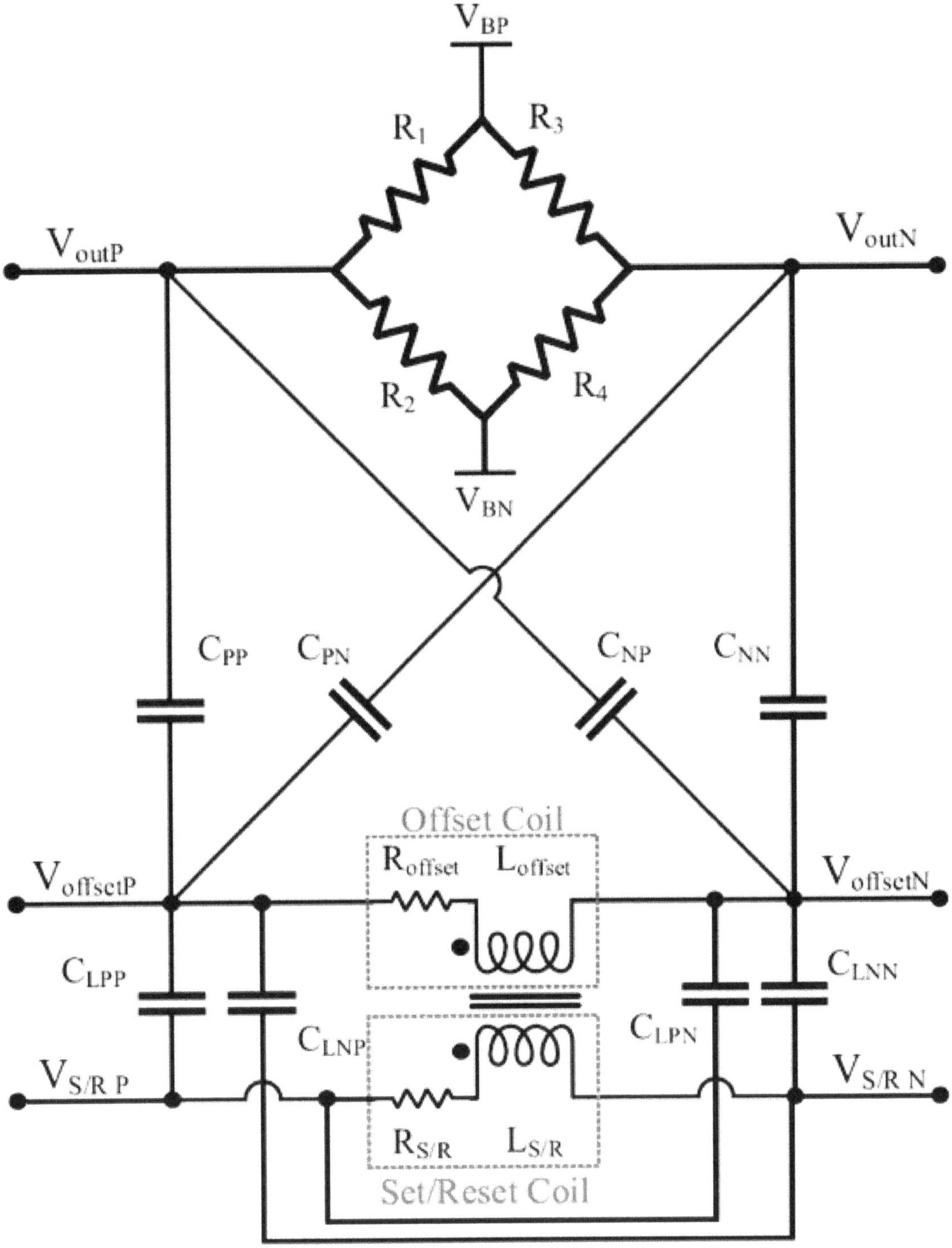

Spin valves are a particular configuration of a sandwich structure. In spin valves, an additional antiferromagnetic (pinning) layer is added to the top or bottom part of the structure, as shown in Figure 8 (right part). In this sort of structure, there is no need for an external excitation to get the antiparallel

Figure 6. Typical sensing operation system diagram of an AMR sensor (Hadjigeorgiou & Sotiriadis, 2020).

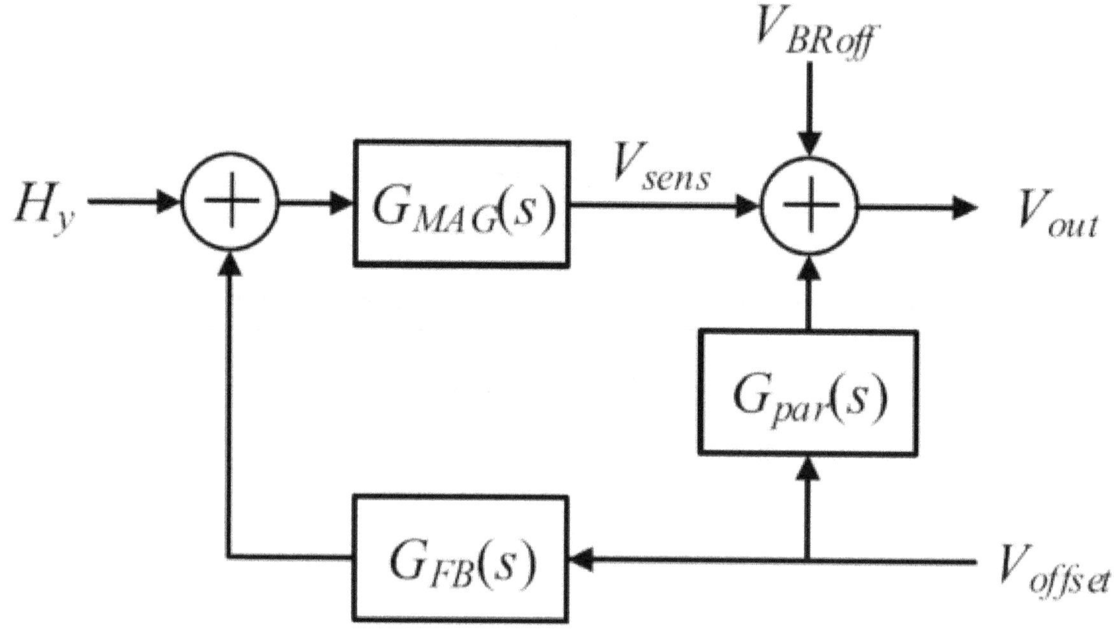

alignment. Despite this, the pinned direction (easy axis) is usually fixed by raising the temperature above the knee temperature (at which the antiferromagnetic coupling disappears) and then cooling it within a fixed magnetic field. Devices based on this phenomenon, have a temperature limitation below the knee temperature. Typical values displayed by spin valves are an MR of 4–20% with saturation fields of 0.8–6 kA/m. For linear applications without excitation, pinned (easy axis) and free layers are preferably arranged in a crossed axis configuration (at 90°), as depicted in Figure 8.

The mathematical expression of the change in resistance for the structure in Figure 8 is:

$$\Delta R = \frac{1}{2} i \left(\frac{\Delta R}{R} \right) R_{\blacksquare} \frac{W}{h} \cos\left(\Theta_p - \Theta_f\right) \tag{27}$$

Figure 7. A Basic GMR structure and its typical response (Grosz et al., 2017).

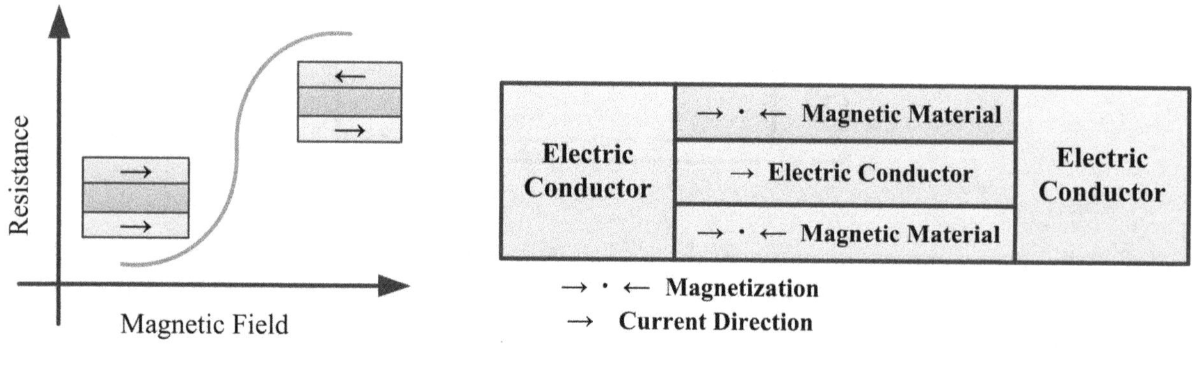

Figure 8. Basic spin value scheme for a multilayer structure in crossed axis configuration (Grosz et al., 2017).

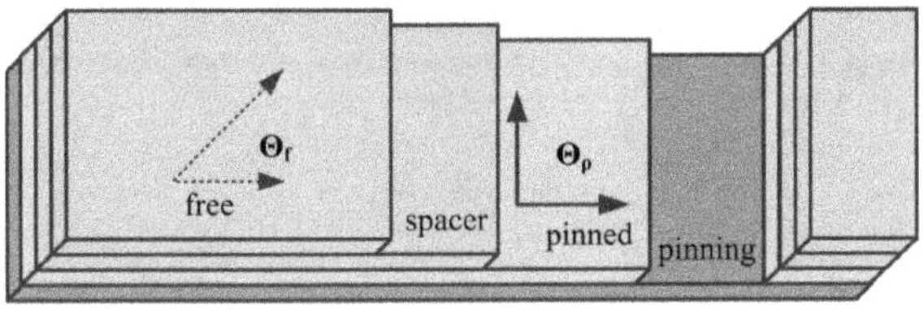

where $\Delta R/R$ is the maximum MR level (*5-20%*), $R_{\blacksquare}$ is the sensor sheet resistance ($\Omega\blacksquare^{-1}$), W is the width (*m*), h is the thickness (*m*), i is the current passing through the sensor (*A*), and Θ_p and Θ_f are the angles of the magnetization of the pinned and free layer (*rads*), respectively (Freitas, 2007).

Both structures that are presented in Figure 7 and Figure 8 can be parts of a Wheatstone bridge. To further improve the sensor's behavior, proper biasing is needed. Usually, a voltage source is needed to bias the sensor bridge. However, the theoretical analysis and the experiments show that is better to use a constant current source for feeding the bridge instead of a constant voltage. This technique minimizes the temperature drifts of the spin values. Furthermore, the use of AC current for biasing improves the sensor's performance in terms of hysteresis, linearity, offset, and noise.

Giant Magneto-Impedance (GMI)

Giant Magnetoimpedance (GMI) effect consists of the huge change of both real and imaginary parts of the impedance upon the application of a static magnetic field. The relative change of impedance can reach ratios up to around 600%, with extremely large sensitivities in the very low magnetic field region. The first reports dealing with GMI appeared several decades ago, in 1935 by researchers E. P. Harrison et al. (Harrison, 1935). The observation and research about the GMI phenomenon were received with less enthusiasm, probably because of the envisaged lower technological expectations and of the apparent lack of intrinsically new magnetic effects related to its origin. This idea became obsolete and now the interpretation requires a deep understanding of the micromagnetic characteristics of the soft magnetic materials and its dependence on dynamic magnetism. Moreover, the effect has attracted great scientific and technological interest in the last years due to the great impedance changes observed in well-suited samples and optimum experimental configurations. Many research teams all over the world investigate GMI and its technological applications. GMI has created a new branch of research merging the micro-magnetics of soft magnets with classical electrodynamics (Grosz, 2017).

Physical Model

The magneto-impedance effect is referring to the complex impedance found in a wire when it is excited with alternating current. The complex impedance is described with the equation:

$$Z = R + jX \tag{28}$$

Figure 9. Scheme to the definition of impedance.

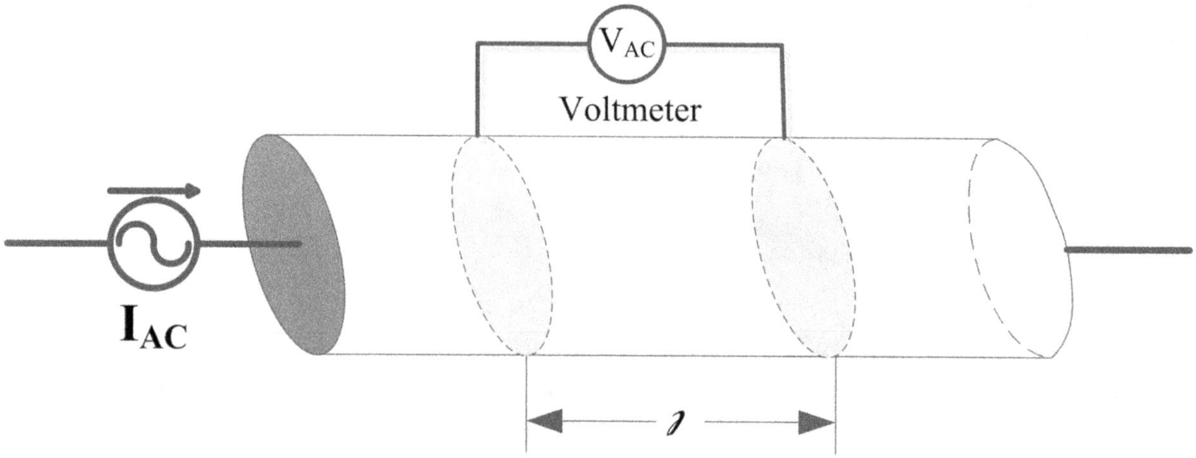

where R is the real part or resistance and X is the imaginary resistance (reactance), as shown in Figure 9.

The analytical impedance for a conductor can be calculated by simultaneously solving the deduced Maxwell equation

$$\nabla^2 H - \frac{\mu_0}{\rho}\dot{\mathrm{H}} = \frac{\mu}{\rho}\dot{\mathrm{M}} - grad\ div\ M \tag{29}$$

and the Landau-Lifshitz equation for the motion of the magnetization vector:

$$M = \gamma M \times H_{eff} - \frac{a}{M_S} M \times \dot{M} - \frac{1}{\tau}(\mathrm{M} - \mathrm{M}_0) \tag{30}$$

where γ is the gyromagnetic ratio, MS the saturation magnetization, $M0$ the static magnetization, He_{ff} the effective magnetic field and a is the damping parameter. The expression of the impedance is given by:

$$Z = R_{DC} kt \frac{J_0(kr)}{2J_1(kr)} \tag{31}$$

Where J_0 and J_1 are the Bessel functions of the first kind, r (m) is the radius of the wire, $R_{DC}(\Omega)$ is the electrical resistance for a DC and $k=(1+j)/\delta m$ with imaginary unit j. δm (m) is the penetration depth in a magnetic medium, with relative circumferential permeability $\mu\varphi$:

$$\delta_m = \frac{c}{\sqrt{4\pi^2 f \sigma \mu_\varphi}} \tag{32}$$

Figure 10. Domain wall configuration in a multilayer structure. The magnetization direction is shown in every domain. The AC current flows in the sandwiches metallic layer, producing an AC flux in the surrounding magnetic layers (Grosz, 2017).

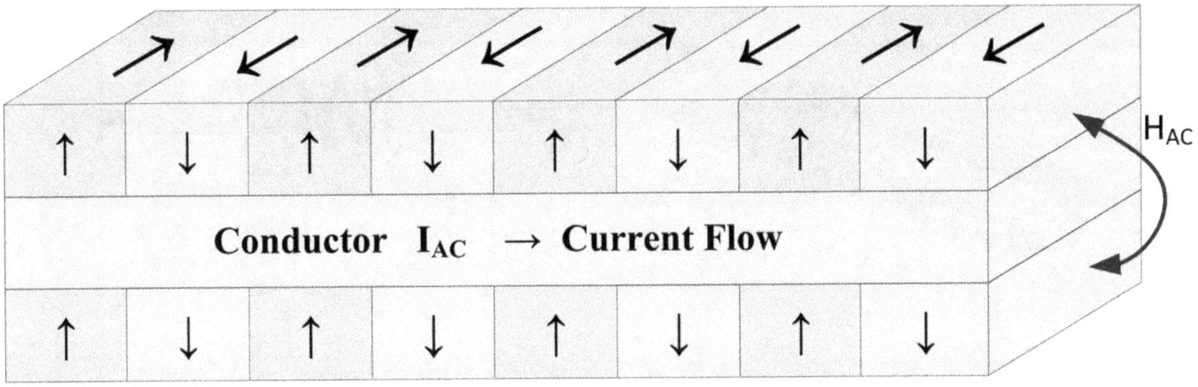

where c (ms-$^{1)}$ the speed of is light, σ (Ωm) the electrical conductivity, and f (Hz) is the frequency of the IAC (A) flowing across the sample (Jordan, 1968). There are two types of magnetic permeability, the longitudinal and the circumferential; in a ribbon, circumferential is referred to as transverse. Figure 10 shows the magnetic domain of a ribbon and Figure 11 shows the longitudinal and the transverse impedance.

There are many theoretical models, each one for a different frequency range, which can be found in the bibliography in more detail. These models are the Quasistatic model, the Eddy current model, the Domain wall model, the Electromagnetic model, and the Exchange-conductivity model.

Figure 11. MI field dependencies measured in narrow NiFe/Au/NiFe sandwich films with (a) the longitudinal anisotropy and (b) the transverse anisotropy. Dashed curves show the reversed magnetic field behavior which can extract the hysteresis of the material (Grosz, 2017).

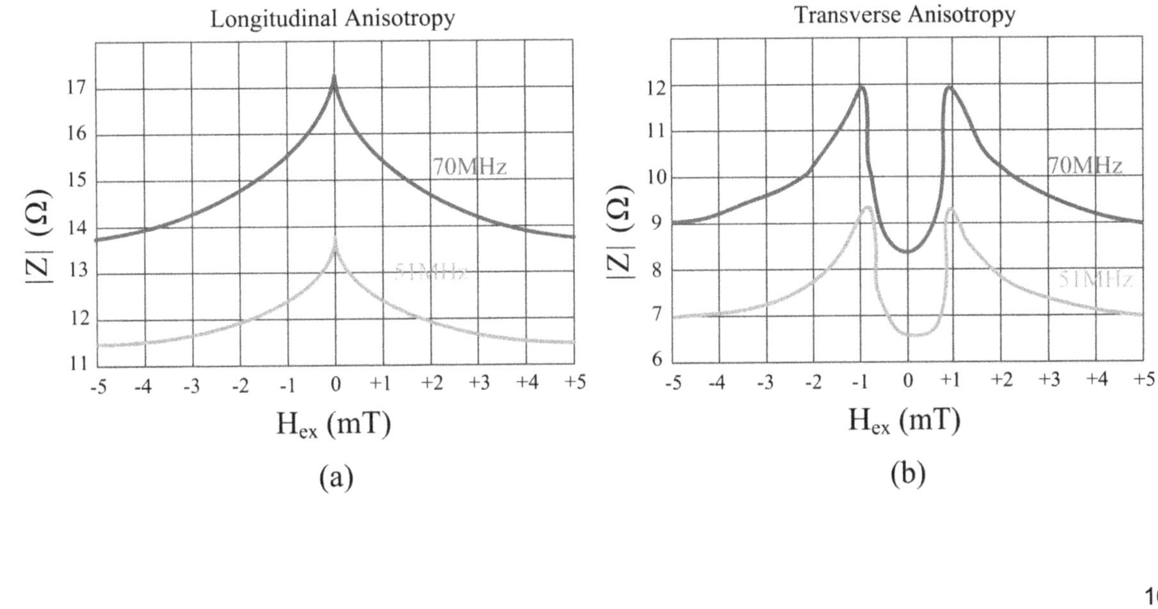

Figure 12. A Parallel Fluxgate on the left (Ripka, 2007) and an Orthogonal Fluxgate on the right (Grosz et al., 2017).

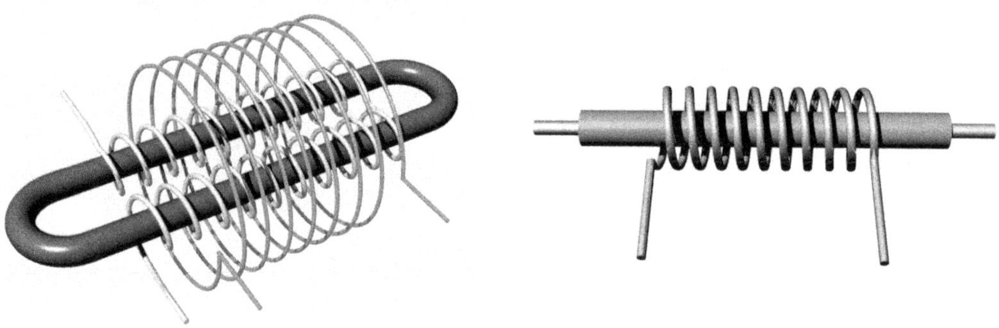

Flux-Gate Sensors

The fluxgate magnetometers received their name from magnetic flux-gating curves. The fluxgate magnetometer is dated back to 1930. It was widely used during WWII as a trigger in magnetic ship mines. Fluxgate sensors' core is coming in many shapes such as rob, ring, and race-track. The design of the core geometry, construction, and magnetic material are crucial factors, for low noise performances. There are two types of fluxgate; the parallel and the orthogonal fluxgate. The difference between these types is that the excitation and pick-up coils are parallel to the fluxgate. On the other hand, the orthogonal fluxgate has a direct excitation to the magnetic core and the pick-up coil is orthogonally oriented to the direction of the excitation current. Figure 12 depicts a parallel fluxgate and an orthogonal fluxgate.

Physical Model

The fluxgate sensor is based on two physical phenomena, the induction law, and the magnetic core saturation. The output voltage of the pick-up coil in Figure 12a is calculated according to:

$$U_i = -NS\left(\frac{dB_E}{dt} + K\mu_0\mu_r\frac{dH_M}{dt} + K\mu_0 H_M\frac{d\mu_r}{dt}\right) \qquad (33)$$

where N is the number of turns of the pick-up coil, S (m^2) is the core cross-sectional area, H_M (Am^{-1}) is the measured external magnetic field intensity, B_E (T) is the alternating excitation flux density in the ferromagnetic core due to the excitation field, $\mu 0$ (Hm-$^{1)}$ is the permeability of vacuum and K is a dimensionless coupling coefficient of the core to the field HM (Am-$^{1)}$. Finally, μr $_i$s the relative permeability of the magnetic core varying with the external magnetic field.

The parallel fluxgate sensors operate based on the following principle. A current pass through the excitation coil, creating a magnetic field HE which encircles the magnetic core. The magnetic field is high enough to force the magnetic core to enter the magnetization saturation region Ms Figure 13a shows the magnetic flux of one half of the core without an external magnetic field while Figure 13c presents

Figure 13. Fluxgate B-H loop without (a), with (b) external magnetic field. At the bottom the excitation magnetic field and the output voltage. (Gordon et al., 1965).

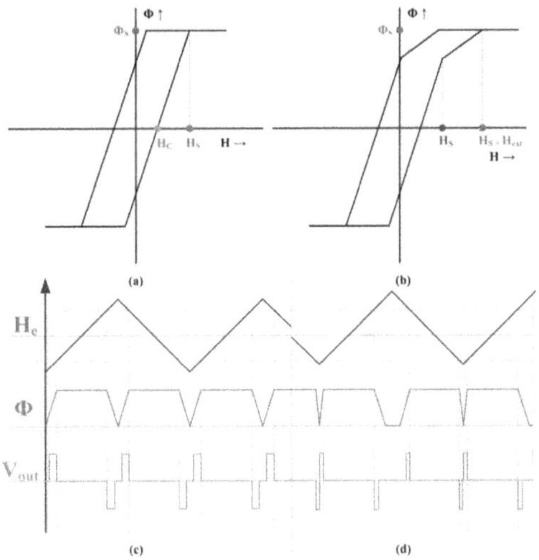

the output voltage. When an external magnetic field He_{xt} is applied, the magnetic flux is different in the two halves of the core as shown in Figure 13b. The output signal is given in Figure 13d.

If the excitation of the core is sinusoidal, H_e=Asin(ωt), then the magnetic flux Φ of the core is,

$$\Phi = 2S\left(aH_{ext} - bH_{ext}^3 - \frac{3}{2}bA^2H_{ext} + \frac{3}{2}bA^2H_{ext}\cos\left(2\omega t\right)\right) \tag{34}$$

where $a>0(Vm/A)$, $b>0(Vm^3/A^3)$, are coefficients, A (Am^{-1}) is the amplitude of the excitation magnetic field, H_{ext} (Am^{-1}) is the external magnetic field and ω (rad/s) is the rotating frequency. Considering equation 34, it is clear that the simplest way to extract the external magnetic field ($He_{xt)}$ is the second harmonic. In reality, there are higher-order harmonics present in the output signal, due to the nature of the B-H loop (hysteresis, approach to saturation). Another factor is the non-sinusoidal excitation waveforms which have higher harmonics (Hadjigeorgiou et al., 2015). These effects are sometimes taken into account but most fluxgate magnetometers use only the second harmonic.

Superconducting Quantum Interference Device (SQUID)

Superconducting Quantum Interference Device (SQUID) is a magnetometer that measures the magnetic flux passing throw it. It may be the most sensitive magnetometer that has ever been created. There are two types of SQUID; the low temperature which is using liquid Helium for cooling, and the high temperature which is using liquid Nitrogen. Both are based on the Josephson effect (Josephson, 1962). They consist of a superconducting loop interrupted by two resistively shunted Josephson tunnel junctions. The low-temperature variation, operating at 4.2 K, is almost noiseless in the order of $\mu\Phi / \sqrt{Hz}$ or $fT / \sqrt{Hz}$. They have large bandwidth and a flat frequency response ranging from DC to GHz (Drung et al., 2007). They are excellent for a wide variety of applications, such as e.g. biomagnetism, magne-

Figure 14. The SQUID ring is shaped as a square washer with a hole in the center and a slit. The Josephson junctions (purple colure) are located at the outer edge of the slit, as indicated. On the right, an electrical equivalent of the SQUID is presented (Grosz et al., 2017).

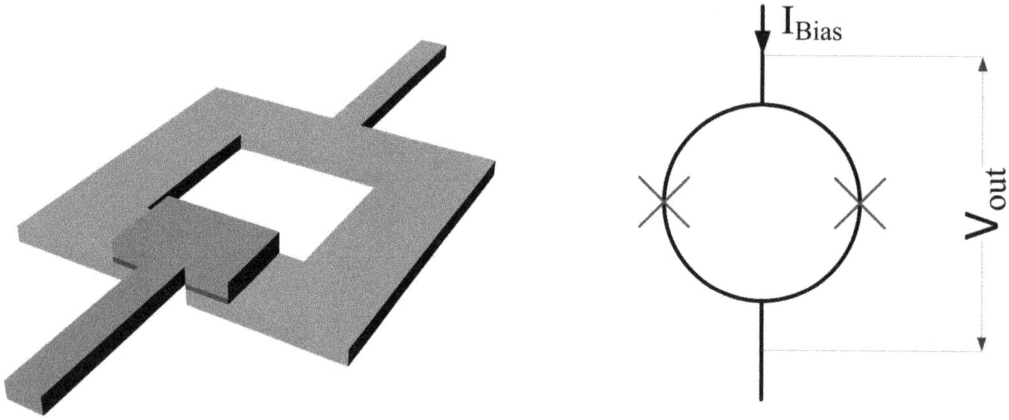

toencephalography (MEG), calibration of other systems, and geophysical exploration. Nevertheless, the cryogenic conditions needed to operate properly are expensive including not only the installation costs but also the maintenance costs. (Ripka, 2001; Tumanski, 2011).

Physical Model

SQUID converts magnetic flux into a voltage across the device. The operation of SQUID is based on two physical phenomena: flux quantization in a closed superconducting loop in units of the flux quantum $\Phi_0=h/2e=2.07\times 10fTm^2$, and Josephson tunneling. The Josephson Junction (JJ) also called Cooper pairs, has zero spin and follows boson statistics. Figure 14 depicts a superconductive ring with a Josephson junction at one end.

Both superconductive branches are in the same quantum state and can be described by a collective superconducting wave function $\Psi=\Psi_0 exp(i\phi)$, with $\phi(x,t)$ being the time and space-dependent phase and ns=$_|\Psi|2$ 'he Cooper pair density. Whenever two superconductive elements are connected through a thin insulator (weak couple), there is a current exchange. In the Josephson tunnel junction (SIS), the thin insulating barrier (I) is placed between two superconductors (S). The current passing from the Josephson junction is described by the first Josephson equation IC=$_I C,_{0si}$n(φ), with IC,0_{th}e maximum critical current and φ being the difference across the junction. When the maximum critical current is exceeded, the phase difference across the junction increases over time and a DC voltage across the junction appears. It is described by the second Josephson equation:

$$\frac{\partial\varphi}{\partial t}=\frac{2e}{\hbar}V_{DC}=\frac{2\pi}{\Phi_0}V_{DC} \quad (35)$$

Figure 15. The Current-Voltage characteristics of a Josephson tunnel junction (Grosz et al., 2017).

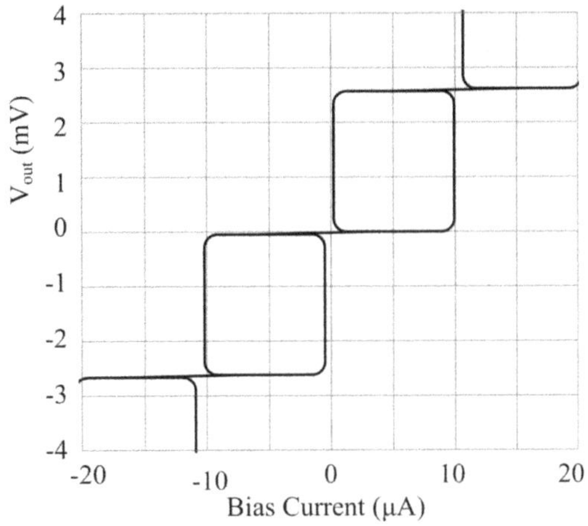

where $\hbar = 1.055\times10^{-34}$ Js is the reduced Plank's constant. The current across the Josephson junction oscillates with the Josephson frequency $2\pi V D_{C/}\Phi 0$. Figure 15 depicts this oscillation that can be measured (Josephson, 1962).

In practice, there is a feedback coil that helps to the linearity of the closed-loop system and the output voltage. However, the electronics needed to be implemented for a linear behavior, introduce new noise sources much higher than the sensor itself.

ADVANCES IN MAGNETIC SENSORS FOR SPACE MEASUREMENTS

This chapter, so far, has presented the different sensor types and the physics principles they operate on. Extensive reviews on this subject can be found in the work of Mahdi et al., (2003), Lenz, (1990; 2006), and Tumanski, (2007). Magnetometers in general have been used for measurements in space since the early 1950s. A detailed review of the magnetic sensors used in space measurements can be found in Acuña, (2002), Díaz-Michelena, (2009) and Hospodarsky, (2016). Moreover, a detailed instrument classification for space applications can be found in Brown, (2019). Finally, a review of optical fiber sensors based on magnetic fluid can be found in Alberto et al., (2018). Overall, as science and time advances, different ideas and implementations tailored to specific needs have emerged. In this section, the authors will attempt to briefly present some of them.

In Griffin et al., (2012) a compact low-noise magnetic gradiometer is designed and calibrated. The gradiometer was developed as a technology pathfinder for a diagnostics payload to characterize magnetic disturbances around space instruments susceptible to magnetic fields and gradients such as the candidate Cosmic Vision mission LISA/NGO (Laser Interferometer Space Antenna/ New Gravitational-wave Observatory) under ESA Contract AO/1-6085/09/NL/AF. The gradiometer key design requirements are tabulated in Table 2. The gradiometer is capable of determining both the local magnetic flux vector and the gradient tensor matrix. It achieves this with a combination of five pairs of fluxgate magnetometers mounted within a block of Macor™. Each fluxgate sensor within the gradiometer is aligned along with one of the three orthogonal axes, and thus the magnetic flux vector can be inferred from the output

Table 2. Gradiometer design requirements (Griffin et al., 2012).

Specifications	Requirement
Magnetic measurement quantities	Magnetic flux vector: B Magnetic flux gradient: ∇B
Noise floor	< 30 nT/m/$\sqrt{\text{Hz}}$ at 0.1 mHz < 10 nT/m/$\sqrt{\text{Hz}}$ at 1 mHz < 1 nT/m/$\sqrt{\text{Hz}}$ at 1 Hz
Working environment	Terrestrial (±60 μT) and space (in the range of a few to several 10s nT)
Measurement Range	±50 μT/m
Residual gradient offset	< 10 nT / m
Magnetic cross field rejection	> 40 dB
Volume	< 40 cm3 and < 4.5cm tip-tip max. dimension
Mass	< 100 g

of the sensors with reference to the axis of interest. Mechanical reference features are machined onto three faces of the gradiometer head to permit accurate alignment of the head with respect to the EUT. The gradient can be determined from the difference in the output of the sensor pairs and the physical separation of the two sensors. Since the gradient tensor matrix is symmetric and the trace is zero, only five of the nine gradient terms are needed to characterize the complete matrix. The selection of the five particular gradient terms to be measured is not unique. The terms measured in the gradiometer head are; $\partial By/\partial y$, $\partial Bz/\partial z$, $\partial Bx/\partial y$, $\partial Bx/\partial z$, and $\partial Bz/\partial y$. Within the mathematical constraints, these terms were selected on the basis of:

- minimizing the overall volume of the gradiometer head,
- maximizing the physical separation of the sensors to have the largest possible measurement baseline within the allocated volume,
- avoiding configurations that have sensors in close physical proximity.

Fluxgate magnetometers were selected for use in the gradiometer due to the following reasons

- accurate vector flux measurements with low cross-field sensitivity can be achieved with careful design and assembly of the sensors,
- the individual sensors can be made to be very compact (the individual sensors are composed of a pair of sense/excitation coils 18 mm long and 2 mm diameter spaced apart by 6 mm in the Förster configuration),
- the noise performance of fluxgate magnetometers is compatible with the low-noise requirements of the gradiometer and,
- the power consumption and complexity of the bias and readout electronics are relatively low.

The sensors are read out by a differential Phase Sensitive Detector circuit with feedback on the sense coil. The bandwidth of the sensor (DC – 5 Hz) is defined by a series of low-pass filters in the output circuit. A 5 m harness is used to interface the gradiometer head with the readout electronics. The output

Figure 16. Combination of sensor outputs (Han et al., 2012).

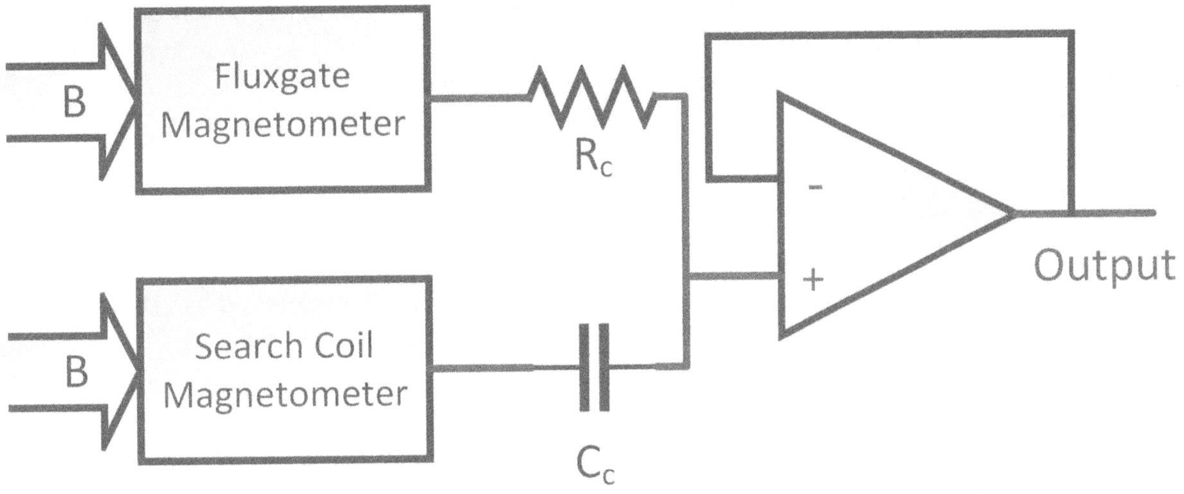

of the individual sensors is buffered by low-noise amplifiers mounted within the gradiometer head to improve the immunity of the system to EMC and to eliminate the effect of the length of the harness from the calibration of the sensor itself. The system was designed to operate with a 24-bit sampling of the sensor outputs which provides a small margin on the required sampling digitization. The bias and readout electronics can be powered by either an external 12 V DC power supply or from internal lead-acid rechargeable batteries. The noise performance at the lower end of the frequency range ($\leq 10^{-2}$ Hz) was found to be non-compliant concerning the requirement. An investigation was made into the primary noise source in this frequency range. It was found that the noise performance of the Operational Amplifiers in the buffer circuit located at the front end and in the Phase Sensitive Detector limited the noise performance of the overall gradiometer. The noise performance of the system can be improved by selecting lower noise, but higher dissipation devices. The current selection is based on the relative weighting of the competing noise and power dissipation requirements (Griffin et al., 2012).

A hybrid sensor combining a fluxgate and search coil was proposed in Han et al., (2012). The hybrid magnetometer presents a wide operational frequency region, ranging from dc to over 30 kHz. Also, it presents a very high resolution. The hybrid sensor combines a pT-level fundamental mode orthogonal fluxgate and a search coil magnetometer. It is assembled in a coaxial manner; the thin fluxgate sensor head is partially inserted into the inner bore of the cylindrical magnetic core of the search coil in the coaxial structure. The search coil magnetometer consists of a cylindrical magnetic core (5 cm long) made of a 50 mm width amorphous tape wound (10 turns) around a plastic pipe of an outside diameter of 10 mm. The primary coil is 800 turns, and the feedback coil is 10 turns. The influence of the self-resonance of the SCM is suppressed since the magnetic flux changes in the core are reduced through the feedback magnetic flux. The fluxgate sensor head employs a U-shaped wire core to minimize magnetic interference. The fluxgate sensitivity is increased by a factor of 1.8 because of the flux concentration effect of the cylindrical magnetic core of the SCM. Authors of Han et al., (2012) report a fluxgate cut-off frequency of 6 kHz and that of the search coil beyond 30 kHz.

An R-C divider combines the two outputs. The output of the Fluxgate (FG) which detects the magnetic field from dc to low frequencies, and the output of SCM which detects the magnetic field of high frequency, are summed after adjusting the gains of both magnetometers to be equal. The addition is made over the frequency where two magnetometers have substantial overlap, typically two decades. The center frequency of the overlapping region is called the crossover frequency, $f_{co}=1/2\pi Cc_{R}c$. The method is depicted in Figure 16 and equation 36 expresses the output fusion.

$$V_{sum} = \frac{\left(V_{SCM}R_c + V_{FG} / j\omega C_c\right)}{\left(R_c + 1/ j\omega C_c\right)} \tag{36}$$

Near the crossover frequency where $V_{SCM}=V_{FG}$, $V_{sum}=V$.The reported resolution is $8pT/Hz$ at 1 Hz, $4pT/Hz$ from 10–200 Hz, and the noise decreases as the frequency increases. Furthermore, the frequency range of magnetic field measurement is very wide from dc to 33 kHz, with a phase delay shift of 50 degrees (Han et al., 2012).

Authors of Brown et al., (2014) designed and developed a vector magnetometer called MAGIC (MAGnetometer from Imperial College), that is based on an anisotropic magnetoresistive hybrid sensor. The magnetometer has been developed for a three CubeSat constellation mission called TRIO-CINEMA (CubeSat for Ions Electrons and Neutrals). There was a specific requirement to accommodate a magnetometer sensor on an ultra-light 1 m stacer boom, with a maximum harness mass of only 10 g, to place the sensor away from spacecraft magnetic fields produced by ferromagnetic material and current loops. One of the reasons an AMR sensor was selected is that it offers the capability to modulate the direction of the easy axis and this provides a mechanism to track and compensate for the sensor bridge offset and maintain the sensor in its lowest noise operating state even in the presence of a large background ambient field. The modulation (commonly referred to as flipping) is achieved by driving bipolar current signals into on-chip coils that are aligned with the easy axis direction of the sensing element and integrated into the manufactured AMR package. However, the length of the harness (needed due to the boom) will introduce line losses that will attenuate the flipping current and eliminate the benefits of a driven closed-loop operation of the sensor. The authors dealt with this issue by migrating the flip current generation next to the AMR sensors. So, the only flipping related signals transmitted down the harness to the sensor are a voltage bias and drive clock. This solution saves power and by decoupling the efficiency of the flipping drive for the harness length enables the use of a long boom. The hybrid device combines an orthogonal triad of commercial anisotropic magnetoresistive (AMR) sensors (three single-axis Honeywell HMC1001) with a totem pole HBridge drive on a ceramic substrate. Each single axis sensor is operated in a driven closed loop with first harmonic detection of the DC magnetic field and the feedback voltage sampled by an 8 channel 24-bit delta-sigma Analogue to Digital Converter (ADC). The magnetometer has a sensitivity of better than 3 nT in a 0–10 Hz band and a total mass of 104 g. (Brown et al., 2014)

Laboratoire de Physique des Plasmas (LPP) has designed and built the Search Coil Magnetometer (SCM) (Le Contel et al., 2016) for the MMS(Magnetospheric Multiscale) mission to measure the 3D magnetic field fluctuations in the 1 Hz–6 kHz frequency range. The design was based on a long heritage of instruments that functioned nominally in the solar system; onboard the Cassini, Cluster, and THEMIS missions (Le Contel et al., 2016). The MMS SCM was designed to reduce mass, as compared to THEMIS for instance, without reducing too much the sensitivity. Therefore, the SCM requirements were a tradeoff between mass and sensitivity. The SCM consists of a tri-axial set of magnetic sensors with its associated preamplifier box. The core of the magnetic sensor is made of ferromagnetic material. Its characteristics (composition and shape) are optimized to obtain a high magnetic amplification with a low mass. The core length is 10 cm and the diameter 4 mm. A primary winding with more than ten thousand turns is implemented to collect the voltage induced by the time variation of the magnetic flux. A secondary winding with a smaller number of turns provides flux feedback. This feedback allows the removal of the resonance and flattens the frequency response of the antenna. Furthermore, flux feedback removes the phase variations associated with temperature variations. The SCM sensor including the tri-axis structure has a mass of 214 g and is very compact (100 mm^3). Its sensitivity is less or equal than $2pT/\sqrt{(Hz)}$ at 10 Hz, $0.18pT/\sqrt{(Hz)}$ at 100 Hz and $0.025pT/\sqrt{(Hz)}$ at 1 kHz according to the MMS requirement (Le Contel et al., 2016).

A miniature optically pumped magnetometer based on the rubidium isotope ^{87}Rb is presented in Korth et al., (2016) for operation in space. Optically pumped magnetometers operate on one of the several modes linked to the atomic spin system. Three variables (M_x, M_y, M_z) of Bloch's spin system equations describe the behavior of the atomic spin system (Bloch, 1946). This atomic scalar instrument design implements M_x as well as M_z mode operation. The dual-mode prototype combines the high sampling rates of M_x magnetometers and the accuracy of M_z magnetometers. The miniature rubidium magnetometer selectively uses M_x or M_z mode magnetic field detection and leverages a novel microelectromechanical system (MEMS) fabricated vapor cell and a custom silicon-on-sapphire (SOS) complementary metal-oxide-semiconductor (CMOS) integrated circuit. The vapor cell has a volume of only 1mm^3 so that it can be efficiently heated to its operating temperature by a specially designed, low-magnetic-field-generating resistive heater implemented in multiple metal layers of the transparent sapphire substrate of the SOS-CMOS chips. The SOS-CMOS chip also hosts the Helmholtz coil and associated circuitry to stimulate the magnetically sensitive atomic resonance and temperature sensors. The prototype instrument has a total mass of fewer than 500 g and uses less than 1 W of power, while maintaining a sensitivity of $15pT/\sqrt{Hz}$ at 1 Hz, comparable to present state-of-the-art absolute magnetometers. The reported measurements from the two modes agree better than one part in 10^{-3} in a 2000 nT applied field (Korth et al., 2016).

Authors of Miles et al., (2016) designed and developed a low-power, miniature fluxgate magnetometer aiming at cube satellite (CubeSat) magnetospheric constellation missions with scientific goals. They attempted to overcome the innate challenges and difficulties of the development of specialized instruments with suitable dimensions and mass as well as power requirements to be fitted in a CubeSat without limiting the scientific utility. This magnetometer was primarily intended as a variometer in order to study the perturbations of the background field, rather than absolute field measurements, for the of Ex-Alta 1 3U CubeSat of the University of Alberta. The assembled instrument was a complete, flight-ready scientific fluxgate magnetometer of sufficient fidelity, mounted on a novel 60 cm deployable boom. The sensor was a double-wound ring-core design, similar to that of the Cassiope/e-POP design, using two custom miniature ferromagnetic ring-cores and occupying a volume of 36 × 32 × 28 mm. The sensor base and mount were manufactured from 30% glass-filled polyetheretherketone (PEEK) plastic to provide a robust, thermally stable mount for the sense windings. The miniature sensor had a mass of 47 g (121 g, sensor with boom). The consumption of the electronics was reported to be 400 mW, a feat achieved by using digital processing instead of traditional analog signal conditioning. A single 10 × 10 cm CubeSatKit-compliant PCB housed the electronics. Finally, the deployable boom occupied a single 10 × 10 cm exterior panel, stowed along the outer surface of the spacecraft, and deploy the fluxgate sensor to a distance of 60 cm, reducing the noise from the onboard reaction wheel to less than 1.5 nT at the sensor. The sensor was designed to achieve a minimum magnetic resolution of 0.1 nT, a maximum magnetic noise of $50\,pT/\sqrt{Hz}$, a nominal cadence of 1 sps, and the capability to burst sample to above 10 sps. This resolution needed to be achievable in the full strength of the Earth's magnetic field at LEO, yielding a target magnetic range of at least ±60 μT. Specific science goals and constellation geometry will determine future mission requirements; however, axis orthogonality within 0.5° (and measured to 0.005°), thermal gain stability of 50 ppm/°C or better, and drift of less than 500 pT/d seemed likely reasonable targets. The reported $150-200\,pT/\sqrt{Hz}$ at 1 Hz noise floor of the fabricated sensor was higher than the targeted value of $50\,pT/\sqrt{Hz}$. Also, the reported noise level was significantly larger than that of other low-mass sensors such as THEMIS ($10\,pT/\sqrt{Hz}$ but volumetrically larger) or SMILE ($30\,pT/\sqrt{Hz}$). However, the performance should be sufficient to demonstrate the utility and initial feasibility of magnetic measurements from miniaturized fluxgate sensors on a CubeSat platform. Work is ongoing to produce lower noise cores for future fluxgate sensors by optimizing their manufacturing processes and the heat treatment of the permalloy foil. At these noise levels, other technologies such as AMR sensors provide competitive performance and are well suited for CubeSat applications. However, with new core development work, authors believe that fluxgate sensors with significantly lower noise will be available for future applications. The instrument produced 100 sps of vector magnetic data at 35 pT resolution over full Earth's field of up to ±65 μT (Miles et al., 2016).

Authors of Dias et al., (2020) developed a MEMS magnetometer based on the Lorentz-force, in the frame of European Space Agency contract "LS-MKT-MMG-0001 - Miniaturization of a Magnetometer based on Micro Technology". Recent MEMS magnetometers advances focus mainly on Lorentz-force-based devices with capacitive readouts, using either amplitude or frequency modulated approaches and/or devices exploring internal thermal-piezoresistive vibration amplification. In this work, authors aimed for a ±64 μT operation range, a minimum bandwidth (BW) of 50 Hz, as well as noise below $3nT/\sqrt{Hz}$ and a minimum electrical capacitance variation of 1 aF for a reference magnetic field of 20 nT (the target resolution at the 50 Hz BW). These were the necessary characteristics for a low-cost capacitive magnetometer for space applications such as LEO (Low Earth Orbit) and MEO (Medium Earth Orbit) spacecrafts (altitude range of 500- 2000 km). The developed magnetometer is designed for off-resonance operation in vacuum, enabling larger bandwidths at the expense of some sensitivity loss. Authors implemented a spring configuration lever-like suspension springs which present over tenfold lower electrical resistance (thus allowing higher currents) than conventional guided-beam springs with the same compliance. The higher operation currents, therefore, generate proportionally higher Lorentz forces. This geometrical improvement reduced the ratio of mechanical stiffness to electrical resistance. By combining multiple Lorentz force-generating bars and off-resonance operation in vacuum (1 mbar), these devices can yield: a noise of $2.65nT/\sqrt{Hz}$ (theoretical) with a bias current of 4 mA in each of the 6 Lorentz bars ($63.6nT.mA/\sqrt{Hz}$) and sensitivity of 50.5 aF/μT (1 aF.μT^{-1}.mA^{-1}; 102 μV/μT with the electronic readout), corresponding to a 20 nT resolution (1 aF capacitance variation), over a 50 Hz bandwidth. This resolution figure over such large bandwidth greatly improves upon amplitude-modulated as well as frequency-modulated Lorentz force magnetometers performance (Dias et al., 2020).

In Novotny et al., (2020) an AMR magnetometer with digital feedback for space applications is proposed. The authors designed and developed a novel concept of a digitally compensated, low-noise magnetometer based on anisotropic magnetoresistance sensors. They aim to use modern microcontrollers and their digital signal processing capabilities in order to replace several analog components. Moreover, the cost was a major parameter in the design in order to satisfy the low-budget requirements for most CubeSat missions. So, parts were limited to commercial off-the-shelf components. STM32F334 is selected as the main controller due to its high-resolution PWM (pulse-width-modulation) capability used for digital compensation. After analog-to-digital conversion, the signal processing is executed with software. The AMR sensor HMC1021 for the magnetic field is operated as a closed-loop compensation system using the built-in offset coil. With feedback compensation, the sensor is used only as a zero indicator, and the amplitude of the compensation current that is fed to the offset coils is the measurement output. This indirect method of measurement enables higher performance in linearity, stability, and hysteresis. after demodulation carried out in the firmware, the offset of the magnetic measurement is given only by the magnetic offset of the sensor itself. There is no contribution of the signal conditioning chain, so there is no reason to use precise analog parts like OPAMPS or ADC converters. Also, due to this design, there are no low-frequency noise requirements on the signal chain. The flipping frequency is set to approximately 216 Hz as a compromise between the useful bandwidth of the magnetometer and the digital signal processing speed. This frequency was also selected to ensure minimal interference with the local 50 Hz disturbing signal and its harmonic frequency. For the design of the sensor preamplifier, only noise at this frequency and the bandwidth of the magnetometer are relevant. Most op-amps have 1/f corner on lower frequencies, so only the white noise is important. In a practical implementation, this means that there are only a very few requirements of the design of the electronics (Novotny et al., 2020):

- stability and noise of the main voltage reference
- stability of the sensing resistor
- white noise (at 180–250 Hz) of the bridge amplifier
- stability and low-frequency (LF) noise of op-amps in the voltage-to-current converter and the voltage reference buffer.

OPA2210 is selected as the low-noise preamplifier, due to its low voltage and current noise at a flipping frequency and its low-voltage, rail-to-rail operation. A low voltage rail-to-rail ADA4805 was selected in place of every other op-amp, where low LF noise was needed. The developed prototype exhibited the following characteristics ±100 μT range, ±50 ppm nonlinearity, noise density @1 Hz equal to $250 pTRMS / \sqrt{Hz}$, an integral noise @0.1-10 Hz 780 pTRMS, a 3 dB bandwidth of 30 Hz, a sample rate of 126 Sa/s and dimensions of 35x90x7 mm (Novotny et al., 2020).

Magnetic Cleanliness

Magnetic cleanliness, in general, is the minimization (ideally zero) of the magnetic field in a region. Magnetic shielding is the process of blocking the unwanted magnetic field in a specific area. The magnetic cleanliness can be achieved in what is called the Gauss chamber which is a special form of Faraday cage for the magnetic field. The Gauss chamber can be passive or active or both.

A magnetic field is much harder to prevent from entering a space compared to an electric field. There is no cheap solution that can control the magnetic field in a space, especially for very demanding applications such as very low magnetic field at low frequencies. The magnetic field can penetrate, even metals, but it depends on the frequency of the magnetic field. A metal cage (Gauss chamber) will drop the amplitude of a magnetic field B_0, exponentially. The thickness of each metal needed to reduce the

Figure 17. Gauss chamber for instrument calibration (Bartington Inc., (c), 2020).

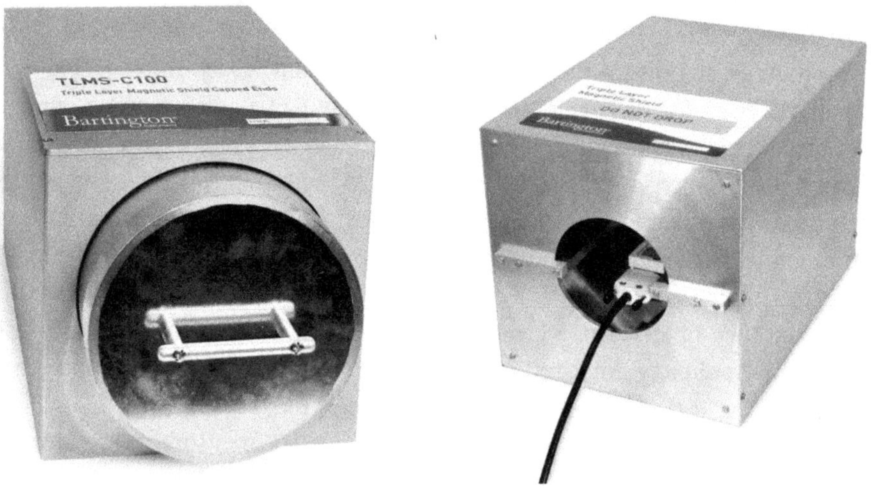

magnetic field to $1/e$ of the original field is called skin depth (δ). The skin depth can be calculated with the following mathematical expression:

$$\delta = \sqrt{\frac{\rho}{\pi f \mu_r \mu_0}} \sqrt{\sqrt{1+\left(2\pi f \rho \varepsilon_r \varepsilon_0\right)^2} + 2\pi f \rho \varepsilon_r \varepsilon_0} \tag{36}$$

where ρ (Ωm) is the resistivity, μ_r is the relative magnetic permeability, ε_r is the relative permittivity of the sheet material, *f (Hz)* is the frequency of the magnetic field, μ_0 (Hm^{-1}) is the magnetic permeability of vacuum and ε_0 (Fm^{-1}) is the electrical permittivity of vacuum (Jordan, 1968).

The best way to block the magnetic field is to use a superconductor. The reason for that is the property of the superconductor to have zero resistivity. Consequently, even the thinnest superconductors can block the magnetic field independently of the frequency (Ripka, 2001). However, this type of insulation is extremely expensive not only to build but also to maintain due to the need for cryogenic conditions. A more practical construction of a Gauss chamber is made from metal alloys with high magnetic permeability called Mu-Metal. Mu-Metals have a relative permeability of the order of 20000 to 200000 compared to pure iron, which exhibits a permeability limited to 5000 (Grosz et al., 2007; Tumanski, 2011). Due to the exponential absorption of the magnetic field, the Gauss chamber should be constructed by multiple layers of Mu-Metal instead of a thick sheet. Gauss champers are used in Magnetic Resonance Imaging (MRI's) rooms and Magnetoencephalography (MEG's) rooms (Ishiyama et al., 1991). A magnetic shield's shape is an important parameter in the design phase. For example, a spherical chamber is 4/3 or 134% more effective than a cylindrical one (Mager, 1970). However, some geometries are not so useful in practice, so every design has its own trade-offs. For a cylindrical Gauss chamber to achieve a 60 *dB* reduction of the magnetic field at 50 *Hz*, a three-layer Mu-Metal is required and each layer needs to be at least 1 mm thick. Figure 17 presents an example of a Mu-metal Gauss chamber.

A passive Gauss chamber sometimes is not enough to yield the necessary result, and the use of active magnetic shielding is required. Active shielding is the creation of an intended vector field, opposite

Figure 18. Helmholtz coils for controlling the magnetic field in 3D axes. (a) circular structure, (b) square structure (Bartington Inc. (a), 2020; Bartington Inc. (b), 2020).

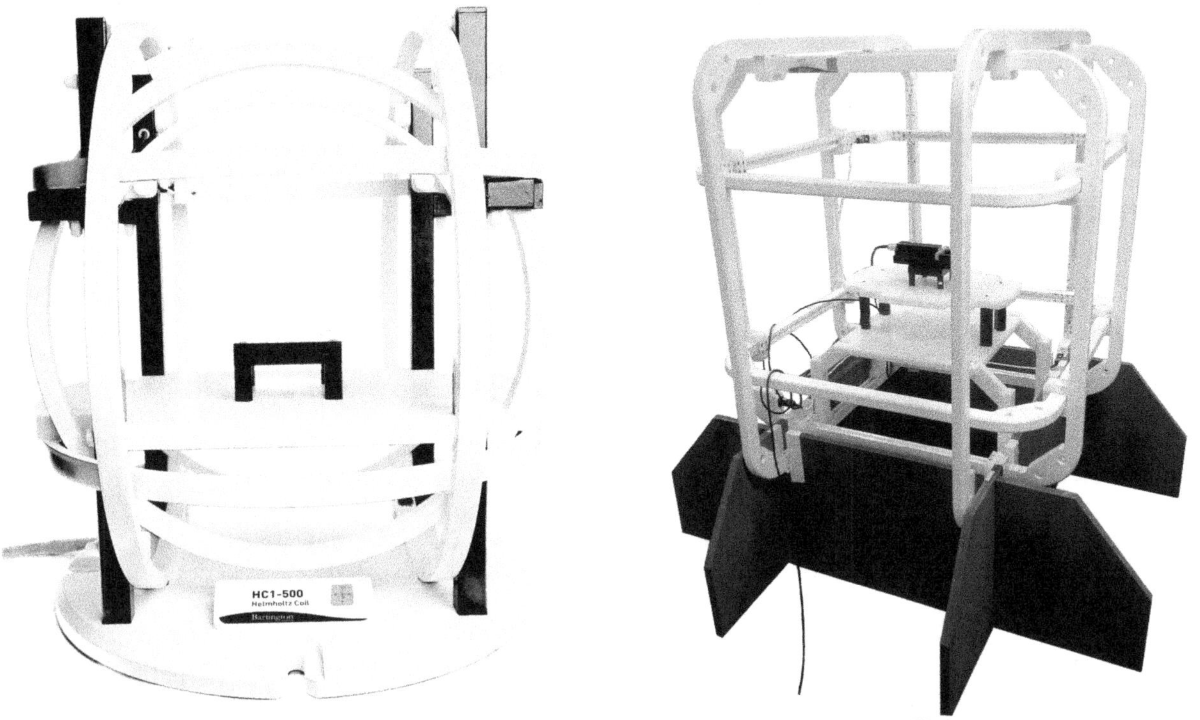

(a) **(b)**

to the one needed to minimize (nullify). The vector sum of the two fields ideally goes to zero. Usually, and for practical 3-dimensional problems, the active shielding equipment is composed of three Helmholtz coils, which are orthogonally oriented to each other (Tumanski, 2011). Maxwell coils can also be utilized, but they are rarely used due to the higher construction costs and other functional limitations. Figure 18 depicts two types of Helmholtz coils manufactured by Bartington Inc. ©, one with circular coils and one with square coils.

CONCLUSION

This chapter presents a basic overview of the field of magnetic field acquisition. A variety of sensors appropriate for magnetic field measurements is reviewed covering the majority of field applications, with a special focus on space missions. These sensors have the advantages that they can easily operate under harsh environmental conditions and elevated temperatures, they are very reliable due to the absence of moving parts and they can be embedded inside building materials. Magnetic sensors are categorized primarily based on their sensing principle, which impacts directly the performance characteristics like measurement range, resolution, frequency response, working temperature, and manufacturing cost.

Magnetic Sensors are commonly used for laboratory measurements and satellite magnetic measurements. Understanding the various types of magnetometers and their limitations is crucial for the interpretation of the satellite magnetometer results. For space use, magnetic sensors must have some unique characteristics. Deep knowledge of the sensor's characteristics leads to an optimal selection of the magnetic sensor in accordance with the application requirements. Finally, the chapter concludes with a brief discussion about magnetic cleanliness equipment.

REFERENCES

Acuña, M. H. (2002). Space-based magnetometers. Review of Scientific Instruments. doi:10.1063/1.1510570

Alberto, N., Domingues, M. F., Marques, C., André, P., & Antunes, P. (2018). Optical fiber magnetic field sensors based on magnetic fluid: A review. Sensors (Switzerland). doi:10.339018124325

Bartington Inc. (2020a). *Operation Manual for HC1 Helmholtz Coils*. Retrieved September, 2020, from https://www.bartington.com/wp-content/uploads/pdfs/operation_manuals/HC1_OM3226.pdf

Bartington Inc. (2020b). *Operation Manual for HC2 Helmholtz Coils*. Retrieved September, 2020, from https://www.bartington.com/wp-content/uploads/pdfs/operation_manuals/HC2_OM3342.pdf

Bartington Inc. (2020c). *Operation Manual for TLMS-O Open and TLMS-C Capped Magnetic Shields*. Retrieved September, 2020, from https://www.bartington.com/wp-content/uploads/pdfs/operation_manuals/Magnetic_Shields_OM1938.pdf

Bloch, F. (1946). Nuclear induction. *Physical Review*, *70*(7-8), 460–474. Advance online publication. doi:10.1103/PhysRev.70.460

Brown, N. F. (2019). Space Scientific Instrument Taxonomy (SSIT). *IEEE Aerospace Conference Proceedings*. 10.1109/AERO.2019.8741740

Brown, P., Whiteside, B. J., Beek, T. J., Fox, P., Horbury, T. S., Oddy, T. M., Archer, M. O., Eastwood, J. P., Sanz-Hernández, D., Sample, J. G., Cupido, E., O'Brien, H., & Carr, C. M. (2014). Space magnetometer based on an anisotropic magnetoresistive hybrid sensor. *The Review of Scientific Instruments*, *85*(12), 125117. Advance online publication. doi:10.1063/1.4904702 PMID:25554336

Daniel, C. M. (1988). *The Theory of Magnetism I: Statics and Dynamics*. Springer.

Dias, R. A., Moreira, E. E., Alves, F. S., Mesquita, D., Gaspar, J., & Rocha, L. A. (2020). Design Optimization of a Lorentz Force, Amplitude Modulated, MEMS Space Magnetometer. *Proceedings of the IEEE International Conference on Micro Electro Mechanical Systems (MEMS)*. 10.1109/MEMS46641.2020.9056126

Díaz-Michelena, M. (2009). Small magnetic sensors for space applications. Sensors. doi:10.339090402271

Drung, D., Abmann, C., & Beyer, J. (2007). Highly Sensitive and Easy-to-Use SQUID Sensors. *IEEE Transactions on Applied Superconductivity, Volume*, *17*(2), 699–704. doi:10.1109/TASC.2007.897403

Foner, S. (1959). Versatile and Sensitive Vibrating-Sample Magnetometer. *The Review of Scientific Instruments, 30*(7), 548–557. doi:10.1063/1.1716679

Fraden, J. (2010). *Handbook of Modern Sensors Physics, Designs, and Applications*. Springer-Verlag. doi:10.1007/978-1-4419-6466-3

Fratantuono, L. (2015). *A Reading of Lucretius' De Rerum Natura*. Lexington Books.

Freitas, P. P., Ferreira, R., Cardoso, F., & Cardoso, F. (2007). Magnetoresistive sensors. *Journal of Physics Condensed Matter, 19*(16), 165221. doi:10.1088/0953-8984/19/16/165221

Gordon, D., Lundsten, R., & Chiarodo, R. (1965). Factors affecting the sensitivity of gamma-level ring-core magnetometers. *IEEE Transactions on Magnetics, Volume, 1*(4), 330–337. doi:10.1109/TMAG.1965.1062987

Griffin, D. K., Masseglia, O., Hall, M., Trougnou, L., Hewitson, M., Howe, C., Poyntz-Wright, O., Leopoldi, M., Ding, L., Turner, S., & Harmon, S. (2012). Design and calibration of a compact low-noise magnetic gradiometer. *Proceedings of the 2012 ESA Workshop on Aerospace EMC 2012*.

Grosz, A., Haji-Sheikh, J. M., & Mukhopadhyay, C. S. (2017). *High Sensitivity Magnetometers*. Springer. doi:10.1007/978-3-319-34070-8

Hadjigeorgiou, N. (2014). Experimental Study Noise of HMR2003 Sensor with or without the Use of Amplifier AMP04. *Key Engineering Materials, 605*, 629–632. doi:10.4028/www.scientific.net/KEM.605.629

Hadjigeorgiou, N., Hristoforou, E., & Sotiriadis, P. P. (2017). Closed-loop current-feedback, signal-chopped, low noise AMR sensor with high linearity. *IEEE Modern Circuits and Systems Technologies, 6*.

Hadjigeorgiou, N., Konstantopoulos, C., & Masxas, D. (2015). Fourier Analysis for Orthogonal and Parallel Fluxgate. *Key Engineering Materials, 644*, 270–273. doi:10.4028/www.scientific.net/KEM.644.270

Hadjigeorgiou, N., Tsalikidou, A.C., Hristoforou, E., & Sotiriadis, P. P. (2017). Highly Linear and Low Noise AMR Sensor Using Closed Loop and Signal-Chopped Architecture. *World Academy of Science, Engineering and Technology, 11*(8).

Hadjigeorgiou, N. G., & Sotiriadis, P. P. (2020). Parasitic Capacitances, Inductive Coupling, and High-Frequency Behavior of AMR Sensors. *IEEE Sensors Journal, 20*(5), 2339–2347. Advance online publication. doi:10.1109/JSEN.2019.2953351

Han, F., Harada, S., & Sasada, I. (2012). Fluxgate and search coil hybrid: A low-noise wide-band magnetometer. *IEEE Transactions on Magnetics, 48*(11), 3700–3703. Advance online publication. doi:10.1109/TMAG.2012.2196762

Harrison, E. P., Turney, G. L., & Rowe, H. (1935). Electrical Properties of Wires of High Permeability. *Nature, 135*(3423), 961. doi:10.1038/135961a0

Honewell Inc. (2011). *Hall effect sensing and application*. Retrieved September 2017, from www.honeywell.com/sensing

Hospodarsky, G. B. (2016). Spaced-based search coil magnetometers. In *Journal of Geophysical Research*. Space Physics. doi:10.1002/2016JA022565

Hristoforou, E., Chiriac, H., & Neagu, M. (1997). A New Magnetic Field Sensor Based on Magnetostrictive Delay Lines. *IEEE Transactions on Instrumentation and Measurement, Vol, 46*(2), 632–635. doi:10.1109/19.571942

Hristoforou, E., & Reilly, R. E. (1991). Nonuniformity in amorphous ribbon delay lines after stress and current annealing. *Journal of Physics, 69*, 5008.

Ishiyama, A., & Hirooka, H. (1991). Magnetic Shielding for MRI Superconducting Magnets. *IEEE Transactions on Magnetics, Volume, 27*(2), 1692–1695. doi:10.1109/20.133514

Jiles, D. (2015). *Introduction to Magnetism and Magnetic Materials*. CRC Press. doi:10.1201/b18948

Jiles, D. C., & Atherton, D. L. (1984). Theory of ferromagnetic hysteresis. *Journal of Applied Physics, 55*(6), 2115–2120. doi:10.1063/1.333582

Jordan, C. E., & Balmain, G. K. (1968). *Electromagnetic Waves and Radiation Systems*. Prentice-Hall, Inc.

Josephson, B. D. (1962). Possible new effects in superconductive tunnelling. *Elsevier Physics Letters, Volume, 1*(7), 251–254.

Korth, H., Strohbehn, K., Tejada, F., Andreou, A. G., Kitching, J., Knappe, S., Lehtonen, S. J., London, S. M., & Kafel, M. (2016). Miniature atomic scalar magnetometer for space based on the rubidium isotope 87Rb. *Journal of Geophysical Research. Space Physics, 121*(8), 7870–7880. Advance online publication. doi:10.1002/2016JA022389 PMID:27774373

Le Contel, O., Leroy, P., Roux, A., Coillot, C., Alison, D., Bouabdellah, A., Mirioni, L., Meslier, L., Galic, A., Vassal, M. C., Torbert, R. B., Needell, J., Rau, D., Dors, I., Ergun, R. E., Westfall, J., Summers, D., Wallace, J., Magnes, W., … de la Porte, B. (2016). The Search-Coil Magnetometer for MMS. Space Science Reviews. doi:10.100711214-014-0096-9

Lenz, J., & Edelstein, A. S. (2006). Magnetic sensors and their applications. *IEEE Sensors Journal, 6*(3), 631–649. Advance online publication. doi:10.1109/JSEN.2006.874493

Lenz, J. E. (1990). A Review of Magnetic Sensors. *Proceedings of the IEEE*. 10.1109/5.56910

Lowrie, W. (2007). *Fundamentals of Geophysics*. Cambridge University Press. doi:10.1017/CBO9780511807107

Mager, J. A. (1970). Magnetic Shields. *IEEE Transactions on Magnetics, Volume, 6*(1), 67–75. doi:10.1109/TMAG.1970.1066714

Mahdi, A. E., Panina, L., & Mapps, D. (2003). Some new horizons in magnetic sensing: High-Tc SQUIDs, GMR and GMI materials. *Sensors and Actuators. A, Physical, 105*(3), 271–285. Advance online publication. doi:10.1016/S0924-4247(03)00106-7

Merrill, R. T., & McElhinny, M. W. (1983). The Earth's magnetic field: Its history, origin and planetary perspective. San Francisco: Academic Press.

Metglas Ind. (2011). *Magnetic Alloy*. Retrieved October, 2017, from www.metglas.com

Miles, D. M., Mann, I. R., Ciurzynski, M., Barona, D., Narod, B. B., Bennest, J. R., Pakhotin, I. P., Kale, A., Bruner, B., Nokes, C. D. A., Cupido, C., Haluza-DeLay, T., Elliott, D. G., & Milling, D. K. (2016). A miniature, low-power scientific fluxgate magnetometer: A stepping-stone to cube-satellite constellation missions. *Journal of Geophysical Research. Space Physics*, *121*(12), 11,839–11,860. Advance online publication. doi:10.1002/2016JA023147

Novotny, D., Petrucha, V., Dressler, M., & Platil, A. (2020). AMR magnetometer with digital feedback for space applications. *I2MTC 2020 - International Instrumentation and Measurement Technology Conference, Proceedings*. 10.1109/I2MTC43012.2020.9129039

Prudenziati, M. (1999). *Handbook of Sensors and Actuators. Thick Film Sensors*. Elsevier.

Ripka, P. (1993). Race-track fluxgate sensors. *Sensors and Actuators A. Physical, Volume, 37-38*, 417–421.

Ripka, P. (2001). *Magnetic Sensors and Magnetometers*. Artech House Publishers.

Ripka, P., & Tipek, A. (2007). *Modern Sensors Handbook*. ISTE. doi:10.1002/9780470612231

Tumanski, S. (2007). Induction coil sensors - A review. *Measurement Science & Technology*, *18*(3), R31–R46. Advance online publication. doi:10.1088/0957-0233/18/3/R01

Tumanski, S. (2011). *Handbook of Magnetic Measurements*. CRC Press.

William, G. (1991). *De Magnete*. Dover Publications.

KEY TERMS AND DEFINITIONS

Anisotropic Magnetoresistance (AMR): Based magnetometers are used in devices as varied as global positioning systems to provide dead reckoning capability and in automotive ignition systems to provide crankshaft rotational position.

Giant Magneto-Impedance (GMI): Is a physical effect that expresses the large variation in the electrical impedance that occurs in some materials when subject to an external magnetic field. It should not be confused with Giant Magnetoresistance that is a totally different physical phenomenon.

Giant Magnetoresistance (GMR): Is a quantum mechanical magnetoresistance effect observed in multilayers composed of alternating ferromagnetic and non-magnetic conductive layers.

Superconducting Quantum Interference Device (SQUID): Is a very sensitive magnetometer used to measure extremely subtle magnetic fields, based on superconducting loops containing Josephson junctions.

Chapter 7
Modeling Antenna Radiation Using Artificial Intelligence Techniques:
The Case of a Circular Loop Antenna

Theodoros N. Kapetanakis
https://orcid.org/0000-0002-6975-1334
Department of Electronic Engineering, Hellenic Mediterranean University, Greece

Ioannis O. Vardiambasis
Department of Electronic Engineering, Hellenic Mediterranean University, Greece

Melina P. Ioannidou
Department of Information and Electronic Engineering, International Hellenic University, Greece

Antonios I. Konstantaras
Department of Electronic Engineering, Hellenic Mediterranean University, Greece

ABSTRACT

The forward and the inverse problem of a thin, circular, loop antenna that radiates in free space is modeled and solved by using soft computing techniques such as artificial neural networks and adaptive neuro fuzzy inference systems. On the one hand, the loop radius and the observation angle serve as inputs to the forward model, whereas the radiation intensity is the output. On the other hand, the electric field intensity and the loop radius are the input and output, respectively, to the inverse model. Extensive numerical tests indicate that the results predicted by the proposed models are in excellent agreement with theoretical data obtained from the existing analytical solutions of the forward problem. Thus, the employment of artificial intelligence techniques for tackling electromagnetic problems turns out to be promising, especially regarding the inverse problems that lack solution with other methods.

DOI: 10.4018/978-1-7998-4879-0.ch007

INTRODUCTION

Artificial intelligence incorporates a vast family of biologically inspired techniques that attempt to imitate human cognitive skills. The artificial neural networks (ANNs) and the fuzzy logic systems, such as the adaptive neuro fuzzy inference system (ANFIS), belong to the aforementioned family of techniques. The former simulate the complex networks of the neurons that are found in the human central nervous system (Haykin, 1999), whereas the latter combine the learning abilities of neural networks with fuzzy logic (Jang, 1993).

During the last few decades, a great variety of electromagnetic (EM) problems have been treated by applying methods based on artificial intelligence (Zhang & Gupta, 2000), (Christodoulou & Georgiopoulos, 2001), (Mishra, 2001). The number of pertinent works is huge. Indicative applications include the modeling and design of microwave components and circuits (Devabhaktuni et al., 2001), (Zhang et al., 2003), as well as the design and optimization of antennas (Choudhury et al., 2015). The prediction of the performance of various types of antennas by using ANNS and/or ANFIS has drawn strong attention, recently. For example, the resonant frequencies of L-shaped compact microstrip antennas have been determined by applying ANNs and ANFISs (Kayabasi et al., 2014), the return-loss characteristics and the radiation patterns of pyramidal and corrugated horn antennas have been obtained by implementing ANFIS-based models (Pujara et al., 2014) and the return-loss performance of a planar inverted-F antenna has been predicted by using an ANFIS (Gehani & Pujara, 2015). Furthermore, the radiation characteristics of a short dipole array, by applying radial basis function (RBF) ANNs (Mishra et al., 2015), and those of a circular loop antenna by implementing multilayer perceptron (MLP) ANNS and ANFISs (Kapetanakis et al., 2012a), (Kapetanakis et al., 2018a), have been calculated.

A more challenging task may be the implementation of artificial intelligence techniques in order to solve inverse EM problems. The latter focus on estimation of the properties of the scatterer or the radiator from information contained in the EM field and obtained either from measurements or from analytical/numerical calculations. ANNs and the Finite Element Method (FEM) have been combined in (Low & Chao, 1992) to treat inverse EM problems, whereas an approach based on Hopfield ANNs has been presented in (Elshafiey et al., 1995). Inverse ANN modeling has been applied for the purpose of microwave filter design (Kabir et al., 2008) and for the inversion of a transmitarray database, i.e., the estimation of the element parameters from the transmission coefficients of the antenna array (Gosal et al., 2016). Moreover, several ANN configurations (Kapetanakis et al., 2018b) and an ANFIS (Kapetanakis et al., 2012b) have been implemented in order to model and solve the inverse EM problem of a thin, circular, loop antenna that radiates in free space.

The radiating circular loop antenna is used herein as an example in order to investigate the potential of certain artificial intelligence techniques to treat EM problems. Although the first efforts to solve the equations that describe the behavior of loop antennas occurred many decades ago, the study of such antennas draws strong interest even nowadays. The motivation may be attributed to the fact that there exist several problems that have not been solved completely yet, for example those related to the radiation of thick loops. Moreover, the emerging meta-material theory was a boost for the evolution and application of nano-scaled rings (McKinley, 2019).

The basis for the study of loop antennas was established more than one century ago by Pocklington, who obtained a solution, in the form of a Fourier series, for the current on a closed loop excited by a plane wave (Pocklington, 1897). Several years later, Hallen considered a driven loop but his Fourier series solution for the current contained a singularity; thus, the results of his method were limited to small loops (Hallen, 1938). Since then, a lot of great scientists have contributed to the formulation and solution of the analytical equations that govern the radiation of a loop antenna. In the 1950s and 1960s several researchers, by complementing one another, managed to find a well behaved solution for the current and the impedance of a driven thin, circular loop in the form of a convergent, infinite Fourier series (Storer, 1956), (Wu, 1962), (King, 1969). Presentation of a detailed analytical history of loops is beyond the scope of this chapter; a comprehensive essay on this subject may be found in (McKinley, 2019).

The analytical study of a driven, thin, electrically small loop, i.e., with circumference smaller than half-wavelength, has received wide attention mainly because of its simplicity. The current distribution may be assumed constant around the periphery of such a loop; thus the radiated EM field may be determined by retracing the steps of the analysis for an infinitesimal magnetic dipole (Balanis, 1982). On the contrary, a large loop can handle many resonant modes and it is not that simple to find which of them is dominant. Several mathematical approaches have been proposed over the years for the determination of the EM field around a thin, circular, loop antenna of arbitrary radius carrying an arbitrary current. Some of them are briefly outlined in the next section, whereas an extensive review is presented in (McKinley, 2019). The complexity of the problem increases considerably if the wire radius of the loop antenna is small but finite. The application of the method of moments to solve the integral equations for a circular loop with a small wire radius, has been investigated by several researchers (Anastassiu, 2006), (Fikioris et al., 2008), (Fikioris et al., 2010).

Loop antennas are easy to construct and inexpensive. Especially the small ones are poor radiators, thus they have been used mainly for reception in radio communications, for direction finding and as probes in magnetic sensing applications (Balanis, 1982), (D.S. Smith, 2007). More recently, printed, integrated loop antennas have appeared, like the circularly polarized square-loop presented in (Fakharzadeh & Mohajer, 2014) for mm-wave short range applications.

At the turn of the 21st century, loop antennas have received special attention due to the rapid development of nanotechnology and metamaterials. A lot of loop-like structures, with negative index of refraction, operating in the microwave band have been presented in the literature; indicative works include the design and fabrication of nested, split ring resonators (Pendry et al., 1999), (D.R. Smith et al., 2000) and configurations with copper square loop resonators printed on fiber glass circuit boards (Shelby et al., 2001). The formulation and analysis of loop antennas in radio frequencies may be extended to describe the radiation properties of nanoloops operating in the Terahertz (THz), infrared (IR) and optical regimes. In order to accomplish this, the dispersion and loss of the antenna material should be taken into account. Analytical representations of the radiation characteristics of nanoloops operating in the aforementioned regimes have been reported in the last decade (McKinley et al., 2013a), (Lu et al., 2017); methods of computing the mutual coupling between such antennas have, also, been proposed (Nagar et al., 2017).

The recent interest in nanoloops may be justified by the large number of applications that have appeared in the last decade. An optical nanoantenna array, composed of plasmonic square loops, has been constructed for enhancing the radiation characteristics and controlling the beam pattern (Ahmadi & Mosallaei, 2010). Besides, broadband superdirective radiation has been obtained from a metallic nanoloop in the far-infrared regime by choosing the proper material composition and loop circumference (Pantoja et al., 2017). The coupling between laterally paired THz metamaterial split-ring resonators has been exploited for accomplishing passive tuning of the strength and frequency of the inductive-capacitive (LC) resonances (Chowdhury et al., 2011). A transmit array antenna, consisting of plasmonic, concentric, square-loop scatterers, has been designed to enable light concentration, at a desired distance, in the near IR-region (Memarzadeh & Mosallaei, 2011). Nanoloop arrays are, also, very promising for their potential application in solar cells for light-trapping (McKinley et al., 2013b; Simovski et al., 2013), a perspective from which space technology might benefit.

The objective of this chapter is to demonstrate the applicability of artificial intelligence techniques in modeling and solving EM problems, with emphasis on the inverse ones that are, generally, difficult to deal with. For this purpose, the example of a loop antenna is adopted. On the one hand, the forward problem of the radiating loop is well established and several analytical solutions exist in the literature. Thus, it may serve as benchmark for the results obtained from the artificial intelligence techniques. On the other hand, the corresponding inverse problem lacks analytical and/or numerical solution. It is treated, herein, by applying ANNs and the results are validated with data produced by the aforementioned existing analytical solutions of the forward problem. In the first part of this chapter, a short literature review is provided regarding the radiation of the circular loop antenna. In addition, a very short description of certain artificial intelligence techniques is presented. In the second part of the chapter, the modeling procedure based on ANNs and ANFISs is described and discussed. Indicative numerical results are presented to manifest the potential of the proposed methodology.

BACKGROUND

Analytical Solutions of the Loop Antenna Radiation Problem

The geometry of a current-carrying, thin, circular loop antenna, with arbitrary radius a, is illustrated in Figure 1. The antenna is assumed to radiate in free space and it is referred to as loop hereafter. It is placed symmetrically on the x-y plane at z=0; S $(r' = a, \theta' = 90^\circ, \varphi')$ is a source point and P(r,θ,φ) is an observation point in spherical coordinates. The distance R *b*etween S and P is given by:

$$R = \sqrt{r^2 + a^2 - 2ar\sin\theta\cos(\varphi - \varphi')} \tag{1}$$

The forward problem consists in determining the electromagnetic (EM) field around the loop. The most common analytical formulation starts from the vector potential $\vec{A}$, which is related to the EM field components by the well-known relationships (Balanis, 1982):

$$\vec{H} = \frac{1}{\mu_0}\nabla \times \vec{A} \tag{2}$$

$$\vec{E} = \frac{1}{j\omega\varepsilon_0}\nabla \times \vec{H} = \frac{1}{j\omega\mu_0\varepsilon_0}\left[\nabla\left(\nabla \cdot \vec{A}\right) + \beta^2\vec{A}\right] \tag{3}$$

where $\mu 0$, $\varepsilon 0$ stand for the magnetic permeability and dielectric permittivity of free space, respectively, β is the wavenumber and ω is the circular frequency; time dependence exp(jωt) is used throughout. A representation of the vector potential, in spherical coordinates, may be obtained through integral equations that involve the loop current I(φ):

Figure 1. Circular loop antenna placed symmetrically on the x-y plane

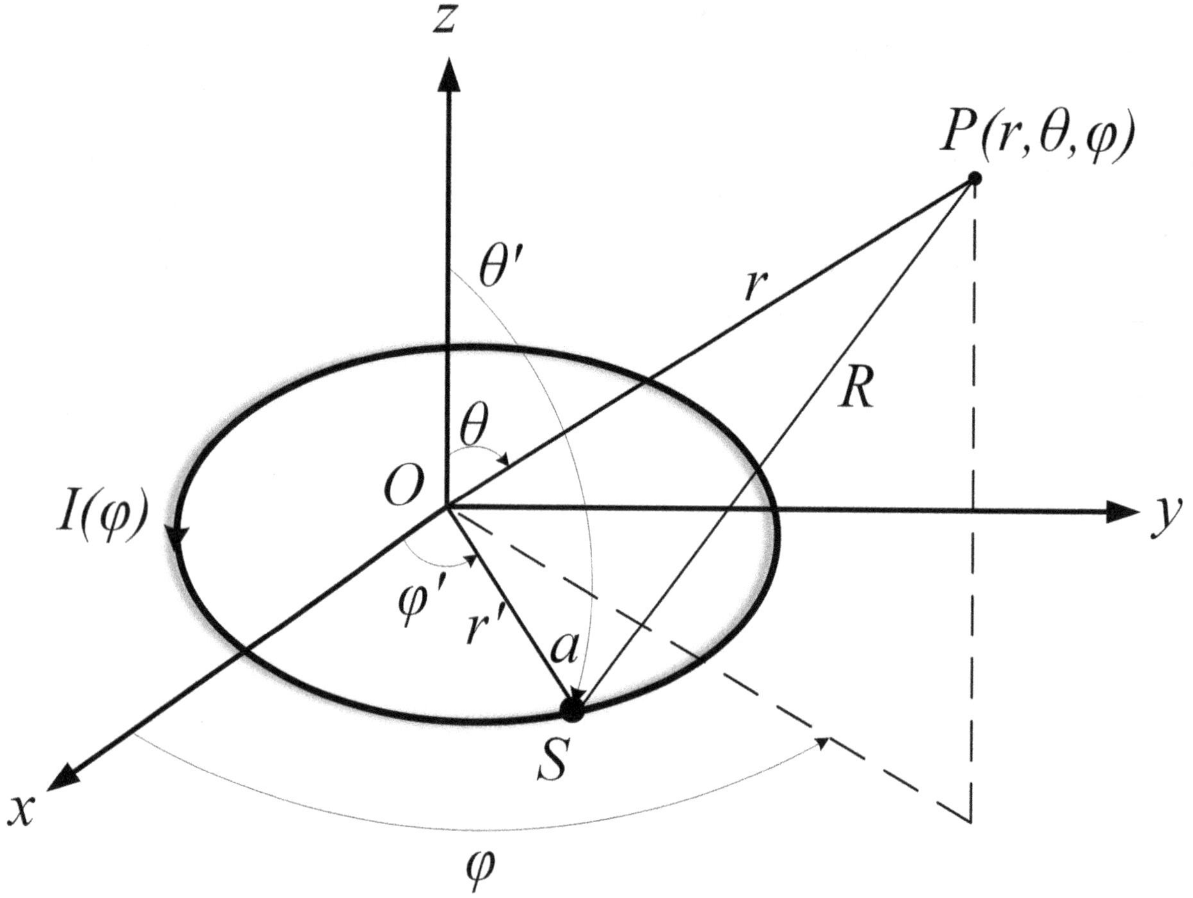

$$A_r(r,\theta,\varphi)=\frac{\mu_0 a\sin\theta}{4\pi}\int_0^{2\pi} I(\varphi')\sin(\varphi-\varphi')\frac{e^{-j\beta R}}{R}d\varphi' \quad (4)$$

$$A_\theta(r,\theta,\varphi)=\frac{\mu_0 a\cos\theta}{4\pi}\int_0^{2\pi} I(\varphi')\sin(\varphi-\varphi')\frac{e^{-j\beta R}}{R}d\varphi' \quad (5)$$

$$A_\varphi(r,\theta,\varphi)=\frac{\mu_0 a}{4\pi}\int_0^{2\pi} I(\varphi')\cos(\varphi-\varphi')\frac{e^{-j\beta R}}{R}d\varphi' \quad (6)$$

The approach summarized below was proposed by Werner (Werner, 1996) and is based on recognizing that the three components of $\vec{A}$ in equations (4)-(6) comprise a common term:

$$A_r(r,\theta,\varphi) = -\frac{\mu_0}{2j\beta r}\frac{\mathrm{d}}{\mathrm{d}\varphi}\Im(r,\theta,\varphi) \tag{7}$$

$$A_\theta(r,\theta,\varphi) = -\frac{\mu_0}{2j\beta r\tan\theta}\frac{\mathrm{d}}{\mathrm{d}\varphi}\Im(r,\theta,\varphi) \tag{8}$$

$$A_\varphi(r,\theta,\varphi) = \frac{\mu_0}{2j\beta r\cos\theta}\frac{\mathrm{d}}{\mathrm{d}\theta}\Im(r,\theta,\varphi) \tag{9}$$

where $\Im(r,\theta,\varphi)$ stands for the integral of the current distribution as follows:

$$\Im(r,\theta,\varphi) = \frac{1}{2\pi}\int_0^{2\pi} I(\varphi')e^{-j\beta R}d\varphi' \tag{10}$$

Consequently, cos(βR) and sin(βR), that constitute the term $e^{-j\beta R}$ in the integrand of equation (10), are expanded in series of spherical Bessel and Hankel functions of the first and second kind (Abramowitz & Stegun, 1972), respectively. By applying Euler's identity and by performing the term by term integration in equation (10), the latter results in:

$$\Im(r,\theta,\varphi) = G_0\, e^{-j\beta\sqrt{r^2+a^2}} + \sum_{m=1}^{\infty}\frac{2^m}{m!}G_m(\varphi)B_m(r,\theta)F_{m-1}(r) \tag{11}$$

The abbreviations used in equation (11) are defined in Table 1 where $h_m^{(2)}(\cdot)$ is the spherical Hankel function of the second kind of order *m*.

By expanding the current distribution in the following Fourier cosine series

$$I(\varphi) = \sum_{n=0}^{\infty} I_n\cos(n\varphi) \tag{12}$$

the series representation in equation (11) evaluates to:

$$\Im(r,\theta,\varphi) = I_0\, e^{-j\beta\sqrt{r^2+a^2}} + \sum_{\substack{m=1\\ m-n=2k\\ k=0,1,\ldots}}^{\infty}\sum_{n=0}^{m} C_{mn}\cos(n\varphi)B_m(r,\theta)F_{m-1}(r) \tag{13}$$

The coefficient C_{mn} is defined in Table 1. The vector potential $\vec{A}$ may then be derived from equations (7)-(9) by substituting equation (13) into equations (7)-(9) and performing the differentiations. Finally, the EM field components follow from equations (2)-(3) and are given below:

$$H_r(r,\theta,\varphi)=\frac{\beta^3a^2\cos\theta}{8j}\sum_{m,n}(m-n)(m+n)C_{mn}\cos(n\varphi)B_{m-2}(r,\theta)F_{m-1}(r) \tag{14}$$

$$H_\theta(r,\theta,\varphi)=\frac{(\beta a)^2\sin\theta}{8j}\sum_{m,n}C_{mn}\cos(n\varphi)B_{m-1}(r,\theta)\left[m\beta^3r^2F_m(r)-(m-n)(m+n)F_{m-1}(r)\right] \tag{15}$$

$$H_\varphi(r,\theta,\varphi)=-\frac{\beta^3ar\cos\theta}{4j}\sum_{m,n}nC_{mn}\sin(n\varphi)B_{m-1}(r,\theta)F_m(r) \tag{16}$$

$$E_r(r,\theta,\varphi)=\frac{\eta\beta^2a\sin\theta}{4}\sum_{m,n}nC_{mn}\sin(n\varphi)B_{m-1}(r,\theta)\left[(\beta a)^2F_{m+1}(r)-(m+1)F_m(r)\right] \tag{17}$$

$$E_\theta(r,\theta,\varphi)=\frac{\eta\beta^2a\cos\theta}{4}\sum_{m,n}nC_{mn}\sin(n\varphi)B_{m-1}(r,\theta)\left[mF_m(r)-F_{m-1}(r)\right] \tag{18}$$

$$E_\varphi(r,\theta,\varphi)=\frac{\eta\beta^2a}{4}\sum_{m,n}\cos(n\varphi)B_{m-1}(r,\theta)\left[n^2C_{mn}F_m(r)-mC_{mn}F_{m-1}(r)\right] \tag{19}$$

η in equations (17)-(19) stands for the wave impedance in free space. The notation $\sum_{m,n}$ has been adopted in equations (14)-(19) instead of the double summation $\sum_{\substack{m=1\\ k=0,1,\ldots}}^{\infty}\sum_{\substack{n=0\\ m-n=2k}}^{m}$, for the sake of brevity.

Far-zone approximations, i.e., for βr>>1 and R≃r, may be obtained by applying the asymptotic expression of the spherical Hankel function of the second kind for large arguments (Abramowitz & Stegun, 1972). The resulting, non-zero, components of the magnetic field intensity are (Werner, 1996):

$$H_\theta(r,\theta,\varphi)\simeq\frac{\beta a}{2}\frac{e^{-j\beta r}}{r}\sum_{n=0}^{\infty}j^nI_n\cos(n\varphi)J_n'(\beta a\sin\theta) \tag{20}$$

Table 1. Abbreviations used in equations (11), (14)-(19)

Symbol	Definition
G_0	$\frac{1}{2\pi}\int_0^{2\pi} I(\varphi')d\varphi'$
$G_m(\varphi)$, m>1	$\frac{1}{2\pi}\int_0^{2\pi} I(\varphi+\varphi')\cos^m(\varphi')d\varphi'$
$B_m(r,\theta)$	$\left(\frac{\beta^2 ar\sin\theta}{2}\right)^m$
$F_m(r)$	$\frac{h_m^{(2)}\left(\beta\sqrt{r^2+a^2}\right)}{\left(\beta\sqrt{r^2+a^2}\right)^m}$
C_{mn}	$\frac{I_n}{[(m-n)/2]![(m+n)/2]!}$

$$H_\varphi(r,\theta,\varphi) \simeq -\frac{\cot\theta}{2}\frac{e^{-j\beta r}}{r}\sum_{n=0}^{\infty} nj^n I_n \sin(n\varphi) J_n(\beta a\sin\theta) \tag{21}$$

where $J_n(\bullet)$ stands for the Bessel function of the first kind (Abramowitz & Stegun, 1972) and the prime in $J_n'(\cdot)$ denotes differentiation with respect to its argument. The non-zero, components of the electric field intensity are simply: $E_\theta = \eta H_\varphi$ and $E_\varphi = -\eta H_\theta$.

The method outlined above, for the determination of the EM field around a current-carrying loop, may be extended to include the more general case of a loop with a travelling-wave current distribution of the form: $I(\varphi) = I_0 e^{-j\gamma\varphi}$, where the parameter γ is complex with a negative imaginary part (Werner, 2000). The analysis starts from equations (4)-(6), where the expression $I(\varphi) = I_0 e^{-j\gamma\varphi}$ is substituted. Thus, the three components of $\vec{A}$ result in containing the integral: $G_m(\varphi) = (I_0/2\pi)\int_0^{2\pi} e^{-j\gamma\varphi}\cos^m(\varphi-\varphi')d\varphi'$. Analytic evaluation of the latter (Gradshteyn & Ryzhik, 1980) may lead to closed-form expressions for the vector potential components and, consequently, by using equations (2)-(3), for the EM field around the loop. The detailed methodology may be found in (Werner, 2000).

The marginal case of uniform current distribution of the loop, may be easily obtained since $\vec{A}$ simplifies to a single component, i.e., equation (6) with $I(\varphi') = I_0$. An exact integration may then be achieved if the observation point is restricted to $r>a$ and $0 \le \theta \le \pi$. The end result of the analysis (Overfelt, 1996) is the following expression for the vector potential:

$$A_\varphi(r,\theta) = \frac{a\mu_0 I_0}{\pi r\tau}\sum_{m=0}^{\infty}\frac{(-j\beta r)^m}{m!}(\sigma L_m - L_{m+2}) \tag{22}$$

where τ=2a sinθ/r and σ=1+a2/r^2. ᴸm m_ay be calculated from the recurrence relation:

$$L_m = \frac{2(m-2)\sigma L_{m-2} - (m-3)\left(\sigma^2 - \tau^2\right)L_{m-4}}{m-1}, \quad m \geq 3 \tag{23}$$

with $L_0 = \mathbf{K}(q)/\sqrt{\sigma+\tau}$, L_1=π/2 and $L_2 = \sqrt{\sigma+\tau}\,\mathbf{E}(q)$. The notations $\mathbf{K}(q)$ and $\mathbf{E}(q)$ represent the complete elliptic integrals of the first and second kind (Abramowitz & Stegun, 1972), respectively, with modulus $q = \sqrt{2\tau/(\sigma+\tau)}$. Consequently, the EM field components may be obtained from equations (2)-(3) by taking into account that $\nabla \cdot \vec{A} = 0$, since $\vec{A} = A_\varphi \hat{\varphi}$.

An alternative approach for the calculation of the EM field radiated by a loop antenna may be found in (Conway, 2005). Although the analytical formulation therein starts from equations (2)-(3), the vector potential $\vec{A}$ is expressed in cylindrical coordinates (ρ,φ,z):

$$A_\rho\left(\rho,\varphi,z\right) = \frac{\mu_0 a}{4\pi}\int_0^{2\pi} I(\varphi')\sin(\varphi-\varphi')\frac{e^{-j\beta R}}{R}d\varphi' \tag{24}$$

$$A_\varphi\left(\rho,\varphi,z\right) = \frac{\mu_0 a}{4\pi}\int_0^{2\pi} I(\varphi')\cos(\varphi-\varphi')\frac{e^{-j\beta R}}{R}d\varphi' \tag{25}$$

and A_z(ρ,φ,z)=0. The distance R (Figure 1) between a source point S ($\rho' = a, \varphi', z' = 0$) and an observation point P(ρ,φ,z) is expressed as $R = \sqrt{\rho^2 + a^2 + z^2 - 2a\rho\cos\left(\varphi - \varphi'\right)}$. By expanding the current distribution $I(\varphi')$ in a Fourier series, as in equation (12), substituting in (24)-(25) and applying the symmetry properties of the resulting integrands, the components Aρ ₐnd Aφ ₒf the vector potential end up in comprising integrals of the form:

$$C_m = \int_0^{2\pi} \cos m\varphi' \frac{e^{-j\beta R}}{R}d\varphi' \tag{26}$$

Consequently, the electric field intensity is calculated from equation (3) and it may be written as follows:

$$E_\rho\left(\rho,\varphi,z\right) = \frac{j}{8\pi a\varepsilon_0}\sum_{m=1}^{m_t} I_m \sin m\left(\varphi+\varphi_m\right)\left[2m\frac{\partial J_m}{\partial \rho} - a\beta^2\left(C_{m-1} - C_{m+1}\right)\right] \tag{27}$$

$$E_\varphi\left(\rho,\varphi,z\right) = \frac{j}{8\pi a\varepsilon_0}\sum_{m=1}^{m_t} I_m \cos m\left(\varphi+\varphi_m\right)\left[\frac{2m^2}{\rho}C_m - a\beta^2\left(C_{m-1} - C_{m+1}\right)\right] \tag{28}$$

$$E_z(\rho,\varphi,z)=\frac{j}{8\pi a\varepsilon_0}\sum_{m=1}^{m_t} I_m \sin m(\varphi+\varphi_m)\frac{\partial C_m}{\partial z} \tag{29}$$

Obviously, the components of $\vec{E}$ involve the integral C_m given by equation (26); I_m in equations (27)-(29) are the Fourier coefficients of the current distribution and φm stands for an arbitrary constant phase angle. The components of the magnetic field intensity may result in a similar way; the corresponding expressions can be found in (Conway, 2005).

Thus, the determination of the EM field around the loop becomes equivalent to evaluating the integrals of equation (26). Although these integrals can be derived numerically for small values of m, the oscillating nature of the integrand makes the calculations difficult as m increases. However, it has been shown that it is possible to perform the angular integration analytically; the real part of C_m may be expressed as a series of Legendre functions of the second kind and half-integral order, whereas the imaginary part may be written aa a series of elementary terms. The reader is referred to (Conway, 2005) for the pertinent analysis and expressions. Numerical comparisons of results derived from Conway's and Werner's approaches have been successfully carried out by (Conway, 2005). Moreover, the aforementioned results were found to agree with the corresponding results obtained after direct numerical integration of equation (26).

A somewhat different vector analysis of the EM field radiated by a loop antenna employs the dyadic Green's function (Li et al., 1997). In this regard, the radiated EM field around the loop of Figure 1 may be written, in spherical coordinates, as follows:

$$\vec{E}=j\omega\mu_0\iiint_V \bar{\boldsymbol{G}}_E(\vec{r},\vec{r}')\cdot\vec{J}(\vec{r}')dV' \tag{30}$$

$$\vec{H}=\iiint_V \nabla\times\bar{\boldsymbol{G}}_E(\vec{r},\vec{r}')\cdot\vec{J}(\vec{r}')dV' \tag{31}$$

V in equations (30)-(31) stands for the volume occupied by the circular loop and $\bar{\boldsymbol{G}}_E(\vec{r},\vec{r}')$ represents the dyadic Green's function of the electric kind in free space. $\vec{J}(\vec{r}')$ is the volumetric electric current density which is related to the loop current distribution through:

$$\vec{J}(\vec{r}')=\frac{I(\varphi')\delta(r'-a)\delta(\theta'-\pi/2)}{a}\hat{\varphi} \tag{32}$$

where $\delta(\bullet)$ is the Dirac's delta function.

Subsequently, $\bar{\boldsymbol{G}}_E(\vec{r},\vec{r}')$ is expanded in spherical eigenvectors (Morse & Feshbach, 1953) that involve spherical Bessel and Hankel functions as well as associated Legendre functions (Abramowitz & Stegun, 1972), while the current distribution is expanded in a Fourier series as given by equation (12). A closed form solution for $\vec{E}$ and $\vec{H}$ may be obtained as long as the functions $I(\varphi)\cos(m\varphi)$ and $I(\varphi)\sin(m\varphi)$ are integratable analytically. The detailed methodology is described in (Li et al., 1997) and it has been validated by comparing the resulting closed forms of the far-field components with those presented in (Balanis, 1982) and (Werner, 1996). Moreover, successful comparisons with the results given by (Balanis, 1982), (Werner, 1996) and (Overfelt, 1996) have been carried out, in the case of uniform current distribution of the loop.

More recently, another approach based on expansion in spherical eigenvectors has been proposed for the determination of the EM field around a radiating loop (Hamed et al., 2013). Although the EM field expressions are given in terms of spherical Bessel and Hankel functions and associated Legendre functions, they differ from those presented in (Li et al., 1997), since the analysis by (Hamed et al., 2013) does not involve the dyadic Green's function. The formulation starts from equations (4)-(6), like in most previous treatments. By expanding the term $e^{-j\beta R}/R$ in spherical eigenvectors and performing a simple integration, the following expressions are obtained for the components of the vector potential:

$$A_r(r,\theta,\varphi) = -j\beta\frac{\mu_0 a\sin\theta}{2}\sum_{n=1}^{\infty}(G_{n-1}-G_{n+1})I_n\sin n\varphi \tag{33}$$

$$A_\theta(r,\theta,\varphi) = -j\beta\frac{\mu_0 a\cos\theta}{2}\sum_{n=1}^{\infty}(G_{n-1}-G_{n+1})I_n\sin n\varphi \tag{34}$$

$$A_\varphi(r,\theta,\varphi) = -j\beta\frac{\mu_0 a}{2}\left[G_1 I_0 + \sum_{n=1}^{\infty}(G_{n-1}+G_{n+1})I_n\cos n\varphi\right] \tag{35}$$

The coefficients I_n in equations (33)-(35) originate from the series expansion of the (arbitrary) current distribution which, unlike equation (12), includes real and imaginary terms: $I(\varphi') = \sum_{n=-\infty}^{\infty} I_n e^{jn\varphi'}$. The coefficients G_n incorporate associated Legendre functions $P_m^n(\cos\theta)$, spherical Bessel functions of the first kind $j_m(\bullet)$ and spherical Hankel functions of the second kind $h_m^{(2)}(\cdot)$:

$$G_n = \sum_{m=n}^{\infty}(2m+1)\frac{(m-n)!}{(m+n)!}P_m^n(\cos\theta)P_m^n(0)\begin{cases} j_m(\beta a)h_m^{(2)}(\beta r), & r>a \\ h_m^{(2)}(\beta a)j_m(\beta r), & r<a \end{cases} \tag{36}$$

Substituting equations (33)-(36) into equations (2)-(3) expressions for the EM field components around the loop are obtained. The reader is referred to (Hamed et al., 2013) for a detailed analysis; successful comparisons with the expressions given by (King, 1969) and (Werner, 1996) have been carried out for the far-field approximation.

The approaches outlined above constitute analytical solutions of the forward problem, i.e., the determination of the EM field around a radiating loop antenna. The inverse problem may be defined as the estimation of the loop radius when the EM field intensity is known at specific observation points. This problem is, obviously, much more complex than the forward one and, presently, it lacks analytical treatment. Thus, an alternative approach, based on artificial intelligence techniques, is developed in the next section to address the forward as well as the inverse problem of the radiating loop antenna.

Artificial Neural Networks: A Short Overview

ANNs belong to the broad family of artificial intelligence techniques and were inspired by the principles of the nervous system of living organisms. The human nervous system may be approached with a simplified diagram like the one depicted in Figure 2; it consists of three elements, i.e., the sensors (sense organs), the brain (network of neurons), and the executive organs. The sensors respond to stimuli (electrical signals) and forward the information to the network of neurons, where the decisions are taken and then the appropriate instructions are given to the executive organs.

Figure 2. Simplified block diagram of the human nervous system

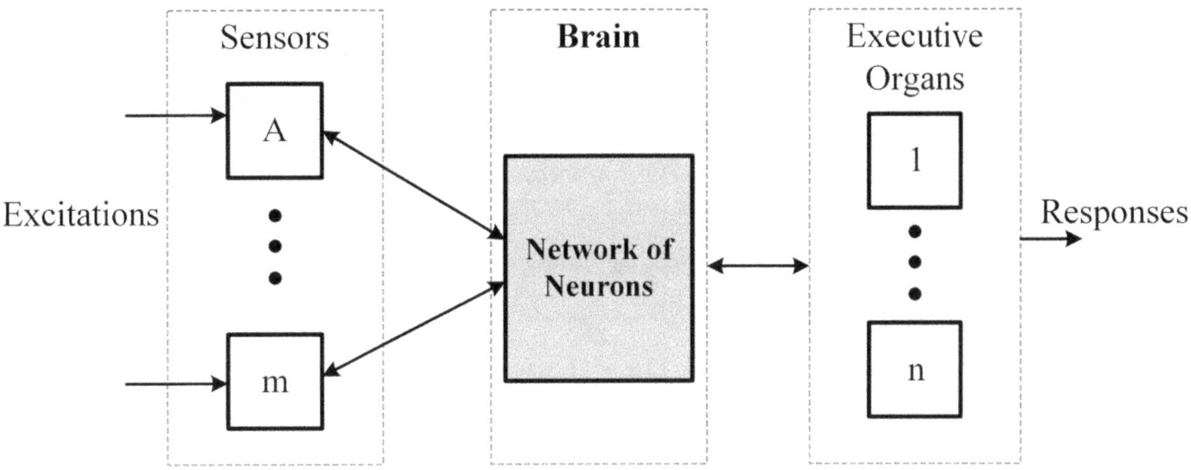

The main functional component of an ANN is the artificial neuron which may be treated as the elementary information processing unit. Figure 3 shows a biological neuron that forms the basis for the design of an artificial neuron. The latter is a simplified form of the biological neuron; it comprises three basic structural elements: the synapses, the adder and the activation function (Haykin, 1999).

With reference to a typical artificial neuron, for example the v-th neuron of Figure 3, each synapsis multiplies the input $x_{m,v}$ by, its own, synaptic weight $w_{m,v}$, m=1,2,…,M; M stands for the total number of synapses that serve as inputs to the adder. Unlike in a biological neuron, the synaptic weight of an artificial synapsis may take both negative and positive values. Consequently, the adder adds the weighted input signals and, then, the activation function Ψ is applied in order to limit the output y_v to a finite value between [-1,1] or [0,1]. The output of the adder may be written as:

$$u_v = \sum_{m=1}^{M} w_{m,v} x_{m,v} \tag{37}$$

Figure 3. Biological neuron and structural diagram of an artificial neuron

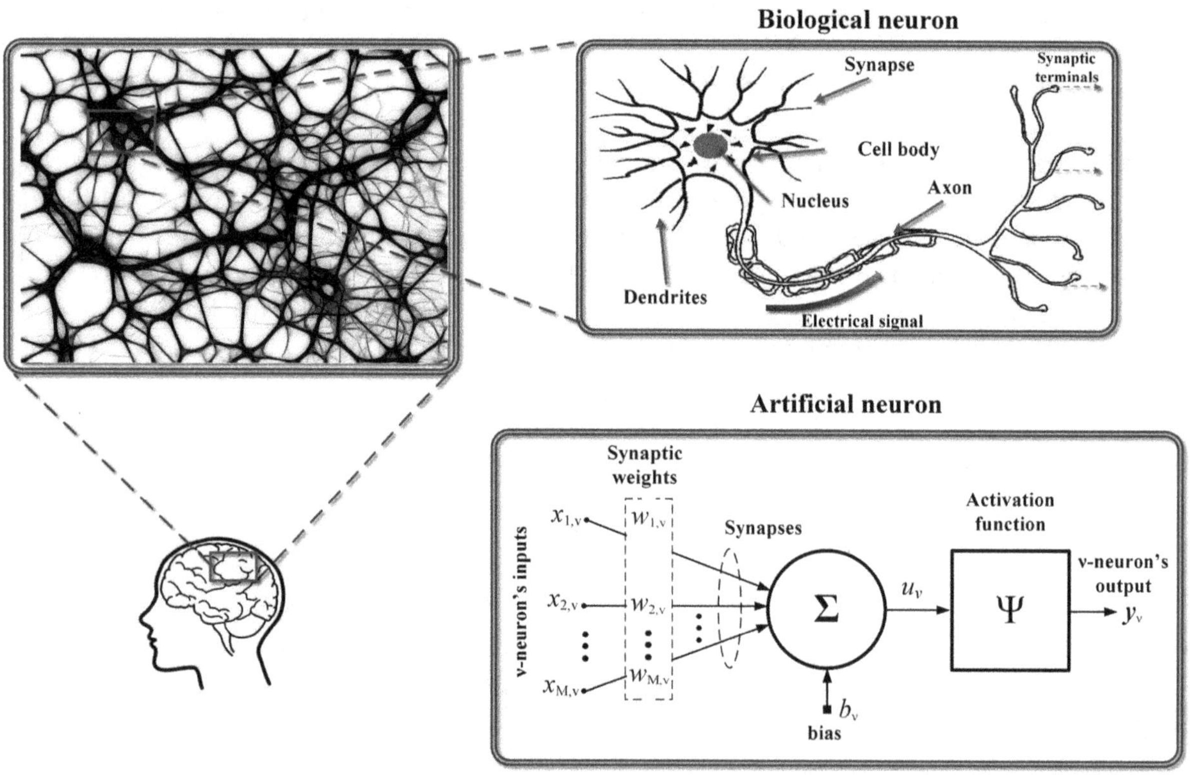

The output of the ν-th neuron is, then, $y_\nu=\Psi(u_\nu+b_\nu)$, where b_ν stands for a bias term used to adjust u_ν (Christodoulou & Georgiopoulos, 2001).

Although there is no restriction on the kind of the activation function, a nonlinear one has proven to be the most suitable for solving EM problems. A popular family of activation functions are the sigmoid ones, such as the logistic: $\Psi_\ell\left(u_\nu+b_\nu\right)=1/\left[1+e^{-\pm(u_\nu+b_\nu)}\right]$ and the hyperbolic tangent: $\Psi_t\left(u_\nu+b_\nu\right)=\tanh\left[\pm(u_\nu+b_\nu)/2\right]$. The former is plotted in Fig. 4(a), whereas the latter is plotted in Fig. 4(b), for various values of the parameter α. A key feature of these functions is that they are continuous and differentiable, which allows for the adjustment of the synaptic weights during training process. Another common activation function is the radial basis one. It is given by: $\Psi_r\left(u_\nu+b_\nu\right)=e^{-(u_\nu+b_\nu)^2}$ and it is widely used with the RBF ANNs, that will be discussed later. The radial basis activation function is plotted in Fig. 4(c) and, evidently, it is similar to a Gaussian function.

The way in which artificial neurons are interconnected to form a network and the flow of information determines the architecture of the ANN. A typical architecture comprises an input layer, one or more intermediate layers and an output layer. The intermediate layers are isolated from the external excitation signals, fed to the input layer, and are referred to as hidden layers. A plethora of architectures may be encountered in the literature, but, generally, they may be classified into two categories. The first one is the feedforward architecture, where information, through the synapses, follows a path without feedback and the output does not depend on the previous condition. The second one is the feedback architecture, where the output of a neuron is fed back either to the input of the same neuron or to neurons of previous

Figure 4. (a) Logistic, and (b) hyperbolic tangent activation functions for different values of α. (c) Radial basis activation function.

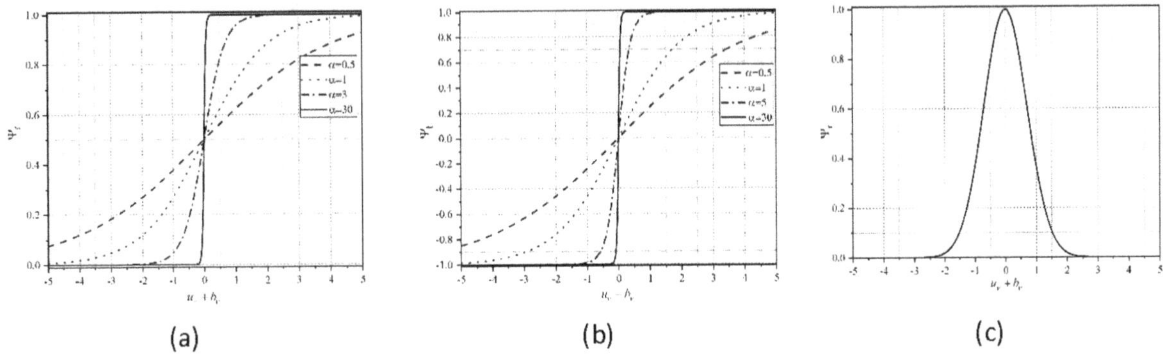

layers (Haykin, 1999). Figure 5 includes the most widely used ANN architectures; the frames bounded by continuous line represent the ANNs applied to solve the problem of the radiating circular loop, in the next section.

Figure 5. Neural network architectures

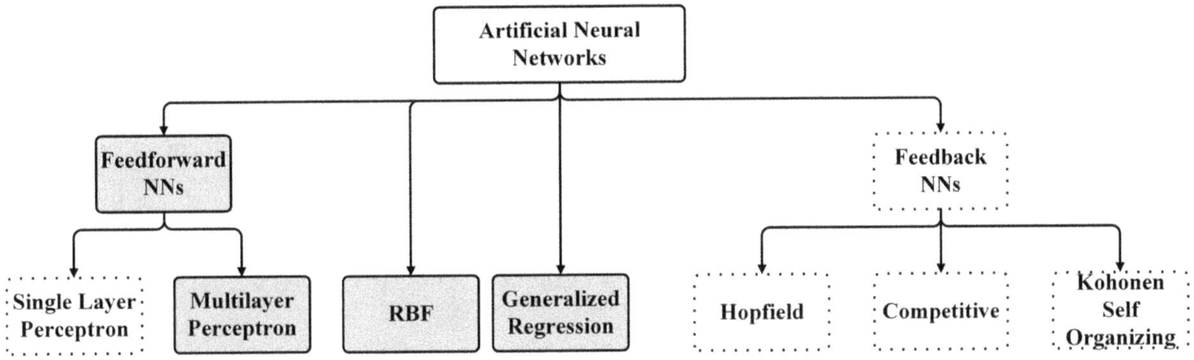

In order to apply an ANN for the solution of a problem, a model should be built from known data. Three distinct datasets are commonly used in different stages of the model development, i.e., the training, the validation and the testing dataset. The first step into the creation of the model is taken by training the ANN. The training dataset consists of sample pairs $\left(\mathbf{x}_k^{\mathrm{tr}}, \mathbf{y}_k^{\mathrm{tr}}\right)$, k=1,2,…,K, where $\mathbf{x}_k^{\mathrm{tr}}$ is a vector that represents the inputs, whereas the vector $\mathbf{y}_k^{\mathrm{tr}}$ represents the desired outputs; K stands for the number of samples that compose the training dataset. One may argue that training is the most important stage in the construction of the model, since during the learning process the synaptic weights, the bias terms and all the intrinsic parameters of the ANN are adjusted. Firstly, they are initialized and during training they are iteratively updated until the end of the training process. Another task that takes place during training and is critical for the achievement of accurate results, is the determination of the number of neurons that compose each hidden layer. This number depends on the degree of nonlinearity and the complexity of the problem. Actually, there is no reliable algorithm or even rule of thumb to calculate the exact number of hidden neurons required in order to train successfully an ANN. The appropriate number of neurons is commonly estimated either empirically or by following a trial and error process (Zhang et al., 2003).

The validation dataset is used to monitor the performance of the ANN, avoid overfitting, and control the termination criterion of the learning process. The testing dataset is an independent set of data used to assess the quality of the ANN model and is kept unseen during training. The datasets used for training, validating and testing the ANN are constructed from results obtained from existing analytical/numerical methods and/or experimental procedures, etc.

The performance of a trained ANN is evaluated in terms of accuracy and generalization (Devabhaktuni et al., 2001); the latter is related to the ability of the ANN to apply the learnt information to "unknown" data. Several criteria and statistical measures are considered in order to assess the performance of an ANN during testing as well as the degree of success of the learning process; specific information about the training and testing errors will be given in the next section.

One of the most popular ANNs is the MLP, which is a layered structure consisting of one input layer, a number of hidden layers and one output layer. MLP may be considered as "universal approximator", since it has been proven that it can estimate any nonlinear, continuous function accurately, provided that it comprises at least two hidden layers with an appropriate number of neurons (Hornik et al., 1989). The detailed description of the MLP architecture and operation may be found in almost all pertinent textbooks. The structure of a MLP ANN is briefly outlined below, based on (Haykin, 1999) and (Christodoulou & Georgiopoulos, 2001).

The typical architecture of a MLP ANN with three layers is shown in Figure 6. The number of neurons included in each layer is denoted by M_n, n=1,2,…,N. The index n is used to identify the layers; the value n=1 corresponds to the input (first) layer, whereas n=N is assigned to the output (last) layer. The vector $\mathbf{x}=\left\{x_1,x_2,\ldots,x_{M_0}\right\}$ constitutes the input to the first layer of the ANN, where the notation M_0 has been reserved for the number of inputs to this layer. The vector $\mathbf{y}^n=\left\{y_1^n,y_2^n,\ldots,y_{M_i}^n\right\}$ denotes the output of the n-th layer. As it may be verified from Figure 6, the MLP ANN adopts a serial structure, i.e., the output of the n-th layer is fed to the input of the (n+1)-th layer. Thus, $\mathbf{y}^n$ may be written as:

$$\mathbf{y}^n=\Psi^n\left(\mathbf{y}^{n-1}\mathbf{w}^n+\mathbf{b}^n\right) \tag{38}$$

where $\mathbf{b}^n$ stands for the vector of bias terms of the n-th layer and the matrix $\mathbf{w}^n$ contains the synaptic weights $w_{i,j}^n$ with $i=1,2,\ldots,M_{n-1}$ and j=1,2,…,M_n. The synaptic weights are adjustable parameters, their adjustment being responsibility of the training algorithm. The selection of the latter, among a plethora of algorithms that exist in the literature, depends on the nature of the problem. For example, when fast convergence is needed, without any other restriction, the Levenberg-Marquardt (LM) algorithm may be the most appropriate choice (Beale et al., 2016). The performance of several training algorithms will be evaluated in reference to the loop antenna radiation problem, presented in the next section.

RBF ANNs comprise three layers of neurons: the input, the hidden and the output layer. The hidden layer is characterized by the radial basis activation function (shown in Figure 4c), which is applied to the input data (Powell, 1987). It has been proven that RBF ANNs are able to approximate any function with an arbitrary small error as long as they have a sufficient number of neurons in the hidden layer

Figure 6. Feedforward MLP ANN with one hidden layer

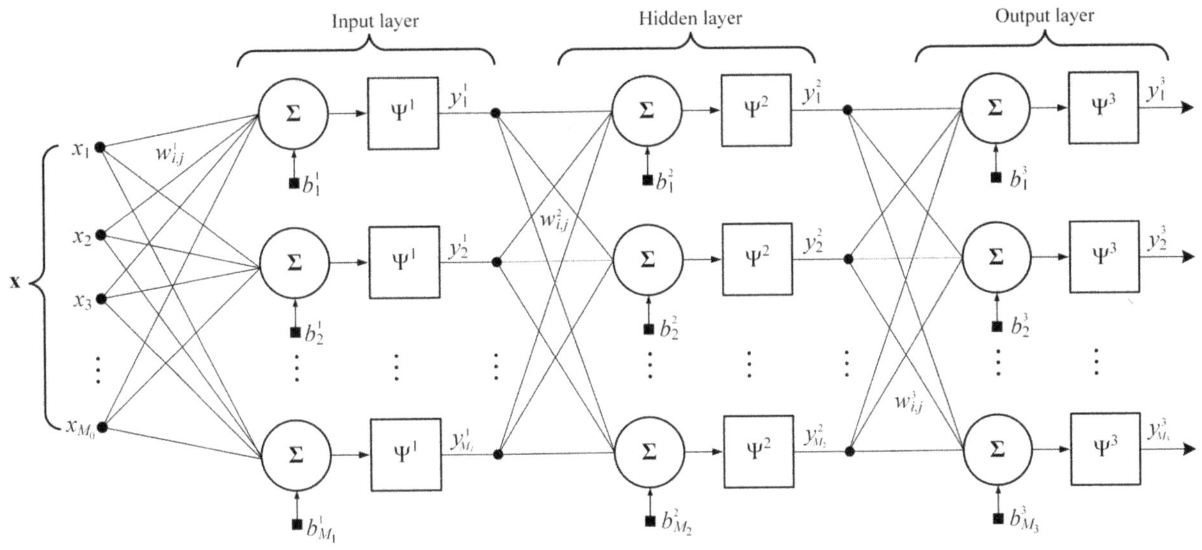

(Park & Sandberg, 1991) and the standard deviation of the (gaussian) activation function along with all synaptic weights are properly adjusted (Hartman et al., 1990). The basic characteristics of a RBF ANN are summarized and compared with the characteristics of a MLP ANN in Table 2.

Table 2. Comparison between RBF and MLP ANNs

RBF ANN	MLP ANN
One hidden layer.	Multiple hidden layers.
The hidden and the output layers consist of different type of neurons. The former is nonlinear, whereas the latter is linear.	Usually, all layers have the same type of neurons and they can be nonlinear.
The appropriate number of hidden neurons results from an iterative procedure.	The number of neurons in all layers is defined before the training process.
The activation function calculates the Euclidean distance between the input and the center vector.	The activation function calculates the product of the input and the weight vector.

Figure 7 depicts the structure of a typical RBF ANN. It is composed of three layers, with the input layer consisting of M_0 nodes; M_0 is actually the length of the input vector. Unlike in MLPs, the input layer in RBF ANNs is connected directly to the hidden layer i.e., without the presence of weights. The number M_b of neurons in the hidden layer results from an iterative procedure. The algorithm starts from a minimum number of neurons; this number is gradually increased until either a predetermined convergence criterion is achieved, or the length of the training data vector is reached. The activation function of the hidden layer, termed as radial basis function, is given by:

$$\Psi_m(\mathbf{x}) = \Psi\left(\left\|\mathbf{x} - \mathbf{c}_m\right\|\right) = \exp\left(-\frac{1}{2\sigma_m}\left\|\mathbf{x} - \mathbf{c}_m\right\|^2\right) \tag{39}$$

where the index $m=1,2,\ldots,M_b$ identifies the hidden neurons and σm stands for the standard deviation of the m-th activation function. The latter is a function of the Euclidean distance between the input and the center vector **cm** which consists of the means of the Gaussian functions of the hidden neurons. The output layer commonly consists of a single neuron and each hidden neuron is connected to the output neuron by its own synaptic weight; thus, only one index, i.e. m=1,2,…,Mb is needed to identify the synaptic weights wm The output of the RBF ANN may be expressed as (Christodoulou & Georgiopoulos, 2001):

$$y(\mathbf{x}) = \sum_{m=1}^{M_b} w_m \Psi_m(\mathbf{x}) = \sum_{m=1}^{M_b} w_m \exp\left(-\frac{1}{2\sigma_m}\left\|\mathbf{x} - \mathbf{x}_m\right\|^2\right) \tag{40}$$

Generalized Regression (GR) ANNs have been introduced by Specht (1991) and fall into the category of probabilistic ANNs. The brief outline that follows is based on (Specht, 1991) and (Christodoulou & Georgiopoulos, 2001).

The structure of a GR ANN is depicted in Figure 8; it comprises three layers: the input, the hidden (pattern), and the output layer. The neurons of the hidden layer incorporate a Gaussian activation function, as follows:

$$\Psi_m(\mathbf{x}) = \exp\left[\frac{-\left(\mathbf{x} - \mathbf{c}_m\right)^{\mathrm{T}}\left(\mathbf{x} - \mathbf{c}_m\right)}{2\sigma^2}\right] \tag{41}$$

where the index $m=1,2,\ldots,M_g$ identifies the hidden neurons and σ is the spread factor; its value is taken to be equal to the common width of all Gaussian functions. The superscript T in Equation (41) is the transpose operator. Unlike the MLPs, the GR ANNs are trained by a single pass and they achieve generalization with less training data than that needed by the corresponding MLP ANNs (Kapetanakis et al., 2018b).

The GR ANN estimates the joint probability density function of the input vector **x** and the output y by summing the Gaussian probability densities centered at the location of each sample pair (x_m, y_m). It approximates the best mean square estimate of y in terms of **x** by using the following expression:

$$\hat{y} = \frac{\sum_{m=1}^{M_g} y_m \exp\left(-\frac{D_m^2}{2\sigma^2}\right)}{\sum_{m=1}^{M_g} \exp\left(-\frac{D_m^2}{2\sigma^2}\right)} \tag{42}$$

D_m in Equation (42) stands for the Euclidean distance between **x** and x_m.

Figure 7. Structure of a typical RBF ANN

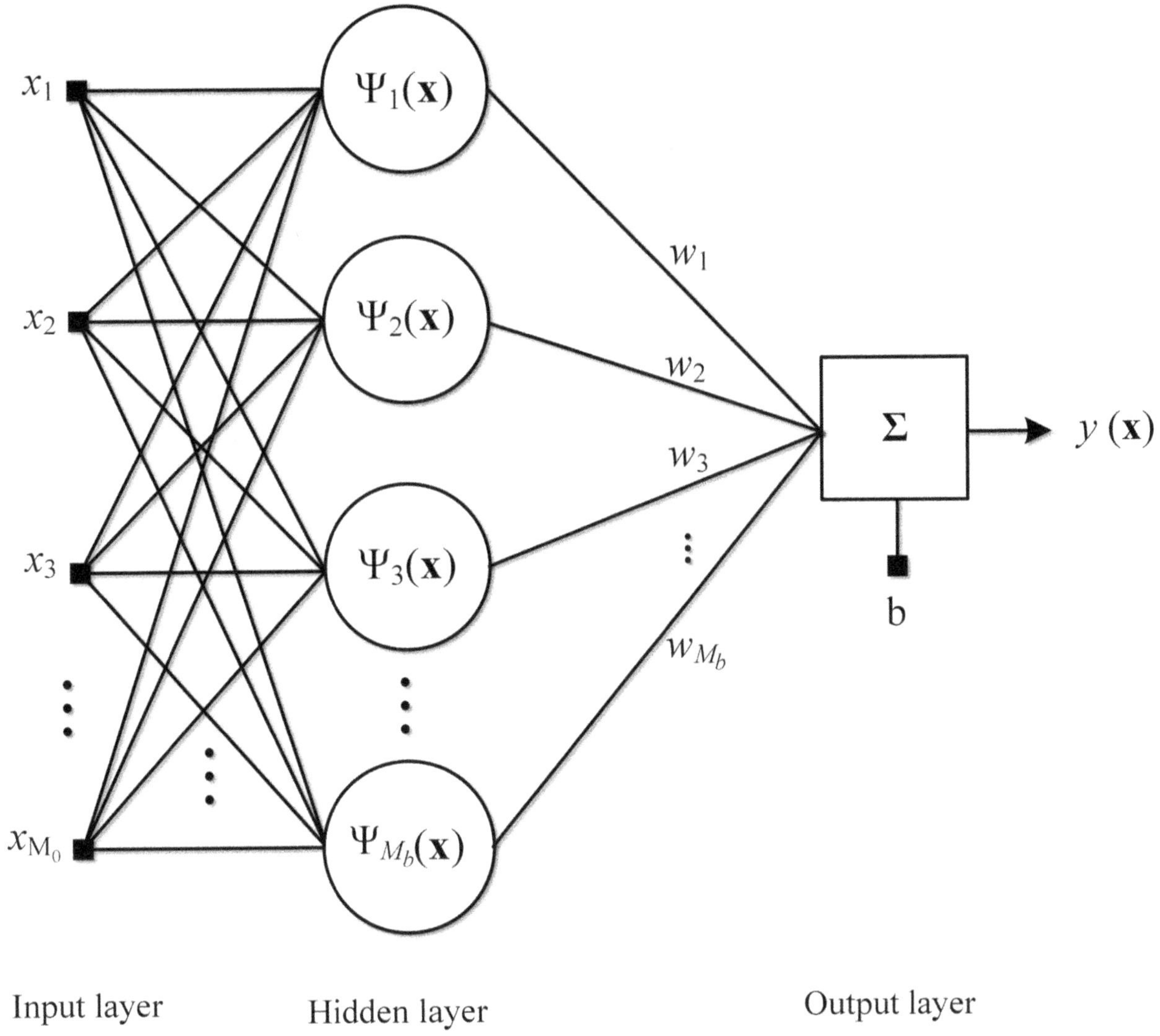

The ANFIS belongs to the family of artificial intelligence techniques and may be used to solve function approximation problems; it combines the most attractive features of fuzzy systems and ANNs. A brief description of the structure and the basic characteristics of an ANFIS is given below; the outline is based on (Jang, 1993).

The core of an ANFIS is a set of fuzzy "if-then" rules with appropriate membership functions, in an attempt to imitate the "rule of thumb" used by humans. The if-then rules are applied to generate the input-output pairs and may contain linguistic values (or labels), such as "high", "small" etc., as well as nonfuzzy equations of the input variables. Figure 9 depicts the architecture of a 2-input, type-3 ANFIS with 2 rules, termed as ANFIS, hereafter, for simplicity. It comprises five layers; each layer consists of nodes and each node performs a specific operation on incoming signals. A set of parameters may be associated with each node; a square, known as adaptive, node does have parameters, whereas a circle,

Figure 8. Structure of a GR ANN

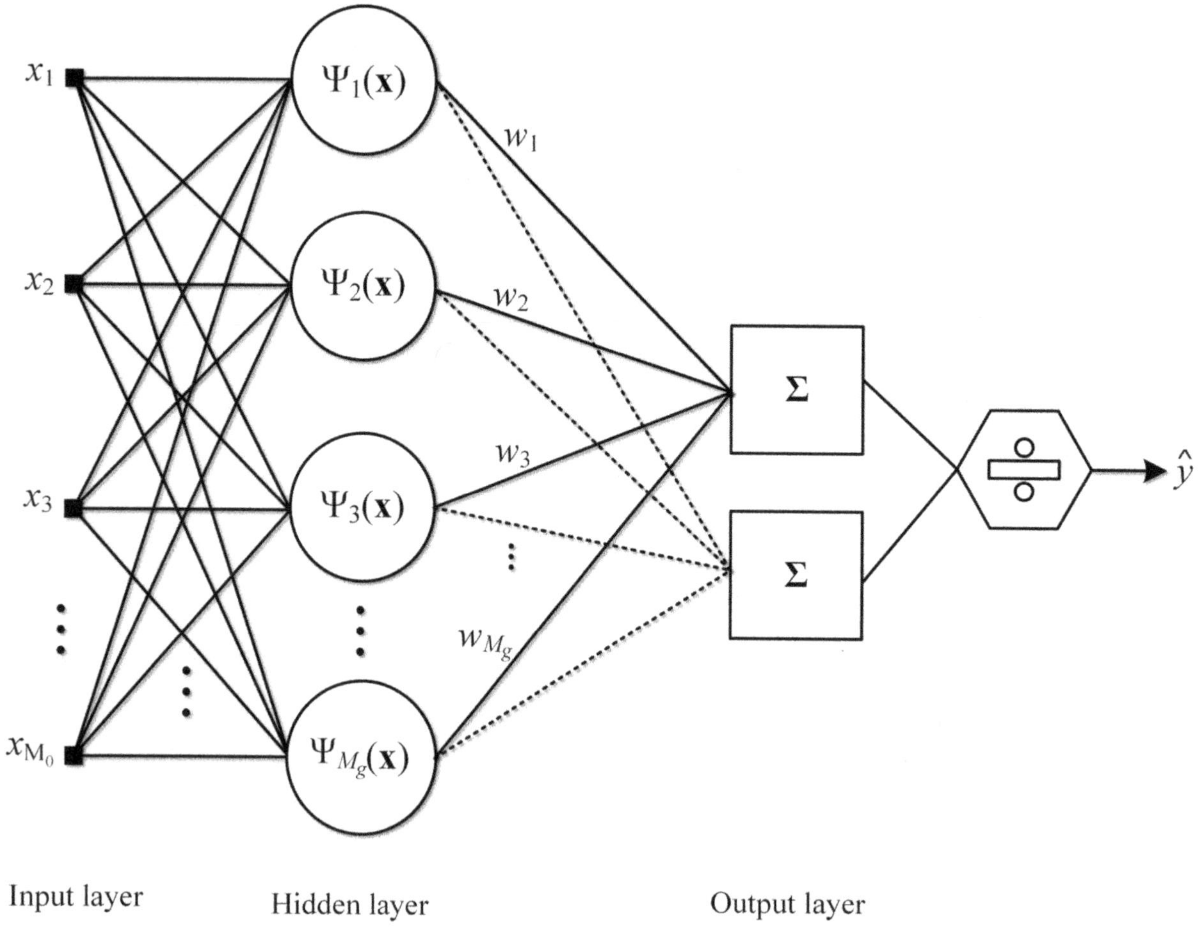

named as fixed, node has no parameters. The name of each layer, as well as the parameters pertaining to it, if any, are shown in Figure 9.

The ANFIS under consideration has two inputs, i.e., x_1, x_2 and one output y. It contains two if-then rules of Tagaki and Sugeno's type, i.e., the rules comprise both linguistic labels and nonfuzzy equations. The former appear at the first part of the rule, i.e., the premise part, whereas the latter constitute the second, or consequent, part. The rules are of the form: "if x_1 is A_m and x_2 is B_m then $y_m = p_m x_1 + q_m x_2 + r_m$", m=1,2. The parameters (p_m,q_m,r_m), are referred to as consequent parameters.

The node function of each layer is termed as O_m^n hereafter; the superscript n=1,2,…,5 identifies the layer, while the subscript m is used to identify the nodes in the n-th layer. The node function of the first layer is a membership function of the label A_m or B_m and it specifies the degree to which the input satisfies this label. Thus, $O_m^1 = \mu_{A_m}(x_m)$ for m=1,2 and $O_m^1 = \mu_{B_{m-2}}(x_{m-2})$ for m=3,4. The second layer multiplies the incoming signals; the symbol "Π" in Figure 9 is, actually, the first letter of the word "product" in Greek. Each node output of this layer represents the firing strength of a rule and is given by:

$$O_m^2 = w_m = \mu_{A_m}(x_1) \cdot \mu_{B_m}(x_2), m=1,2 \tag{43}$$

Figure 9. Layered structure of ANFIS with two inputs and one output

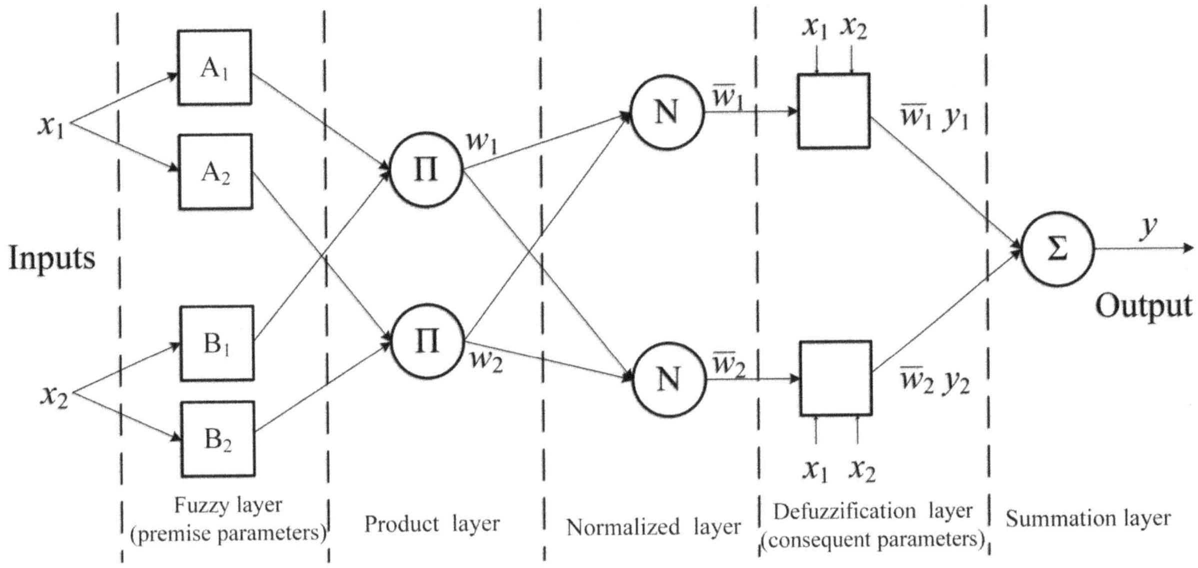

Consequently, the aforementioned firing strengths are normalized according to:

$$O_m^3 = \overline{w}_m = \frac{w_m}{w_1 + w_2}, \; m{=}1{,}2 \tag{44}$$

The defuzzyfication takes place in the fourth layer, where the contribution of the m-th rule to the output of the node is computed, by using the consequent parameters:

$$O_m^4 = \overline{w}_m y_m = \overline{w}_m \left(p_m x_1 + q_m x_2 + r_m\right), \; m{=}1{,}2 \tag{45}$$

Finally, the single, fixed node of the last layer sums all the incoming signals and provides the overall output of the ANFIS:

$$O_m^5 = y = \sum_{m=1}^{2} \overline{w}_m y_m \tag{46}$$

FORWARD AND INVERSE PROBLEM SOLUTION USING ARTIFICIAL INTELLIGENCE TECHNIQUES

Modeling of the Forward Problem

The artificial intelligence techniques, outlined in the previous section, are applied herein in order to model the forward problem of the radiating, circular, thin loop antenna, i.e., the determination of the EM field around the antenna, given the loop radius and the observation angle. The stages required to solve the problem are shown in Figure 10. The definition of the problem relies upon the identification of whether it is a clustering problem, a classification one or a function approximation, with the latter being the case regarding the radiating loop. Then, the inputs and outputs of the models are selected according to availability; moreover, an attempt is made to select those inputs that have the greatest impact on the outputs. The selection of the data range follows the good practice to generate data in the interval $[\mathbf{x}_{min}-\delta, \mathbf{x}_{max}+\delta]$, where $[\mathbf{x}_{min}, \mathbf{x}_{max}]$ is the range that the model is required to perform and δ is used to expand this range in order to increase the reliability of the model when it is forced to operate near the boundaries (Zhang & Gupta, 2000). The datasets may be constructed from results obtained from analytical solutions, simulations, experiments, etc. Herein, the existing analytical solution, given in (Werner, 1996), is used to construct data for training, validating and testing the models. Two different architectures are adopted for the forward problem modeling: the MLP ANN and the ANFIS. Their specific structure, as well as a description of the training process will be presented below. Finally, the accuracy of the results is checked according to predetermined statistical measures and criteria.

All models have two inputs and one output. The inputs are the loop radius a, which is allowed to take values in the range $[0.05\lambda, 5\lambda]$, where λ stands for the wavelength, and the observation angle θ (Figure

Figure 10. Flow chart the solution of the loop antenna radiation problem

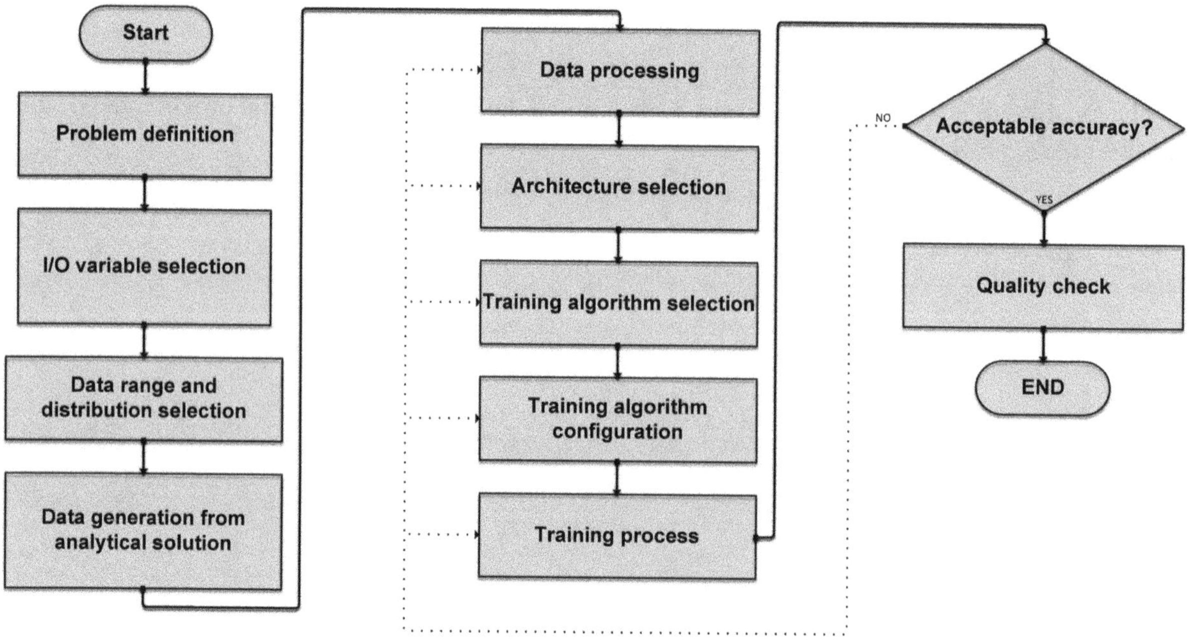

1) which varies in the range [0°, 180°]. The output U is the radiation intensity, in the elevation plane of the far field, given by:

$$U(\theta,\varphi)=\frac{r^2\left|E_\varphi\left(r,\theta,\varphi\right)\right|^2}{120\pi} \tag{47}$$

where $E\varphi$ is the φ-component of the electric field intensity and may be calculated from Equation (19); the loop is assumed to carry uniform current. It is well-known that the radiation intensity may resume the same value for different values of the angle θ. Thus, in order to avoid performance degradation and difficulties during training caused by multivalued data, the range [0°, 180°] is divided in subranges according to the following three schemes: one subrange, i.e., [0°, 180°], three subranges, i.e., [0°, 60°), [60°, 120°), [120°, 180°] and six subranges, i.e., [0°, 30°), [30°, 60°), [60°, 90°), [90°, 120°), [120°, 150°), [150°, 180°]. Accordingly, one, three or six ANNs (or ANFISs), are implemented to solve the problem. Henceforth, the index i=1,3,6 is used to denote the number of the aforementioned subranges; consequently, the notation ANNi (or ANFISi) means that the model comprises i ANNs (or ANFISs), each one handling data in the corresponding subrange (Kapetanakis et al., 2018a).

The structure of the MLP ANNs applied to solve the forward problem follows the general block diagram depicted in Figure 6, albeit with two hidden layers. The hyperbolic tangent activation function (Figure 4(b)) is adopted and the LM algorithm is chosen for training the ANNs. All simulations have been performed in MATLAB environment, where the NN toolbox has been used (Beale et al., 2016). Data generated from Equation (47) are grouped in three distinct datasets and are used for training, validating and testing the models, respectively. The basic characteristics of the MLP ANNs, along with the size of the datasets, are presented in Table 3.

Table 3. Configuration of the ANN models

Model characteristics	ANN_1	ANN_3	ANN_6
Number of ANNs	1	3	6
Number of hidden neurons	50×50	35×35	30×25
Number of samples in the training dataset	6417	3208	3208
Number of samples in the validation dataset	1132	566	566
Number of samples in the testing dataset	1000	1000	1000

The ANFIS models implemented to predict the radiation intensity of the loop are of Takagi and Sugeno type, like the one described briefly in the previous section. However, the proposed models are much more complicated than that depicted in Figure 9. Their characteristics are found after an extensive trial-and-error process and are shown in Table 4. The ANFISs are trained by a hybrid algorithm, which combines gradient descent and least-square methods for parameter identification.

Table 4. Configuration of the ANFISs

Model characteristics	$ANFIS_1$	$ANFIS_3$	$ANFIS_6$
Number of ANFISs	1	3	6
Membership function type	Combination of 2 Gaussian functions	Combination of 2 Gaussian functions	Combination of 2 Gaussian functions
Number of membership functions	16	12	10
Number of samples in the training dataset	5100	2550	2550
Number of samples in the validation dataset	766	766	766
Number of samples in the testing dataset	1000	1000	1000
Number of fuzzy rules	256	144	100
Number of consequent parameters	896	528	380

Results and Discussion

The quality of the ANN and ANFIS models implemented herein is assessed by using an independent set of data, i.e., the testing data which is kept unseen from the models during training. The testing dataset consists of triplets of the form $\left(a_k^{te}, \theta_k^{te}, U_k^{te}\right)$, $k = 1, 2, \ldots, K$ with a_k^{te}, θ_k^{te} being the input values of the loop radius and the observation angle, respectively, and U_k^{te} being the required output of the model during testing; K stands for the number of samples used for testing and has been set equal to 1000 for all cases examined. The statistical measures calculated herein for the evaluation of the models' performance are the absolute error (AE), the mean absolute error (MAE) and the mean relative error (MRE), defined as follows:

$$\mathrm{AE}_{te,k} = \left|U_k^{pr} - U_k^{te}\right| \tag{48}$$

Figure 11. Normalized radiation intensity versus normalized circular loop radius a/λ for θ=60°. (a) ANN1, (b) ANN3, (c) ANN6.

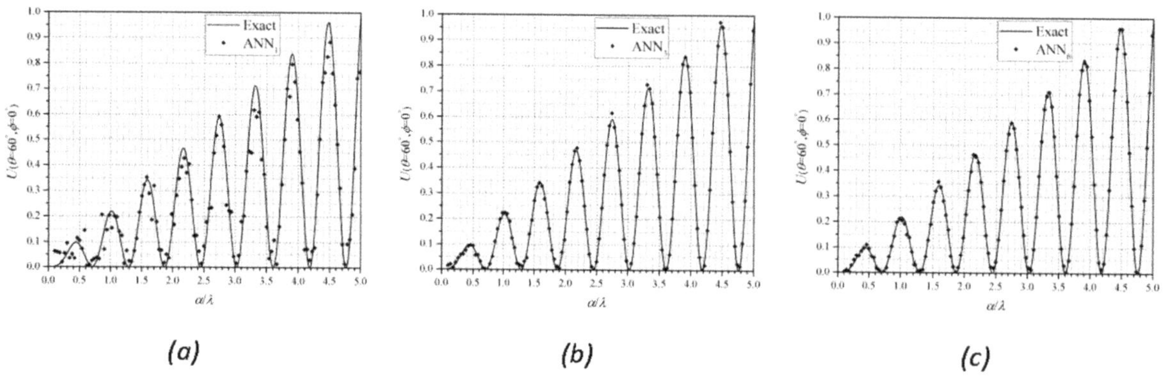

$$\mathrm{MAE}_{te} = \frac{1}{K}\sum_{k=1}^{K}\left|U_k^{pr} - U_k^{te}\right| \tag{49}$$

$$\mathrm{MRE}_{te} = \frac{1}{K}\sum_{k=1}^{K}\frac{\left|U_k^{pr} - U_k^{te}\right|}{U_k^{te}} \tag{50}$$

where U_k^{pr} is the *k*-th predicted value of the radiation intensity by the model (MLP ANN or ANFIS).

A comparison of the radiation intensity predicted from the six models, i.e., ANN_i and ANFIS_i, for i=1,3,6, with theoretical data extracted from (47) is offered by Figures 11-16. The normalized radiation intensity of the radiating loop versus its normalized radius a/λ is plotted in Figures 11-12, for a fixed angle of observation, i.e., θ=60°. The solid line corresponds to U(θ=60°, φ=0°), as calculated from Equation (47), whereas the markers represent the predicted values of the radiation intensity by the model indicated in the inset. It is evident from Figures 11(a) and 12(a) that the ANN1 and ANFIS1 fail to predict accurately the radiation intensity for small values of the loop radius, roughly up to a/λ=*1*. As a/λ increases, the predicted values tend to coincide with the theoretical ones, albeit the ANFIS1 is rather better than the ANN1 with regard to certain peaks of the radiation intensity, such as the one that occurs for a/λ=3.32. However, if the range [0°, 180°] is split and each model handles a smaller range of values of the observation angle, as in Figures 11(b), 11(c), 12(b) and 12(c), the MLP ANNs and ANFISs are able to predict much more accurately the radiation intensity. Still, ANN3 and ANFIS3 exhibit some small failures at certain peaks, which appear to diminish in Figures 11(c) and 12(c), where the results obtained by ANN6 and ANFIS6 are being plotted. This is a reasonable outcome, since the narrower the observation angle data ranges are, the rarer the occurrence of multivalued data becomes and, consequently, the greater the success of the training process and the accuracy of the results is.

The normalized radiation pattern, i.e., U(θ,φ=0°)/U(θ,φ=0°)|max, for various values of the loop radius is plotted in Figures 13-16. The solid line corresponds to the theorical values of the radiation intensity, whereas the markers represent the predicted values. Even a quick glance at Figures 13-16 suggests that they follow roughly the aforementioned trend: the predictions of the radiation intensity by ANN6 and

Figure 12. Normalized radiation intensity versus normalized circular loop radius a/λ, for θ=60°. (a) ANFIS1, (b) ANFIS3, (c) ANFIS6.

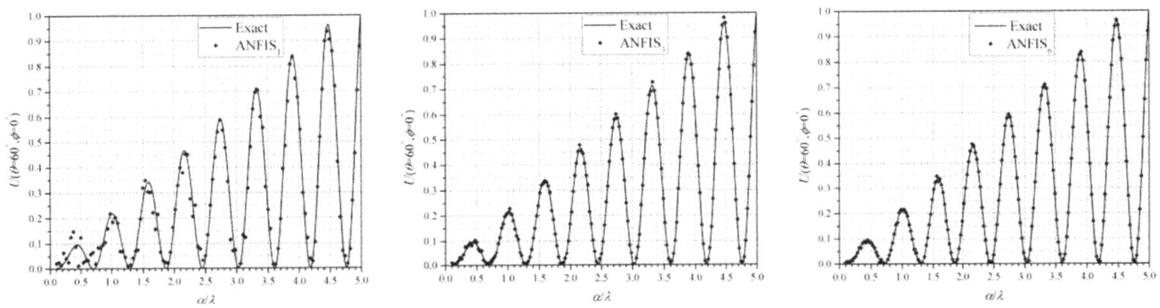

(a) (b) (c)

ANFIS6 (Figures 13(c), 14(c), 15(c) and 16(c)) are much more accurate than those obtained when using a smaller number of θ-subranges. It is remarkable that even for a=$3\lambda/2$, where the radiation pattern consists of several lobes and zeroes (Figures 15 and 16), ANN6 and ANFIS6 manage to maintain their optimum performance regarding the radiation intensity prediction.

Figure 13. Normalized radiation pattern of a loop with radius a=0.2λ. (a) ANN1, (b) ANN3, (c) ANN6.

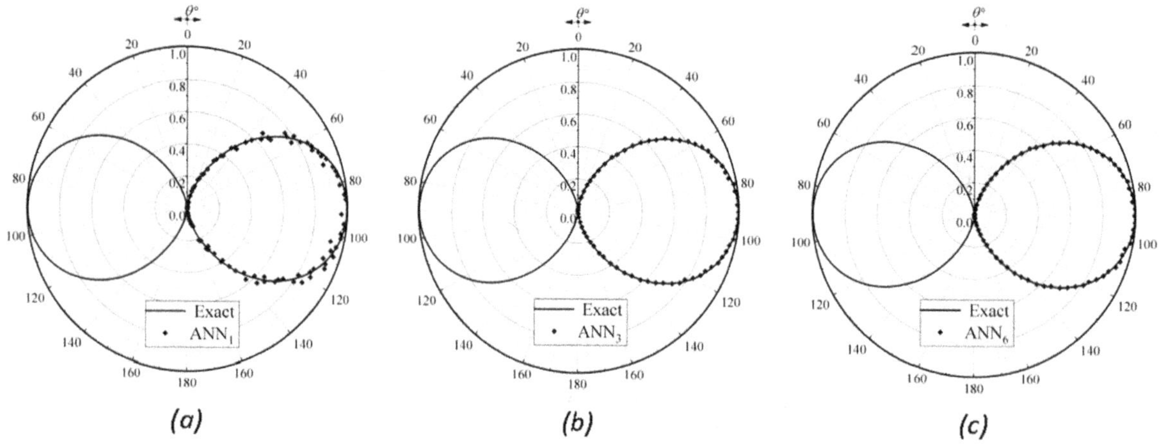

The testing absolute error AEte (minimum and maximum values), the testing mean absolute error MAEte, and the mean relative error MREte, as calculated from Equations (48)-(50), respectively, are given in Tables 5 and 6. The results of Table 5 were produced by using a testing dataset with triplets of the form $\left(a_k^{te}, \theta = 30^\circ, U_k^{te}\right)$, for the first four columns, or $\left(a_k^{te}, \theta = 60^\circ, U_k^{te}\right)$, for the last four columns.

Figure 14. Normalized radiation pattern of a loop with radius a=0.5λ. (a) ANFIS1, (b) ANFIS3, (c) ANFIS6.

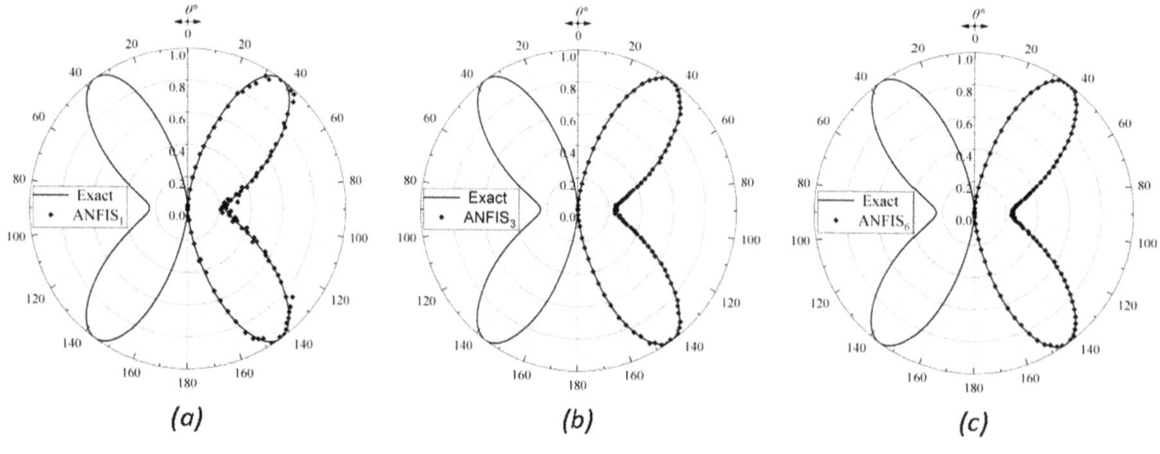

Figure 15. Normalized radiation pattern of a loop with radius a=1.5λ. (a) ANN1, (b) ANN3, (c) ANN6.

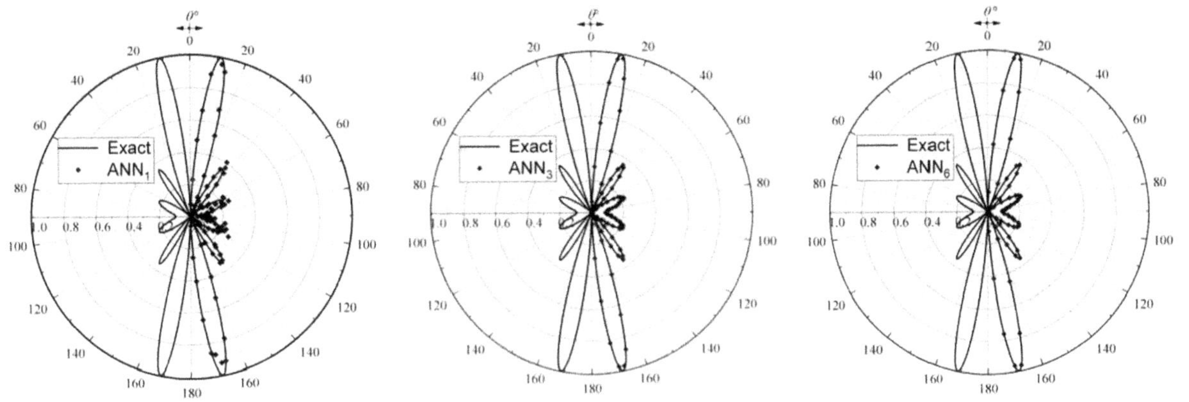

(a) (b) (c)

Figure 16. Normalized radiation pattern of a loop with radius a=1.5λ. (a) ANFIS1, (b) ANFIS3, (c) ANFIS6.

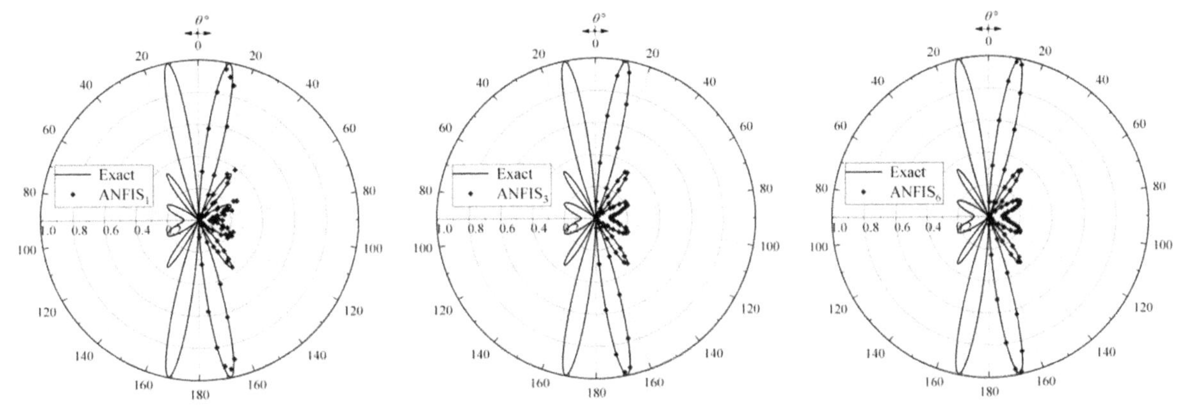

(a) (b) (c)

Table 5. Testing errors of all ANN and ANFISs models (for 1000 samples of a)

	θ=30°				θ=90°			
Model	$AE_{te,max}$	$AE_{te,min}$	MAE_{te}	MRE_{te} (%)	$AE_{te,max}$	$AE_{te,min}$	MAE_{te}	MRE_{te} (%)
ANN_1	1.24×10^{-1}	1.98×10^{-5}	2.60×10^{-2}	9.01	1.79×10^{-1}	6.06×10^{-5}	3.84×10^{-2}	14.19
$ANFIS_1$	1.05×10^{-1}	4.64×10^{-5}	2.03×10^{-2}	7.04	1.56×10^{-1}	1.93×10^{-6}	3.26×10^{-2}	12.05
ANN_3	1.80×10^{-2}	1.32×10^{-5}	4.36×10^{-3}	1.51	4.14×10^{-2}	2.33×10^{-5}	8.48×10^{-3}	3.13
$ANFIS_3$	2.55×10^{-2}	1.18×10^{-5}	5.71×10^{-3}	1.98	2.80×10^{-2}	2.17×10^{-6}	6.81×10^{-3}	2.51
ANN_6	2.05×10^{-2}	4.36×10^{-7}	4.53×10^{-3}	1.57	3.77×10^{-2}	6.62×10^{-6}	7.47×10^{-3}	2.75
$ANFIS_6$	3.39×10^{-2}	1.05×10^{-6}	7.44×10^{-3}	2.58	2.21×10^{-2}	1.18×10^{-5}	4.81×10^{-3}	1.77

Table 6. Testing errors of all ANN and ANFISs models (for 1000 samples of θ)

	a=0.61λ				a=1.5λ			
Model	$AE_{te,max}$	$AE_{te,min}$	MAE_{te}	MRE_{te} (%)	$AE_{te,max}$	$AE_{te,min}$	MAE_{te}	MRE_{te} (%)
ANN_1	1.46×10^{-1}	6.79×10^{-5}	3.24×10^{-2}	8.56	1.71×10^{-1}	6.42×10^{-5}	3.23×10^{-2}	13.41
$ANFIS_1$	1.29×10^{-1}	1.46×10^{-6}	2.82×10^{-2}	7.45	1.20×10^{-1}	3.14×10^{-5}	2.68×10^{-2}	11.13
ANN_3	3.62×10^{-2}	8.44×10^{-6}	9.57×10^{-3}	2.53	2.37×10^{-2}	1.98×10^{-6}	5.51×10^{-3}	2.29
$ANFIS_3$	3.24×10^{-2}	1.34×10^{-6}	8.38×10^{-3}	2.22	2.76×10^{-2}	5.39×10^{-6}	5.91×10^{-3}	2.45
ANN_6	2.85×10^{-2}	1.96×10^{-5}	7.48×10^{-3}	1.98	2.03×10^{-2}	1.25×10^{-5}	4.70×10^{-3}	1.95
$ANFIS_6$	2.40×10^{-2}	1.22×10^{-5}	5.67×10^{-3}	1.50	2.57×10^{-2}	1.14×10^{-5}	4.57×10^{-3}	1.90

Accordingly, the testing datasets for Table 6 were triplets of the form $\left(a=0.61\lambda,\theta_k^{te},U_k^{te}\right)$ or $\left(a=1.5\lambda,\theta_k^{te},U_k^{te}\right)$, k=1,2,...,1000 for all cases examined.

Tables 5 and 6 suggest that the technique of dividing the range [0°, 180°], for the angle θ, and using a different ANN (or ANFIS) to handle each subrange, may improve significantly the accuracy of the results. The familiar remark that the greater the number of the observation angle subranges is the more accurate the prediction of the radiation intensity by the corresponding model becomes, may be readily verified from Tables 5 and 6. For example, the $MAEt_e$ is roughly one order of magnitude smaller for ANN6 and ANFIS6 compared to the corresponding $MAEt_e$ for ANN1 and ANFIS1 regardless of the specific values of θ and a. Furthermore, it may be verified from Table 6, that the MREte $_i$s, generally, greater when a=1.5λ than in the case of a=0.61λ, an outcome that is much more pronounced for the ANN1 an$_d$ ANFIS1. T$_h$is may be attributed to the fact that the radiation pattern of the loop with greater radius, i.e., a=1.5λ, has a greater number of lobes and zeros (Figures 15-16) than the smaller loop (a=0.61λ). Thus, it becomes more difficult for the model (ANN or ANFIS) to capture the more frequent variation of the radiation intensity (Kapetanakis et al., 2018a).

Modeling of the Inverse Problem

Three different ANN architectures are implemented, in MATLAB environment, for modeling the inverse problem of the radiating loop: The MLP, the RBF and the GR ANNs. Data generated from Equation (19), by assuming that the loop carries uniform current are employed to train, validate and test the aforementioned models. Selecting ANNs as the means of solving the inverse problem requires for data to be swapped, so that $|E\varphi|$ v_{al}ues form the inputs of the model, whilst the output is associated to loop radius data (Kapetanakis et al., 2018b). Calculations regarding particular field points Pn(r), *where* r = (*400*λ, θn, φ) *are* performed (Figure 1) for n=1,2,…,N. Th*e* parameters λ and N repre*s*ent the wavelength and the number of observation points which are located around the antenna, respectively. Because of the radiation pattern symmetry, all the selected values of the θn obser$_v$ation angle fall in the range of [0°, 90°]. The observation points' maximum number considered in this work is 12, whilst the optimum number of observation points N is ex*a*mined later in the chapter. The particular locations of every observation point, termed as observers hereafter, rely upon N and *a*re presented in Table 7. The loop radius is allowed to obtain values from the range of [0.1λ, 8λ] including edge values.

Table 7. Observers' number (N) and observation angles (θ_n)

N	θ_n
2	30°, 80°
3	15°, 60°, 85°
4	30°, 70°, 80°, 90°
5	30°, 70°, 75°, 80°, 90°
6	5°, 30°, 60°, 70°, 75°, 90°
7	30°, 45°, 60°, 65°, 75°, 85°, 90°
8	15°, 30°, 45°, 60°, 65°, 75°, 85°, 90°
9	30°, 45°, 55°, 60°, 65°, 70°, 75°, 80°, 90°
10	5°, 30°, 45°, 55°, 60°, 65°, 70°, 75°, 80°, 90°
11	5°, 15°, 30°, 45°, 55°, 60°, 65°, 70°, 75°, 80°, 90°
12	5°, 15°, 30°, 45°, 55°, 60°, 65°, 70°, 75°, 80°, 85°, 90°

In order to solve the inverse problem, the first step is training the ANNs. The dataset used for training comprises $\left(\boldsymbol{E}_m^{tr}, a_m^{tr}\right)$ sample pairs, for m=1,2,...,M, where $\boldsymbol{E}_m^{tr}$ stands for an N-vector representing the inputs to the model, i.e. $|E\varphi(r,\theta_{n,}\varphi)|$ v$_a$lues, for n=1,2,…,*N*, and a_m^{tr} represents the required output shown to the ANN during training that corresponds to the loop radius. The parameter M corresponds to the integer number of data samples that are included in the training data set.

The mean square training error defined as:

$$\text{MSE}_{tr} = \frac{1}{M}\sum_{m=1}^{M}\left(a_m^{pr} - a_m^{tr}\right)^2 \tag{51}$$

is employed to evaluate the performance during the training process, where a_m^{pr} represents the mt*h* value regarding the loop radius predicted by the ANN. Further improvement of the accuracy of the results requires the encompassment of average MSEtr$_w$hich is determined by multiple repetitions, 33 epochs in this particular case, of the training process using the same training data set. Thus, the MSEtr$_n$otation refers to the average MSE mentioned above. The training process ends when MSEtr$_f$alls beneath 10-6· The latter value corresponds to the output of the ANN, i.e. the normalized loop radius, achieving a three-digit accuracy.

To assess the training efficiency of the ANNs, an independent testing dataset is used. This set comprises data samples in the form of $\left(\boldsymbol{E}_k^{te}, a_k^{te}\right)$, *k*=1,2,…,*K* where $\boldsymbol{E}_k^{te}$ and a_k^{te} are the *N*-input vector and the output, respectively, of the aforementioned ANN during the testing process. The parameter *K* represents the overall number of data samples used during the testing process and equals 500 for all the cases that were examined in this particular work. The mean square error MSE_{te} during testing is defined similarly to that of equation (51) should *M* be replaced by *K*, a_m^{tr} by a_k^{te} and a_m^{pr} by a_k^{pr}.

The MLP ANN implemented to solve the radiating loop inverse problem is based on the general configuration scheme shown in Figure 6, albeit it comprises two hidden layers, instead of one, and the output layer is formed by a single neuron that calculates the sum of all the weighted outputs produced by neurons in the intermediate (second) hidden layer. According to the notation adopted in Figure 6, the M_1 number of input neurons does coincide with the number of inputs M_0. The latter is actually the number of observers, thus $M_0=M_1=N$. The number of neurons encompassed in the first and second hidden layer is denoted by M_2 and M_3, respectively, whereas $M_4=1$ corresponds to the single neuron in the output layer.

MLP ANNs training is conducted by exploiting twelve (12) different training algorithms (Beale et al., 2016), as shown in Table 8. Each algorithm's performance can be assessed by referring to the two last columns of Table 8, where the MSE_{tr} and the MSE_{te} are given. Table 8 suggests that the values of MSE_{te} are in general greater than the values of MSE_{tr}, as was expected to be the case. Firstly, the Bayesian Regularization backpropagation (BR) and secondly, the Levenberg-Marquardt (LM) algorithms appear to be superior over the remaining ten (10) training algorithms, regarding the MSEs, as is readily verified by the results shown in Table 8. The MSE_{te} for the BR algorithm is computed to less than 10^{-5}, while at the same time the MSE_{te} for all the other training algorithms, excluding LM, is computed to be greater by 20dB at least. Although not reported here, similar results were obtained for all the examined *N* and *M* values.

Table 8. Algorithms used for training the MLP ANN and number of learning iterations (N=6, M=600)

Abbreviation	Algorithm	MSE_{tr}	MSE_{te}
BR	Bayesian Regularization backpropagation	6.75×10^{-5}	8.00×10^{-5}
LM	Levenberg-Marquardt	3.00×10^{-5}	3.61×10^{-5}
CGB	Conjugate Gradient with Powell/Beale restarts	6.65×10^{-4}	7.94×10^{-4}
BFG	Broyden Fletcher Goldfarb Shanno Quasi-Newton	1.29×10^{-4}	2.63×10^{-4}
GDX	Gradient Descent with momentum and adaptive learning rate backpropagation	1.44×10^{-3}	7.07×10^{-3}
GD	Gradient Descent backpropagation	5.86×10^{-3}	6.96×10^{-3}
GDM	Gradient Descent with Momentum backpropagation	6.26×10^{-3}	5.65×10^{-3}
OSS	One Step Secant	6.04×10^{-4}	5.54×10^{-3}
RP	Resilient backpropagation	2.65×10^{-4}	2.75×10^{-3}
CGF	Conjugate Gradient with Fletcher-Powell updates	1.82×10^{-3}	2.30×10^{-3}
SCG	Scaled Conjugate Gradient backpropagation	5.35×10^{-4}	2.05×10^{-3}
CGP	Conjugate Gradient backpropagation with Polak-Ribiére updates	1.29×10^{-3}	1.57×10^{-3}

In order to specify the optimum number of hidden neurons in the model a lengthy trial and error process was performed. Indicative results are shown in Table 9 regarding the corresponding investigation, for the MLP ANN trained using the LM algorithm, abbreviated as MLP-LM henceforth. The MSE_{tr}, the MSE_{te} and the relative MSE_{te}, denoted as $RMSE_{te}$ and defined as

$$\mathrm{RMSE}_{te}=\frac{1}{K}\sum_{k=1}^{K}\left(\frac{a_k^{pr}-a_k^{te}}{a_k^{te}}\right)^2 \tag{52}$$

are presented therein, for different values of M_2 and M_3. Table 9 indicates that MSE_{tr} varies by 20dB being dependent on the M_2 and M_3 combination used, while the MSE_{te} and the $RMSE_{te}$ variations are smaller by some margin. The investigation's final outcome, which is based on further results as well not shown for the purpose of brevity, is the deployment of M_2=50 and M_3=5 neurons in the first hidden layer and the second hidden layer of the model, respectively. Furthermore, a momentum term (Beale et al., 2016) equal to 0.9 was used with a decrease and increase factor equal to 0.8 and 1.5, respectively, in order to prevent the training algorithm from being trapped into local minima. However, it has been verified that this technique does not produce any significant improvement and thus the results shown in this work don't employ this particular technique.

To select the best suited activation function for either of the two hidden layers required extensive testing upon several functions of which indicative results are being presented in Table 10. The MSE_{tr} and the MSE_{te} are being affected significantly by the choice of each particular activation function. As is indicated by the results in Table 10, the best suited activation function for either hidden layer, first

Table 9. Error metrics for various values of M_2 and M_3 values (MLP-LM of N=6 and M=600)

M_2	M_3	**MSE_{tr}**	**MSE_{te}**	**$RMSE_{te}$**
15	10	2.12×10^{-5}	2.16×10^{-5}	7.09×10^{-4}
15	15	2.52×10^{-5}	6.01×10^{-5}	5.33×10^{-4}
25	5	1.99×10^{-5}	3.32×10^{-5}	1.13×10^{-3}
25	25	3.38×10^{-5}	4.49×10^{-5}	5.37×10^{-4}
30	15	2.68×10^{-5}	3.08×10^{-5}	4.68×10^{-4}
30	30	6.30×10^{-6}	4.81×10^{-5}	6.30×10^{-4}
40	15	2.98×10^{-5}	6.34×10^{-5}	6.04×10^{-4}
40	20	9.03×10^{-6}	1.79×10^{-5}	4.02×10^{-4}
50	5	1.25×10^{-5}	4.42×10^{-5}	1.72×10^{-3}
50	20	1.19×10^{-4}	1.50×10^{-4}	6.62×10^{-4}
60	5	2.01×10^{-5}	2.85×10^{-5}	5.60×10^{-4}
60	45	4.50×10^{-5}	1.35×10^{-4}	1.36×10^{-3}

Table 10. Error metrics for different activation functions (MLP-LM of M_2=50, M_3=5, N=6, M=600)

1st hidden layer	**2nd hidden layer**	**MSE_{tr}**	**MSE_{te}**
Hyperbolic tangent	Hyperbolic tangent	2.98×10^{-5}	3.14×10^{-4}
Logarithmic	Hyperbolic tangent	1.46×10^{-4}	5.53×10^{-4}
Logarithmic	Elliot	1.52×10^{-4}	3.16×10^{-4}
Elliot	Elliot	2.23×10^{-3}	5.74×10^{-3}
Elliot	Linear	2.14×10^{-3}	3.50×10^{-3}
Hyperbolic tangent	Logarithmic	2.76×10^{-3}	8.83×10^{-3}
Linear	Logarithmic	6.25×10^{-2}	6.64×10^{-2}
Linear	Elliot	7.64×10^{-2}	8.74×10^{-2}

and second, is the hyperbolic tangent function (Figure 4b), with the MSE_{tr}(MSE_{tr}) being of the order of 10^{-5} (10^{-4}) in this particular case. Should an inappropriate activation function be selected, the errors mentioned above may increase even up to 30dB.

The architecture of the RBF ANN employed to predict the radius of the loop consists of three layers, as shown in Figure 7. The required number of neurons of the hidden layer M_b is the result of the iterative procedure that was described in the previous section. By applying this procedure, it has been proven that M_b=67 hidden neurons are capable of achieving the predetermined MSE_{tr}, i.e., 10^{-6}.

The GR ANN implemented in order to solve the inverse problem is depicted in Figure 8. It comprises an input layer, a hidden layer and a crisp output layer. The latter performs the summations that appear in the numerator and the denominator of the right-hand side of Equation (42) and calculates the best mean square estimate of the loop radius by dividing the two sums. The appropriate value of the spread factor σ, in Equations (41)-(42), has been proven to be equal to 0.01 (Kapetanakis et al., 2018b).

Results and Discussion

For the evaluation of the performance of ANNs the $RMSE_{te}$ statistical measures were calculated, as given by Equation (52), and the testing mean absolute error MAE_{te}, given by Equation (49), provided that the radiation intensity U is replaced by the loop radius a. Since each input value of the electric field intensity corresponds to various several output values, i.e., loop radii, the inverse problem's solution is multivalued. In effect, any ANN model prediction requires the importation of inputs from two different observers at least. The selection of the location of the observers, outlined in Table 7, were made appropriately so as to avoid performance degradation and compensate for difficulties inflicted by multivalued data during training.

The effect caused by the observers' number N regarding the performance of the ANNs is investigated in Figure 17, by plotting the MAE_{te} (Figure 17a) and the $RMSE_{te}$ (Figure 17b) with respect to N.

Figure 17. (a) MAE_{te} and (b) $RMSE_{te}$ against N for MLP LM and MLP BR setting M at 600, and RBF and GR ANNs setting M at 3000

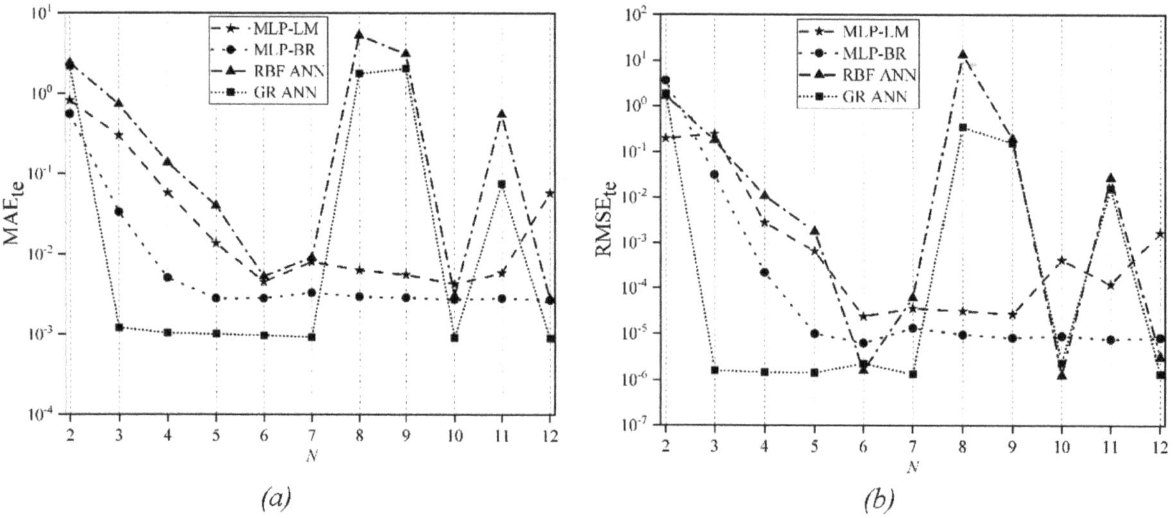

For reference, Table 7 lists the locations of all observers. The architectures of all three ANNs are being considered; training of the MLP was conducted by applying both the BR and LM training algorithms. The training data set size for either ANN used is displayed in the relevant caption. Figure 17 makes evident that the performance of the ANN does not necessarily improve when the number of observers *N* is being increased. This attributes to the fact that the radiation pattern of an antenna consists mainly of multiple lobes and zeroes that are not remaining steady in the initial direction when there is a change of the loop radius. Multiple numerical experiments, not shown in this work for brevity, indicate that it is the observers' location and not the number of observers that affects most significantly the performance of the ANNs.

Figure 17a indicates that the value of the MAE_{te}, for the MLP, reaches its minimum when *N*=6 observers are used. Similarly, the minimum MAE_{te} for RBF ANN is obtained when *N*=10, while the GR ANN's best performance is exhibited upon the deployment of twelve observers. It is quite remarkable, though, that regarding GR ANN, the best performance is almost achieved with as few as three observers and furthermore it does remain nearly constant for multiple *N* values, specifically *N*=3,4,5,6,7,10,12. Similarly, such a behavior is also observed in the case of the MLP-BR, where the required number of observers necessary to approximate closely the best possible performance is *N*=5. The results in Figure 17a verify that the addition of further observers does not necessarily result in a decrease in MAE_{te}. For instance, regarding the RBF ANN, MAE_{te} may increase by as much as 30dB should *N* be increased from six (6) to eight (8). What is further, should these observers be displaced from their actual locations, which are shown in Table 7, then the MAE_{te} appears to deteriorate. These corresponding results are being omitted for the sake of brevity.

It is made apparent from the $RMSE_{te}$ plots with respect to *N* (Figure 17b) that roughly the same trend is followed as with those in the case of MAE_{te}. What differentiates these plots, i.e. MAE_{te} and $RMSE_{te}$, is that MAE_{te} never falls below 10^{-3} whilst in the case of $RMSE_{te}$ the relevant limit is 10^{-6}. It is possible that these limits are due to the restrictions imported during the training stage of the ANN in order to achieve a three-digit accuracy at its output. Results in Figure 17b verify that six (6) observers are capable of producing sufficiently accurate results regardless of the model of the ANN. Yet, the RBF ANN offers slightly better accuracy in spite of the fact that it encompasses greater number of observers, specifically *N*=10. The results in Figure 17b also verify the nearly constant behavior of the GR ANN for various different *N* values.

Figure 18 investigates the dependence of the $RMSE_{te}$ with respect to the training dataset's size *M* for two different *N* values. Table 7 displays the observers' locations. A careful examination of the results in Figure 18 leads to the conclusion that *M* more or less affects the performance of all tested ANNs, for either of the *N* values examined. As a general rule it appears that the $RMSE_{te}$ decreases or it remains nearly constant when *M* is being increased, except in the cases of RBF ANN with *N*=6 (Figure 18a) and MLP-LM with *N*=10 (Figure 18b), for quite small *M* values. For instance, $RMSE_{te}$ appears to deteriorate by nearly one order of magnitude as *M* is being increased in value from six hundred (600) to a thousand (1000), for the RBF ANN with *N*=6 (Figure 18a).

It is also worth noting, regarding Figure 18, that the depicted variation that is undergone by $RMSE_{te}$ appears to be more abrupt for relatively small *M* values of up to a thousand (1000). This variation is pronounced significantly more in the case of RBF ANN with *N*=6 (Figure 18a). The $RMSE_{te}$ value for *M*=300 is greater by approximately 30 dB than the $RMSE_{te}$ obtained for *M*=600. Increasing the size of the training dataset, i.e., $M \geq 1000$, causes the $RMSE_{te}$ variation with respect to *M* to be quite smooth, with the exception of the RBF ANN with *N*=6 (Figure 18a). Figure 18a demonstrates that six (6) ob-

Figure 18 RMSE$_{te}$ for (a) N=6 and (b) N=10 versus M

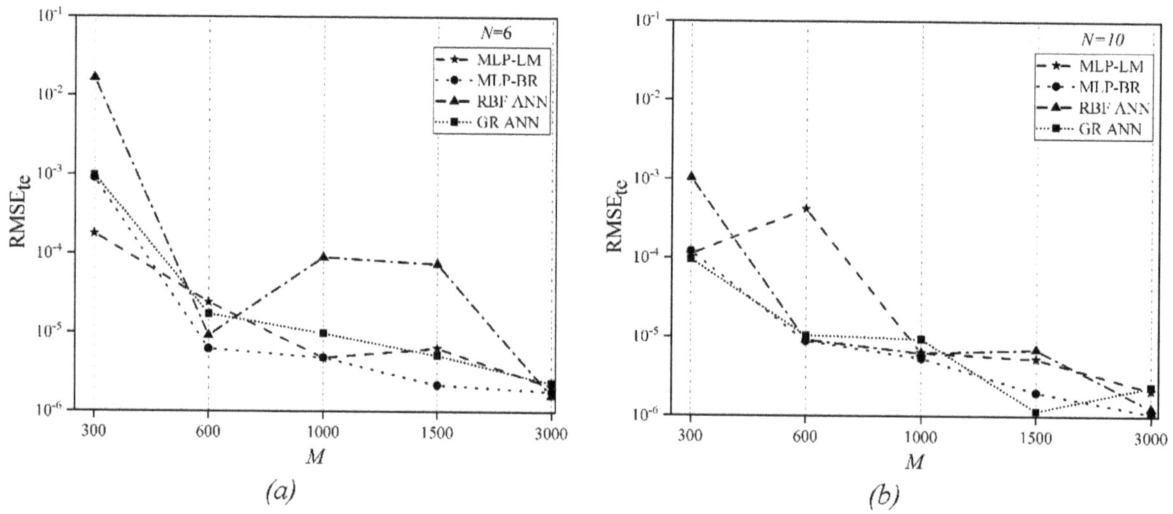

servers and a thousand (1000) training data could suffice in order to achieve RMSE$_{te}$ of less than 10^{-4}. The latter decreases by one order of magnitude, i.e. 10^{-5}, for all the examined ANNs with *N*=10 (Figure 18b). Further increasing of the size of the training dataset may not necessarily result in a significant enhancement of the ANNs' performance.

FUTURE RESEARCH DIRECTIONS

Artificial neural networks have proven to be excellent global approximators. On the short-hand, though, they "suffer" from weaknesses attributed to their extensive dimensionality. Neural networks are of parallel non-linear architectures with significant capacity requirements both in terms of processing speed and memory access times. This problem increases arithmetically with the amount of data to be processed and geometrically should the neural network need to comprise multiple layers and increased number of neurons per layer.

Heterogeneous parallel programming provides a modern state of the art approach in addressing processing times and data access latency issues. Every neuron in a single layer is being operated by a distinct thread organized in a linear block. Multiple blocks of threads cover the neurons' operating requirement of every single hidden layer in the neural network. Thread synchronization ensures that no neural outputs shall be transferred from one layer to another unless all neurons of the former layer have completed the processing of the information fed to them. In that manner, the processing time of the operation of all neurons in a single layer is brought down to the latency caused by the neuron that yielded results last.

Furthermore, heterogeneous parallel processing employees various GPU memories that are quick to channel large amounts of data to multiple threads supported not just by algorithms but by architectural design of the hardware as well. This is achieved via memory management and data transferring in between CPU cores and GPU compute unit back and forth depending on whether a process is serial or parallel, respectively.

Emerging from the advantages enabled by heterogeneous parallel programming, it is possible to devise for the applications described in this work, neural network architectures of increased number of inputs, multiple number of neurons per layers and several hidden layers. The latter leads to deep learning neural network architectures that excel in terms of classical artificial neural networks as they are capable of combining features' extraction and classification capabilities all at once, whilst in the case of plain ANNs the former required domain experts' intervening. In effect, instead of having to train computers how to process and learn from data, with deep learning the neural networks train themselves to process and learn from data. The end result is extensive deep learning neural networks with exceled processing time and memory management capabilities able to process big data both in terms of quantity as well as variability in the types of information fed to the network.

CONCLUSION

The investigation presented in this chapter showed that methods based on artificial intelligence have great potential as alternative techniques for modeling forward and, especially, inverse EM problems. Specifically, the radiation of a circular loop antenna in free space has been modeled successfully by using ANNs and ANFISs. The forward problem consists in determining the radiation intensity by specifying the observation angle and the loop radius, whereas the latter is the required result of the inverse problem; the EM field around the loop serves as input to the model in this case.

The numerical tests have revealed that the selection of the parameters of the models is critical. First and foremost, the data range regarding the observation angle had to be divided in subranges in order to avoid problems caused by multivalued data. Accordingly, one, three or six ANNs (or ANFISs) were implemented to model the forward problem, each one handling data in the corresponding subrange of the observation angle. Thus, the accuracy of the results has been improved significantly, the testing MRE being as low as 1.5-2%, for all six independent ANNs deployed for solving the problem.

Selecting the appropriate configuration of the ANN for solving the inverse loop antenna radiation problem turned out to be a rather complicated task, depending on several parameters such as data-sets' size, hidden neurons' number, the choice of the training algorithm for the MLP ANN and the number/location of the observers. As a general observation, the increase in the observers' number and/or in the training data-set's size does not always enhance the ANNs' performance, whereas a systematic selection of the locations of the observers is of the outmost importance in order to achieve optimum estimations. The testing RMSE may be of the order of 10^{-5}, and sometimes as low as 10^{-6}, if the parameters of the models are appropriately selected.

REFERENCES

Abramowitz, M., & Stegun, I. A. (1972). *Handbook of Mathematical Functions*. Dover.

Ahmadi, A., & Mosallaei, H. (2010). Plasmonic nanoloop array antenna. *Optics Letters*, *35*(21), 3706–3708. doi:10.1364/OL.35.003706 PMID:21042398

Anastassiu, H. T. (2006). Fast, simple and accurate computation of the currents on an arbitrarily large circular loop antenna. *IEEE Transactions on Antennas and Propagation*, *54*(3), 860–866. doi:10.1109/TAP.2006.869929

Balanis, C. A. (1982). *Antenna Theory, Analysis and Design*. John Wiley & Sons.

Beale, M., Hagan, M., & Demuth, H. (2016). *Neural Network Toolbox: User's Guide (version 9)*. The MathWorks Inc.

Choudhury, B., Thomas, S., & Jha, R. M. (2015). Implementation of soft computing optimization techniques in antenna engineering. *IEEE Antennas & Propagation Magazine*, *57*(6), 122–131. doi:10.1109/MAP.2015.2439612

Chowdhury, D. R., Singh, R., Reiten, M., Zhou, J., Taylor, A. J., & O'Hara, J. F. (2011). Tailored resonator coupling for modifying the terahertz metamaterial response. *Optics Express*, *19*(11), 10679–10685. doi:10.1364/OE.19.010679 PMID:21643323

Christodoulou, C., & Georgiopoulos, M. (2001). *Applications of Neural Networks in Electromagnetics*. Artech House.

Conway, J. T. (2005). New exact solution procedure for the near fields of the general thin circular loop antenna. *IEEE Transactions on Antennas and Propagation*, *53*(1), 509–517. doi:10.1109/TAP.2004.838804

Devabhaktuni, V., Yagoub, M. C. E., Fang, Y., Xu, J., & Zhang, Q. (2001). Neural networks for microwave modeling: Model development issues and nonlinear modeling techniques. *International Journal of RF and Microwave Computer-Aided Engineering*, *11*(1), 4–21. doi:10.1002/1099-047X(200101)11:1<4::AID-MMCE2>3.0.CO;2-I

Elshafiey, I., Udpa, L., & Udpa, S. S. (1995). Solution of inverse problems in electromagnetics using Hopfield neural networks. *IEEE Transactions on Magnetics*, *31*(1), 852–861. doi:10.1109/20.364586

Fakharzadeh, M., & Mohajer, M. (2014). An Integrated Wide-Band Circularly Polarized Antenna for Millimeter-Wave Applications. *IEEE Transactions on Antennas and Propagation*, *62*(2), 925–929. doi:10.1109/TAP.2013.2290038

Fikioris, G., Papakanellos, P. J., & Anastassiu, H. T. (2008). On the Use of Nonsingular Kernels in Certain Integral equations for Thin-Wire Circular-Loop Antennas. *IEEE Transactions on Antennas and Propagation*, *56*(1), 151–157. doi:10.1109/TAP.2007.913076

Fikioris, G., Papakanellos, P. J., & Anastassiu, H. T. (2010). Corrections to "On the Use of Nonsingular Kernels in Certain Integral equations for Thin-Wire Circular-Loop Antennas". *IEEE Transactions on Antennas and Propagation*, *58*(10), 3436. doi:10.1109/TAP.2010.2055816

Gehani, A., & Pujara, D. A. (2015). Predicting the return loss performance of a hexa-band PIFA using ANFIS. *Microwave and Optical Technology Letters*, *57*(9), 2072–2075. doi:10.1002/mop.29277

Gosal, G., Almajali, E., McNamara, D., & Yagoub, M. (2016). Transmit array antenna design using forward and inverse neural network modeling. *IEEE Antennas and Wireless Propagation Letters*, *15*, 1483–1486. doi:10.1109/LAWP.2015.2514065

Gradshteyn, I. S., & Ryzhik, I. M. (1980). *Table of Integrals, Series, and Products*. Academic Press.

Hallen, E. (1938). Theoretical investigations into transmitting and receiving qualities of antennae. *Nova Acta Upsaliensis series IV, 11*, 1–43.

Hamed, S. M. A. (2013). Exact field expressions for circular loop antennas using spherical functions expansion. *IEEE Transactions on Antennas and Propagation, 61*(6), 2956–2963. doi:10.1109/TAP.2013.2250241

Hartman, E., Keeler, J., & Kowalski, J. (1990). Layered neural networks with gaussian hidden units as universal approximations. *Neural Computation, 2*(2), 210–215. doi:10.1162/neco.1990.2.2.210

Haykin, S. (1999). *Neural Networks: A Comprehensive Foundation*. Pearson Education.

Hornik, K., Stinchcombe, M., & White, H. (1989). Multilayer Feedforward Networks are Universal Approximators. *Neural Networks, 2*(5), 359–366. doi:10.1016/0893-6080(89)90020-8

Jang, J.-S. R. (1993). ANFIS: Adaptive-Network-Based Fuzzy Inference Systems. *IEEE Transactions on Systems, Man, and Cybernetics, 23*(3), 665–685. doi:10.1109/21.256541

Kabir, H., Wang, Y., Yu, M., & Zhang, Q. J. (2008). Neural network inverse modeling and applications to microwave filter design. *IEEE Transactions on Microwave Theory and Techniques, 56*(4), 2811–2813. doi:10.1109/TMTT.2008.919078

Kapetanakis, T. N., Vardiambasis, I. O., Ioannidou, M. P., & Maras, A. (2018b). Neural network modeling for the solution of the inverse loop antenna radiation problem. *IEEE Transactions on Antennas and Propagation, 66*(11), 6283–6290. doi:10.1109/TAP.2018.2869136

Kapetanakis, T. N., Vardiambasis, I. O., Liodakis, G., & Maras, A. (2012a). Neural network solution of the circular loop antenna radiation problem. In *Proceedings of the 20th Telecommunications Forum* (pp. 1193-1196). 10.1109/TELFOR.2012.6419428

Kapetanakis, T. N., Vardiambasis, I. O., Liodakis, G., & Maras, A. (2012b). Solving the inverse loop antenna radiation problem using a hybrid neuro-fuzzy system. In *Proceedings of the 20th Telecommunications Forum* (pp. 1189-1192). 10.1109/TELFOR.2012.6419427

Kapetanakis, T. N., Vardiambasis, I. O., Lourakis, E. I., & Maras, A. (2018a). Applying neuro-fuzzy soft computing techniques to the circular loop antenna radiation problem. *IEEE Antennas and Wireless Propagation Letters, 17*(9), 1673–1676. doi:10.1109/LAWP.2018.2862939

Kayabasi, A., Toktas, A., Akdagli, A., Bicer, M. B., & Ustun, D. (2014). Applications of ANN and ANFIS to predict the resonant frequency of L-shaped compact microstrip antennas. *Applied Computational Electromagnetics Society Journal, 29*, 460–469.

King, R. W. P. (1969). The loop antenna for transmission and reception. In R. E. Collin & F. J. Zucker (Eds.), *Antenna Theory* (pp. 458–481). McGraw-Hill.

Li, L. W., Leong, M. S., Kooi, P. S., & Yeo, T. S. (1997). Exact solutions of electromagnetic fields in both near and far zones radiated by thin circular-loop antennas: A general representation. *IEEE Transactions on Antennas and Propagation, 45*(12), 1741–1748. doi:10.1109/8.650191

Low, T. S., & Chao, B. (1992). The use of finite elements and neural networks for the solution of inverse electromagnetic problems. *IEEE Transactions on Magnetics*, *28*(5), 2811–2813. doi:10.1109/20.179635

Lu, B. Q., Nagar, J., Yue, T., Pantoja, M. F., & Werner, D. H. (2017). Closed-form expressions for the radiation properties of nanoloops in the Terahertz, infrared and optical regimes. *IEEE Transactions on Antennas and Propagation*, *65*(1), 121–133. doi:10.1109/TAP.2016.2624150

McKinley, A. (2019). *The Analytical Foundations of Loop Antennas and Nano-scale Rings*. Springer. doi:10.1007/978-981-13-5893-7

McKinley, A. F., White, T. P., & Catchpole, K. R. (2013a). Theory of the circular closed loop antenna in the terahertz, infrared, and optical regions. *Journal of Applied Physics*, *114*(4), 044317-1, 044317–10. doi:10.1063/1.4816619

McKinley, A. F., White, T. P., & Catchpole, K. R. (2013b). Designing nanoloop antenna arrays for light-trapping in solar cells. *Proceedings of Photovolt. Specialists Conference (PVSC)*, 1894-1896.

Memarzadeh, B., & Mosallaei, H. (2011). Array of planar plasmonic scatterers functioning as light concentrator. *Optics Letters*, *36*(13), 2569–2571. doi:10.1364/OL.36.002569 PMID:21725482

Mishra, R. K. (2001). An overview of neural network methods in computational electromagnetics. *International Journal of RF and Microwave Computer-Aided Engineering*, *12*(1), 98–108. doi:10.1002/mmce.10009

Mishra, S., Yadav, R. N., & Singh, R. P. (2015). Directivity estimations for short dipole antenna arrays using radial basis function neural networks. *IEEE Antennas and Wireless Propagation Letters*, *14*, 1219–1222. doi:10.1109/LAWP.2015.2399453

Morse, P. M., & Feshbach, H. (1953). *Methods of Theoretical Physics, Part II*. McGraw-Hill.

Nagar, J., Lu, B. Q., Pantoja, M. F., & Werner, D. H. (2017). Analytical expressions for the mutual coupling of loop antennas valid from the RF to optical regimes. *IEEE Transactions on Antennas and Propagation*, *65*(12), 6889–6903. doi:10.1109/TAP.2017.2754411

Overfelt, P. L. (1996). Near fields of the constant current thin circular loop antenna of arbitrary radius. *IEEE Transactions on Antennas and Propagation*, *44*(2), 166–171. doi:10.1109/8.481643

Pantoja, M. F., Nagar, J., Lu, B., & Werner, D. H. (2017). Existence of superdirective radiation models in thin-wire nanoloops. *ACS Photonics*, *4*(3), 509–516. doi:10.1021/acsphotonics.6b00486

Park, J., & Sandberg, I. (1991). Universal approximation using radial-basis-function networks. *Neural Computation*, *3*(2), 246–257. doi:10.1162/neco.1991.3.2.246 PMID:31167308

Pendry, J. B., Holden, A. J., Robbins, D. J., & Stewart, W. J. (1999). Megnetism from conductors and enhanced nonlinear phenomena. *IEEE Transactions on Microwave Theory and Techniques*, *47*(11), 2075–2084. doi:10.1109/22.798002

Pocklington, H. C. (1897). Electrical oscillations in wire. *Cambridge Philosophical Society Proceedings*, *9*, 324.

Powell, M. (1987). Radial basis functions for multivariable interpolation: a review. In J. Mason & M. Cox (Eds.), *Algorithms for Approximation* (pp. 143–167). Clarendon Press.

Pujara, D., Modi, A., Pisharody, N., & Mehta, J. (2014). Predicting the performance of pyramidal and corrugated horn antennas using ANFIS. *IEEE Antennas and Wireless Propagation Letters*, *13*, 293–296. doi:10.1109/LAWP.2014.2305518

Shelby, R. A., Smith, D. R., Nemat-Nasser, S. C., & Schultz, S. (2001). Microwave transmission through a two-dimensional, isotropic, left-handed metamaterial. *Applied Physics Letters*, *78*(4), 489–491. doi:10.1063/1.1343489

Simovski, C., Morits, D., Voroshilov, P., Guahva, M., Belov, P., & Kivshar, Y. (2013). Enhanced efficiency of light-trapping nanoantenna arrays for thin-film solar cells. *Optics Express*, *21*(S4), A714–A725. doi:10.1364/OE.21.00A714 PMID:24104498

Smith, D. R., Padilla, W. J., Vier, D. C., Nemat-Nasser, S. C., & Schultz, S. (2000). Composite medium with simultaneous negative permeability and permittivity. *Physical Review Letters*, *84*(18), 4184–4187. doi:10.1103/PhysRevLett.84.4184 PMID:10990641

Smith, G. S. (2007). Loop antennas. In R. C. Johnson & H. Jasik (Eds.), *Antenna Engineering Handbook* (pp. 5-1–5-24). McGraw-Hill.

Specht, D. F. (1991). A general regression neural network. *IEEE Transactions on Neural Networks*, *2*(6), 568–576. doi:10.1109/72.97934 PMID:18282872

Storer, J. E. (1956). Impedance of thin-wire loop antennas. *AIEE Transactions. Part I Communications and Electronics*, *9*, 606–619.

Werner, D. H. (1996). An exact integration procedure for vector potentials of thin circular loop antennas. *IEEE Transactions on Antennas and Propagation*, *44*(2), 157–165. doi:10.1109/8.481642

Werner, D. H. (2000). Near-field and far-field expansions for traveling-wave circular loop antennas. *Progress in Electromagnetic Research*, *28*, 29–42. doi:10.2528/PIER99100103

Wu, T. T. (1962). Theory of the thin circular loop antenna. *Journal of Mathematical Physics*, *3*(6), 1301–1304. doi:10.1063/1.1703875

Zhang, Q., Gupta, K. C., & Devabhaktuni, V. K. (2003). Artificial neural networks for RF and microwave design: From theory to practice. *IEEE Transactions on Microwave Theory and Techniques*, *51*(4), 339–1350.

Zhang, Q. J., & Gupta, K. C. (2000). *Neural Networks for RF and Microwave Design*. Artech House.

ADDITIONAL READING

Adamidis, G. A., Vardiambasis, I. O., Ioannidou, M. P., & Kapetanakis, T. N. (2019). Design and implementation of single-layer 4×4 and 8×8 Butler matrices for multibeam antenna arrays. *International Journal of Antennas and Propagation*, *2019*, 1–12. doi:10.1155/2019/1645281

Adamidis, G. A., Vardiambasis, I. O., Ioannidou, M. P., & Kapetanakis, T. N. (2020). Design and implementation of an adaptive beamformer for phased array antenna applications. *Microwave and Optical Technology Letters*, *62*(4), 1–5. doi:10.1002/mop.32231

Adekola, S. A. (1983). On the excitation of a circular loop antenna by traveling- and standing-wave current distributions. *International Journal of Electronics*, *54*(6), 705–732. doi:10.1080/00207218308938771

Altshuler, E. E. (1996). A Monopole Loaded with a Loop Antenna. *IEEE Transactions on Antennas and Propagation*, *44*(6), 787–791. doi:10.1109/8.509881

Fikioris, G. (2006). Integral Evaluation Using the Mellin Transform and Generalized Hypergeometric Functions: Tutorial and Applications to Antenna Problems. *IEEE Transactions on Antennas and Propagation*, *54*(12), 3895–3907. doi:10.1109/TAP.2006.886579

Kapetanakis, T. N., & Vardiambasis, I. O. (2016). Radiation performance of satellite reflector antennas using neural networks. *Proceedings of the 3rd International Conference on Mathematics and Computers in Sciences and in Industry (MCSI 2016)*, 85-88. 10.1109/MCSI.2016.026

Kraus, J., & Marhefka, R. (2002). *Antennas for All Applications*. McGraw-Hill.

Liodakis, G., Arvanitis, D., & Vardiambasis, I. O. (2004). Neural network – based digital receiver for radio communications. *WSEAS Transactions on Systems*, *3*(10), 3308–3313.

Prasad, S. M., & Das, B. N. (1970). A circular loop antenna with traveling-wave current distribution. *IEEE Transactions on Antennas and Propagation*, *18*(2), 278–280. doi:10.1109/TAP.1970.1139673

Rayas-Sanchez, J. E. (2004). EM-based optimization of microwave circuits using artificial neural networks: The state of the art. *IEEE Transactions on Microwave Theory and Techniques*, *52*(1), 420–435. doi:10.1109/TMTT.2003.820897

Sergaki, E., Spiliotis, G., Vardiambasis, I. O., Kapetanakis, T., Krasoudakis, A., Giakos, G. C., Zervakis, M., & Polydorou, A. (2018). Application of ANN and ANFIS for detection of brain tumors in MRIs by using DWT and GLCM texture analysis. *2018 IEEE International Conference on Imaging Systems and Techniques (IST)*, 1-6. 10.1109/IST.2018.8577099

Tsalamengas, J. L., Vardiambasis, I. O., & Fikioris, J. G. (1993). TE and TM modes in circularly shielded slot waveguides. *IEEE Transactions on Microwave Theory and Techniques*, *41*(6/7), 966–973. doi:10.1109/22.238510

Vardiambasis, I. O., Tsalamengas, J. L., & Fikioris, J. G. (1995). Hybrid wave propagation in circularly shielded microslot lines. *IEEE Transactions on Microwave Theory and Techniques*, *43*(8), 1960–1966. doi:10.1109/22.402286

Vardiambasis, I. O., Tsalamengas, J. L., & Fikioris, J. G. (1997). Hybrid wave propagation in generalized Goubau-type striplines. *IEE Proceedings. Microwaves, Antennas and Propagation*, *144*(3), 167–171. doi:10.1049/ip-map:19970792

Vardiambasis, I. O., Tsalamengas, J. L., & Fikioris, J. G. (1998). Plane wave scattering by slots on a ground plane loaded with semi-circular dielectric cylinders in case of oblique incidence and arbitrary polarization. *IEEE Transactions on Antennas and Propagation*, *46*(10), 1571–1579. doi:10.1109/8.725291

Volakis, J. (2007). *Antenna Engineering Handbook*. McGraw-Hill.

KEY TERMS AND DEFINITIONS

AE: Absolute error.

ANFIS: Adaptive neuro fuzzy inference system.

ANN: Artificial neural network.

Artificial Intelligence: Machines or techniques that attempt to imitate human cognitive skills.

Gaussian Function: A function that has a characteristic, symmetric graph which resembles the shape of a bell.

GR: Generalized regression.

MAE: Mean absolute error.

Metamaterial: Manmade material with properties that cannot be found in materials encountered in nature, for example material with negative refractive index.

MLP: Multilayer perceptron.

MRE: Mean relative error.

MSE: Mean square error.

Nanotechnology: Techniques, methods and processes that manipulate materials with dimensions of the order of nanometers.

Radiation Intensity: The power radiated from an antenna per unit solid angle, in the far-field.

RBF: Radial basis function.

RMSE: Root mean square error.

Chapter 8
Electromagnetic Transients: EMC Testing and Mitigation Methodologies – Recent Trends

Eleni P. Nicolopoulou
High Voltage Laboratory, National Technical University of Athens, Greece

Panagiotis K. Papastamatis
High Voltage Laboratory, National Technical University of Athens, Greece

Ioannis F. Gonos
High Voltage Laboratory, National Technical University of Athens, Greece

ABSTRACT

This chapter aims at presenting the current international standards and the recent bibliography regarding the transient phenomena associated with space systems, with emphasis on the EMC requirements. The first section of the chapter focuses on the description of the physical mechanisms causing transient electromagnetic phenomena in space. The second section reviews the procedures proposed for testing the immunity of space equipment against these transients, as described in the current space standards and in recent bibliography as non-standardized testing procedures. Finally, the last section investigates possible mitigation approaches and design guidelines against these electromagnetic disturbances, essential for every space system designer.

INTRODUCTION

Spacecraft in synchronous or lower orbits often experience charge build-up on exposed and internal material as a result of geomagnetic storms. The resulting large potential differences can cause incidents of insulation breakdown and electrostatic discharge reaching peak values of several kilovolts. The induced electric current is, in turn, coupled to the terminals of electronic devices, disrupting sensitive digital and analog electronics with possible catastrophic consequences and potential system failure. In the past few decades, various ESD mitigation methods (selection of specific surface materials, grounding techniques,

DOI: 10.4018/978-1-7998-4879-0.ch008

etc.) are being applied by aerospace engineers to modern spacecraft. Nevertheless, equipment-level immunity testing to ESD remains the most reliable method of preventing system failure. International standardization provides test specifications that simulate the flight The ESD phenomena taking place during a flight are reproduced as precisely as possible within the test procedures specified in the international standards.

Apart from the above ESD testing methods, later in this chapter, power bus transients caused by switching loads on and off on the bus will be analyzed. The proposed measurements and immunity test methods of the associated standards are presented and commented on. Some of the Standards presented in this chapter are the Military standard MIL-STD-461 and its latest version (G), the ISO 14302 standard, which specifies the Electromagnetic Compatibility requirements for Space Systems, and the ECSS-E-ST-20-07C European Standard on Electromagnetic Compatibility for space technology of the European Cooperation for Space Standardization. In summary, this chapter aims to present the types and sources of transient electromagnetic phenomena in the space environment, to review and comment on the recent standardization efforts of the various associated bodies, and to propose sufficient protective measures for a robust and electromagnetically compatible space system design.

More specifically, in the "BACKGROUND" section the electromagnetic phenomena expected to occur in the space environment and affect the aerospace systems, as well as their causes are described. The "INTERNATIONAL STANDARDS: TESTING AND MEASUREMENT PROCEDURES" section presents earlier and current testing methods according to different International Standards trying to establish the most accurate simulation and testing of the aforementioned phenomena. Differentiation points as well as the authors' comments on the preferable methods are presented herein. Finally, the "SOLUTIONS AND RECOMMENDATIONS" section includes design guidelines and mitigation techniques against those types of electromagnetic disturbances crucial during the design process of every aerospace system.

BACKGROUND

Charging Effects in Space

The initiation of charging effects in space applications is attributed to the in-flight interaction between the external space plasma environment and the materials and electronic systems in the interior of the spacecraft. The Sun usually acts as the primary source of charge in the form of solar wind, magnetic storms, and ion clouds. During these physical processes the various regions of a spacecraft with different materials and configurations are actually bombarded by space particles and therefore act as a set of electrically floating capacitors that are charged at different rates and different potentials with respect to space plasma potential. The development of these surface potentials is governed by the physical concept of reaching current equilibrium, according to which, on a balance state, the sum of all currents adds up to zero.

Ambient space plasma is a partially ionized gas made up of electrons, protons, and other ions of variable energies. Protons are usually ignored when dealing with spacecraft charging because they are not able to penetrate matter due to their greater mass. On the contrary, electrons and ions penetrate the matter to a certain depth, depending on their energy content and composition of the target material, which leads to the following categories of charging mechanisms and the resulting discharges:

1. **Surface charging:** It is caused by the deposition of charge on the outer surface of the spacecraft, and particularly by lower energy charges that penetrate just below the surface of the exposed structures. The resulting surface discharges affect mainly exposed components, such as external cabling and antenna feeds and to a much smaller extent electrical systems inside the spacecraft, as their transmitted energy is attenuated through the various coupling mechanisms. This deposition of surface charge can lead to unstable conditions and ultimately to spacecraft-to-space breakdowns which are generally similar to dielectric surface failures, but involve only minor discharges in which charge can be lost from the spacecraft into space.
2. **Internal charging:** it is caused by invading energetic particles that accumulate either on dielectric surfaces or on ungrounded internal conductors. This charging process expands over a longer time scale compared to surface charging. The terms "buried charging" or "deep dielectric charging" are also used to describe the process, although the use of the word dielectric can be misleading as ungrounded inner conductors can discharge at a higher peak current and rate of change of current than a dielectric.

In both charging procedures, an intensification of the electric field is observed -due to the high levels of charge concentration and local geometry factors such as the arrangement and thickness of the material- and discharge is triggered when the electric field exceeds the dielectric strength of the material or when the potential difference between adjacent surfaces exceeds a critical value. On a surface discharge event, the ejection of the accumulated charge into space, causes an electrical field pulse in the order of tens of kilovolts per meter in the immediate vicinity of the location. As a result, a replacement current pulse is induced in the spacecraft structure and electronic systems causing transient voltages and currents coupled through wires on the device ports.

Local breakdowns in dielectrics occur when the internal electric field exceeds $2 \cdot 10^7$V/m (Garrett & Whittlesey, 2011) and may severely damage equipment as they are accompanied by rapidly expanding ionization channels that allow electrical conduction and heat. Apart from these direct effects, some common indirect effects are transient currents induced by electromagnetic coupling, permanent deterioration in material properties and leakage currents (European Cooperation for Space Standardization [ECSS], 2008). Transient electrostatic discharge (ESD) effects arising from spacecraft charging are responsible for partial or even total malfunction of electrical and electronic subsystems (such as power, navigation, communications, or instrumentation). In addition, charge concentration increases the content of contaminants thus degrading the thermal properties of surface materials and the performance of optical instruments and solar systems. Contaminants can also distort plasma meters as they can alter particle trajectories (Garrett & Whittlesey, 2011).

Orbits

The extent of exposure to the above-mentioned space charge phenomena, their severity and the extent of their impact strongly depend on the orbit of the spacecraft. Knowledge of the environment of an orbit in terms of electron charging is necessary in order to estimate the relevant ESD threat. Satellites move around the earth in orbits categorized in: medium earth orbits (MEO), low earth orbits (LEO), geosynchronous earth orbits (GEO), and polar earth orbits (PEO). Potentials connected to charging phenomena at altitudes above 400 km may even reach 400-500 V, values that can cause discharges. The surface charge is negative in most cases due to the higher relative velocity of the electrons. However, in the absence of intense magnetic storms, the exposure of the spacecraft to sunlight results in the photoemission current which can easily charge the spacecraft positively (Lai, 2011).

Geosynchronous orbits are of great importance for surface charging as most of the telecommunications satellites are located in this region. In GEO orbits spacecrafts encounter electron particles with energies in the range from 2 eV to 2 keV (ECSS, 2008). Considering that ions and electrons with high energy enter these regions around midnight, the surface charge levels in the GEO latitudes reaching hundreds of volts or even several kV are mainly developed when the spacecraft is not sunlit. As expected, spacecrafts undergo much higher states of charge (typically hundreds of kV) at auroral latitudes (60 ° to 70 °), occurring even at lower altitudes. Deep dielectric charging is enhanced through events such as solar coronal mass ejection or the passage of the spacecraft from the so-called "radiation belts" regions where ions and electrons of very high energy content are predominant (Lai, 2011).

The various loading mechanisms described above, as well as the properties of the selected spacecraft's orbit, are the main parameters for identifying and characterizing the resulting current waveform, presented in the next section and used as a reference for the associated immunity test procedures.

Conventional ESD Transient Current Characterization

The amplitude and duration of the resulting pulse depend on the amount of the charge deposited. In addition, charge accumulation affects primarily the developed charge flow, but also the remaining charge, which is greatly depends on material conductivity and grounding schemes. Materials with resistivity in the order of 10^{12} Ω·cm, when grounded, effectively drive away currents induced in GEO locations that are expected to be on the order of 10^{12} A. Dielectric materials widely applied in contemporary spacecrafts such as Teflon® and Kapton®, flame retardant 4 (FR4) circuit boards and conformal coatings, often have resistivities that can cause problems. In the meantime, the conductive nature of metallic surfaces alone is no guarantee of protection as the electrical isolation should not exceed 10^{12} Ω (Garrett & Whittlesey, 2011).

Based on ESD recordings from telemetry measurements, it is concluded that the waveform of a transient current depends on the location of the ESD event and the properties of the ESD pulse are determined by the following parameters: the rise time by the rate of expansion of the plasma cloud, the pulse width and the peak current by the ESD initiation point and the size of the affected area, while the fall time at the end of the pulse by the remaining charged area at the end of the phenomenon. Typical values of the ESD waveform parameters for various arc initiation points, calculated from incidents mainly on the Galileo and Voyager satellites, are listed in Table 1 (National Aeronautics and Space Administration [NASA], 2013b):

Additional material on the presented spacecraft charging phenomena and the respective mitigation techniques can be found in (Dunn, 2016), (Pisacane, 2016) and (NASA, 2007). Apart from the environmentally induced transient phenomena, such as the electrostatic discharges, the equipment installed in a spacecraft is also exposed to system level transient disturbances, referred to as power bus switching transients, analyzed below.

Power Bus Load-Induced Switching Transients

System-level switching transients have always been an issue requiring special treatment by EMC engineers, especially in the past few decades where the electronic systems with constantly reduced size are

Table 1. Typical values of the ESD waveform parameters for various arc initiation points as calculated from incidents mainly on the Galileo and Voyager satellites (NASA, 2013b)

ESD generation point	Capacitance, C [nF]	Breakdown voltage, V_B [kV]	Energy, E [mJ]	Peak current, I_{PK} [A]	Discharge current rise time, t_r [ns]	Discharge current pulse width, t_D [ns]
Dielectric plate to conductive substrate	20	1	10	2	3	10
Exposed connector dielectric	0.150	5	1.9	36	10	15
Paint on high-gain antenna	550	1	150	150	5	2400
Conversion coating on metal plate	4.5	1	2.25	16	20	285
Paint on optics hood	550	0.360	35000	18	5	600

being co-installed and co-operated with multiple other systems of high energy consumption that lead to the generation of switching transient sources. Sensitive electronics supplied by the same source might be affected by these disturbances, which can lead to disturbing changes in status, resets, incorrect readings or even a complete system failure. Standardization organizations are constantly working on identifying these phenomena, proposing measurement and test methods, and thus ensuring the EMC robustness of electrical and electronic systems.

The generation mechanism of switching transients is based on the turning on and off a heavy load connected to the power bus, or on a possible short-circuit that occurs before the central protection units can react. These transients are divided into two categories: 1) the turn-on or negative-going transients and 2) the turn-off or positive-going transients.

Inrush turn-on transients occur with the EUT in operation when a switch somewhere on the power bus is closed and current is attempting to flow through the load. This change in current is counteracted by the LISN inductance through a voltage drop until the current is balanced. The transition time characteristic depends on the LISN inductance and the RC constant of the load.

Shutdown turn-off transients are a product of the response of the 50μH LISN inductance to switching off a heavy load somewhere on the power bus. This inductance causes an increase of the output voltage in comparison to the input voltage. This voltage spike is limited by the line-to-line capacitance and stray capacitance, while the time parameters of the developed transient are not affected by the impedance of the switched off load (NASA, 1999).

For current-limited buses or fuse-protected loads, in addition to these types of transients that need to be measured and controlled, there are requirements regarding the maximum inrush current observed when the EUT is switched on.

Space agencies and standardization organizations have made great efforts in the past decades to establish robust and reproducible test methods, either for measuring these transients or for testing immunity to these phenomena.

INTERNATIONAL STANDARDS: TESTING AND MEASUREMENT PROCEDURES

Aerospace systems and especially systems that are designed to be installed in unmanned vehicles have always suffered from the difficulty of "ad hoc" maintenance and troubleshooting, while the associated failure events are dramatically more difficult to handle in comparison to ground systems. Therefore, pre-flight testing and compliance with the international emissions and immunity standards is broadly considered the safest and the most cost-efficient precaution available. Space systems, in particular, operating in an electromagnetic environment which is still relatively unpredictable (magnetic storms, charging effects, different orbits etc.) require additional consideration regarding their electromagnetic compatibility robustness.

Electrostatic Discharges

As mentioned earlier, the most reliable mitigation measure against ESD events is pre-flight compliance testing, which allows designers and manufacturers to evaluate in practice the performance of their selected protective measures, some of which are described in a later section. ESD tests in space applications are divided into two main categories based on the source of the discharges:

1. Immunity testing against personnel-borne electrostatic discharges. The simulation of the discharge source is based on the capacitive charging mechanism of the human body and the immunity requirement applies to electrical / electronic devices and systems where human-machine interaction takes place.
2. Immunity testing against machine-induced electrostatic discharges, simulating flight ESD events on an unmanned spacecraft.

Compliance to EMC specifications and the subsequent final installation in a spacecraft system requires execution of both of the above-mentioned immunity testing procedures.

Personnel-Borne Electrostatic Discharges

The standing human body, isolated from the floor by shoes and / or floor coverings, can be viewed as an insulated conductor with a known capacitance value called "Human Body Capacitance" (HBC), measured experimentally in the range of 100-150 pF. The charging process of the HBC, based on friction mechanism, is particularly enhanced by the combination of synthetic fabrics and a dry atmosphere.

Immunity testing against this phenomenon for space applications is described in military standard MIL-Std-461 (Department of Defense: United States of America, 2015) and specifically in section CS118. This method dealing with the effects of human ESD on electrical and electronic equipment, which is a recent addition in the latest version (G) of the standard, is based entirely on the IEC standard 61000-4-2 (International Electrotechnical Commission [IEC]), 2009), the test reference standard for ESD immunity testing used in most EMC testing and certification, and required by most Product Family Standards.

MIL-STD-461G Section CS118

As stated in the above standard, the device under test (EUT) must not exhibit any malfunction or deterioration in performance that is below a specific performance limit or tolerance values specified by the manufacturer. The EUT is subjected to the appropriate test level listed in Table 2 (Department of Defense: United States of America, 2015) which is formed as the discharge of a 150pF capacitor through a 330 Ω resistor with a circuit inductance no greater than 5 μH. Contact discharges are only applied to conductive surfaces, while air discharges are applied to surfaces where contact discharge testing is not possible. This is the first point where MIL-STD-461G and IEC 61000-4-2 disagree. A comparison of the test levels of Table 2 with the test levels of IEC 61000-4-2 shown in Table 3 (IEC, 2009), highlights the omission of contact discharge test levels below 8kV in the military standard thus allowing susceptibility of the tested devices to ESD events of lower level.

Table 2. ESD Test Level according to MIL-STD-461G - CS118

Level	Test Voltage	Method
1	±2kV	Air
2	±4kV	Air
3	±8kV	Contact/Air
4	±15kV	Air

The waveform of the resulting current is presented in Figure 1 (Department of Defense: United States of America, 2015), while in Table 4 and Figure 2 the current parameters for the ±8kV test level and a simplified diagram of an ESD generator are presented accordingly.

Table 3. ESD Test Level according to IEC 61000-4-2

Level	Contact Discharge	Air Discharge
1	±2kV	±2kV
2	±4kV	±4kV
3	±6kV	±8kV
4	±8kV	±15kV

The general test set-up, as presented in Figure 3, shows significant deviations from the IEC 61000-4-2 test up of Figure 4, such as the LISN requirement, the omission of indirect discharges to the horizontal and vertical coupling planes and the strict requirements regarding EUT cabling.

The selection of test points is based on the accessibility of conductive / non-conductive points or surfaces to the operator and personnel. Points of increased vulnerability include: metal screws, seams and openings, LCD displays, connector shells, and exposed connector pins. For each selected test point, 5 positive and 5 negative discharges must be applied according to MIL-STD-461G. For air discharge

Figure 1. Ideal contact discharge current waveform at 8 kV (Department of Defense: United States of America, 2015)

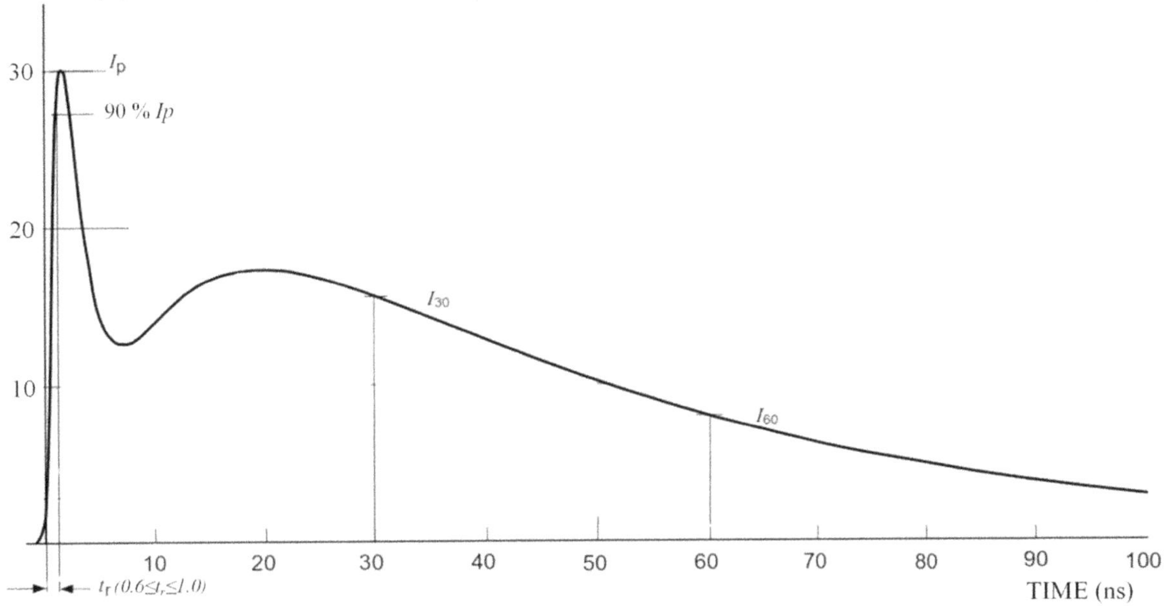

Figure 2. Simplified ESD generator diagram (IEC, 2009)

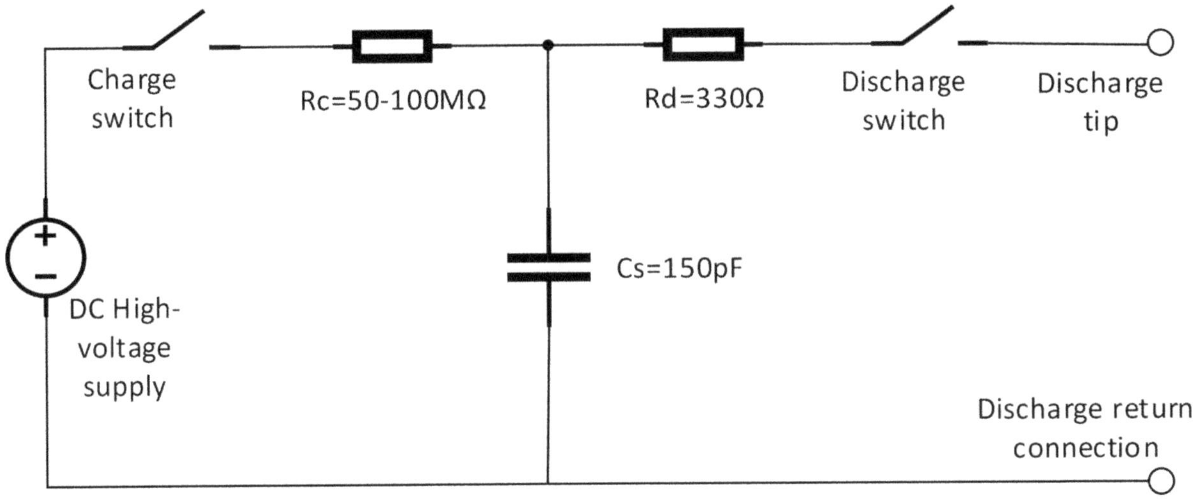

Rc: Charging resistor
Rd: Discharge resistor
Cs: Energy storage capacitor

Table 4. Contact discharge current parameters for 8kV test level (Department of Defense: United States of America, 2015)

Displayed Voltage(kV)	First Current Peak, ±15% (A)	Rise time (ns)	Current at t=30ns, ±30% (A)	Current at t=60ns, ±30% (A)
±8	30	0.6£t_r£1.0	16	8

Figure 3.MIL-STD-461G ESD Test set-up (Department of Defense: United States of America, 2015)

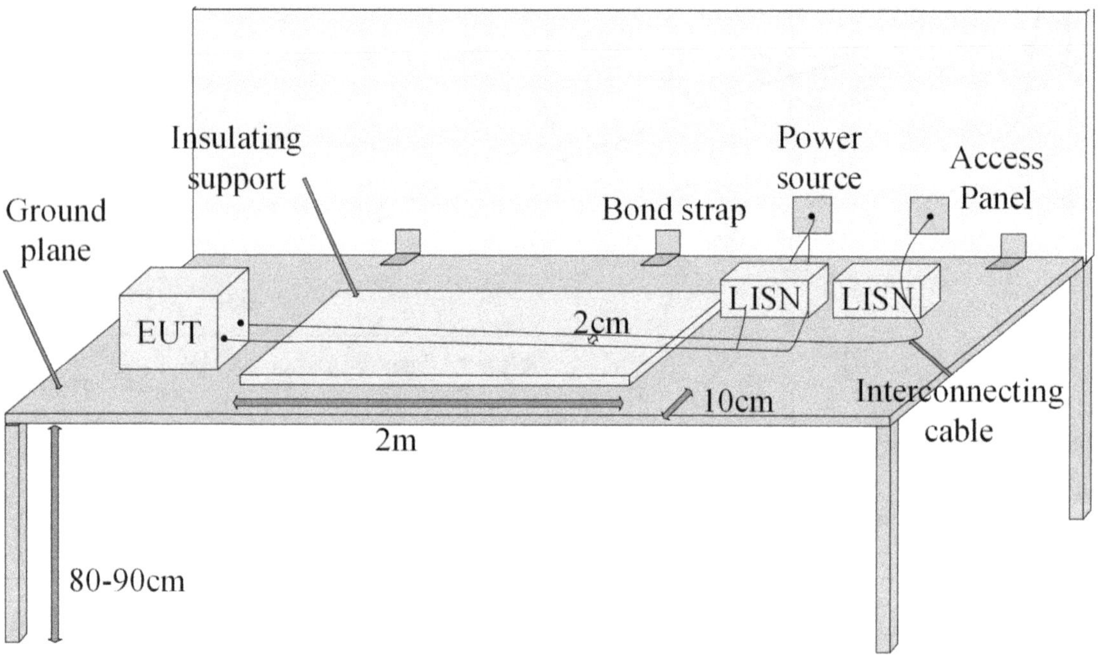

tests where a conductive ground path is not available, the residual charge must be removed with a 1 Ω resistor, an ionizer, or by waiting for the charge to dissipate. It should be noted that the humidity conditions have a strong impact on charge distribution. It is therefore recommended (a strict requirement in the IEC 61000-4-2 standard) that the tests be performed under conditions with relative humidity values in the range of 30% and 60%.

In conclusion, the authors would like to point out that the above test method in relation to human induced ESD immunity tests is a well-established proposal for device-level ESD testing as it is based on a widely used test standard (IEC 61000-4-2) with proven repeatability and reproducibility. On the contrary, the standards for non-human induced electrostatic discharges described in the next section have significantly less coherent and reproducible properties.

Non-Human Induced Electrostatic Discharges

As it has been mentioned before, the reproduction of ESD events in space, contrary to the personnel borne electrostatic discharges described above, is based on a variety of ESD test-setups used in space industry. The methods of ESD application depend mainly on the expected ESD environment and the types

Figure 4. IEC 61000-4-2 ESD Test set-up (IEC, 2009)

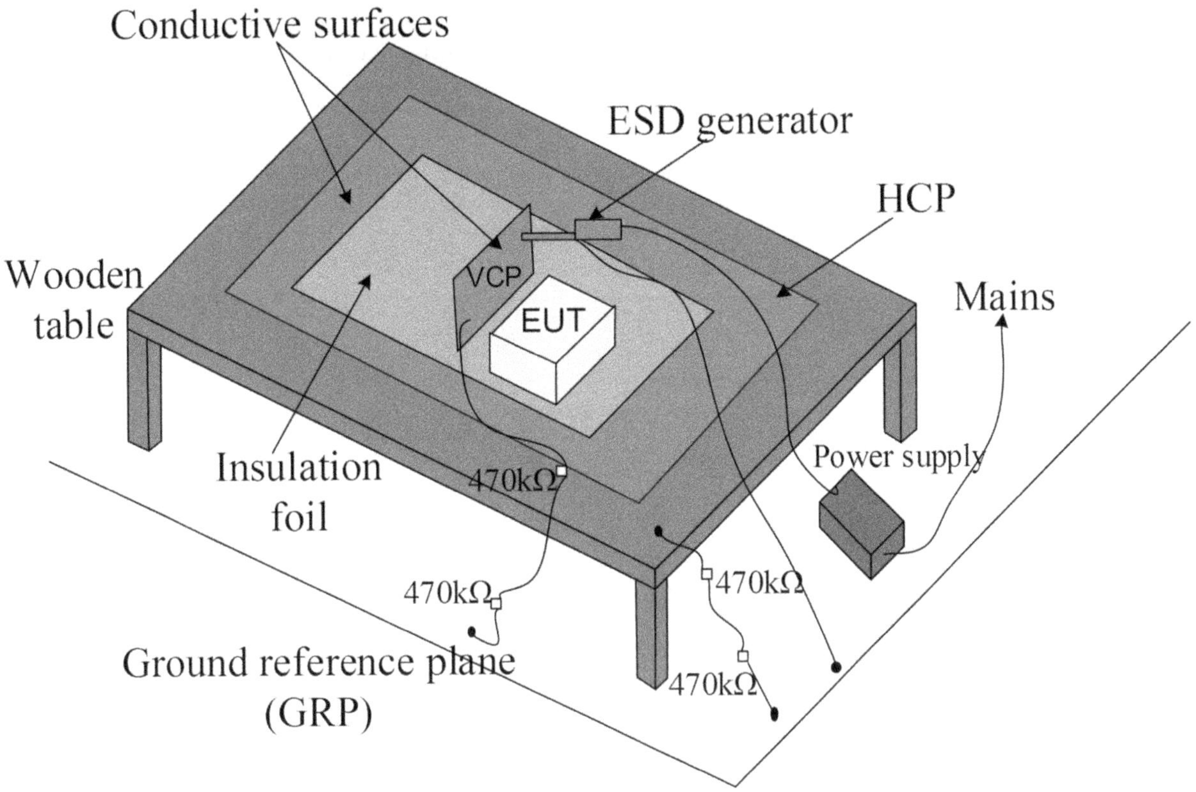

of discharge current each of the parts of the vehicle are subject to. The three most common methods are 1) Radiated Field ESD Testing 2) Single-Point discharges and 3) Structure Current Testing (Garrett & Whittlesey, 2011). More specifically:

- **Radiated field Testing**

During this type of ESD testing, the susceptibility of spacecraft communication and sensor systems to RF interference caused by electrostatic discharge is assessed. Due to the very short rise time of the ESD pulses, the spectral content of this interference can include frequencies of up to several hundred MHz, which are responsible for incorrect measured values or even for the resetting of sensitive system components.

- **Single-Point Discharges**

Local discharge currents are simulated by discharging the arc current on a metallic spacecraft surface having connected the return current wire in close proximity.

- **Structure Current Testing**

Defining the actual replacement current pulses which flow through the structure of the spacecraft and are coupled to the input port of a device due to charge being ejected from its surface is in most cases an impossible task for a space safety assessment. Therefore, testing against these phenomena requires specific analysis that identifies the test currents and test locations. Figure 5 shows some possible test paths for structural current tests.

Figure 5: Suggested test paths for structure current testing

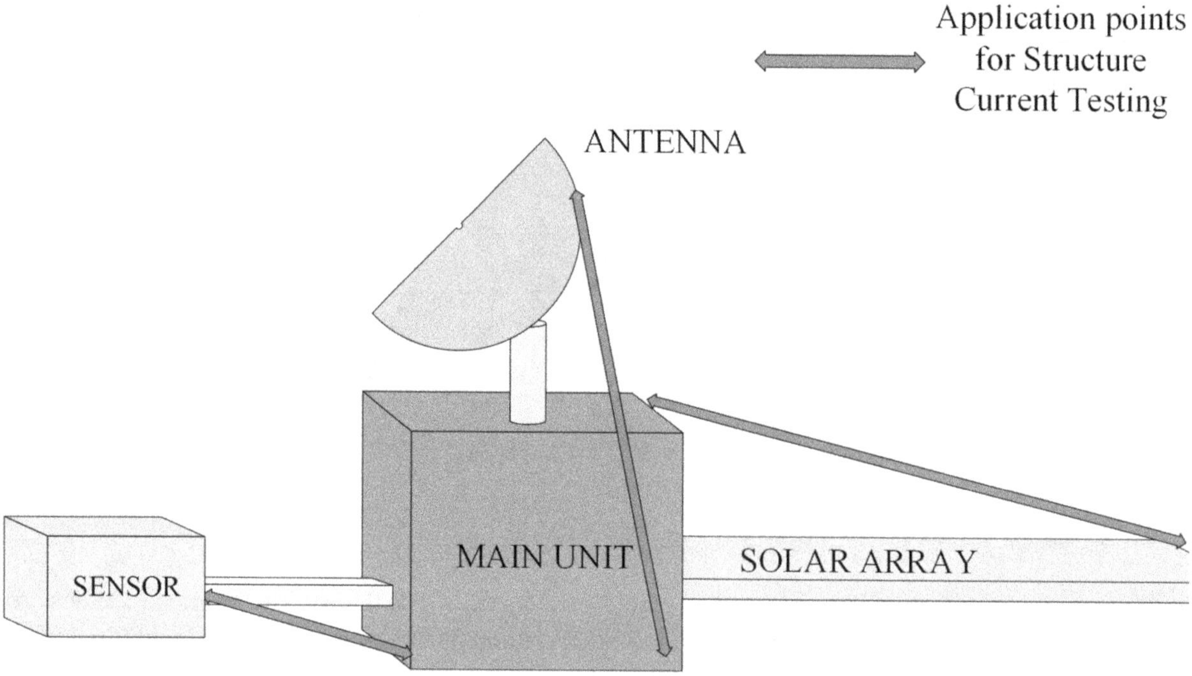

Non-human induced ESD immunity testing has been addressed in the last decades by a variety of international standards, such as MIL-STD-1541A, ISO 14302 and ECSS-E-ST-20-07C, that will be extensively discussed in the next section.

MIL-STD-1541A

The military standard MIL-STD-1541A (Division of the Air Force: United States of America, 1987), has been for decades the main standard for ESD immunity testing proposing a procedure comprising of an indirect, radiation field-based ESD simulation as well as direct discharges to the housing of devices directly exposed to the space environment. This standard distinguishes two required test cases based on the orbit of the spacecraft (synchronous or geosynchronous). In the first case, a 10kV discharge is imposed at a rate of 1 pulse / s for 30 seconds with the discharge electrodes being held at a distance of 30 cm from any exposed surface of the equipment under test (EUT). This process is repeated either by applying direct discharges to each of the top corners of the EUT or by directing the discharge current to the mounting surfaces of the EUT if it is to be installed in a shielded vehicle according to its specifications.

For geosynchronous orbits, the selected test level is determined according to the inter- and intra-system analysis and hazard assessment. MIL-STD-1541A suggests a configuration for the arc discharge source as shown in Figure 6 (Department of the Air Force: United States of America, 1987).

Figure 6. Arc discharge source according to MIL-STD-1541A (Department of the Air Force: United States of America, 1987)

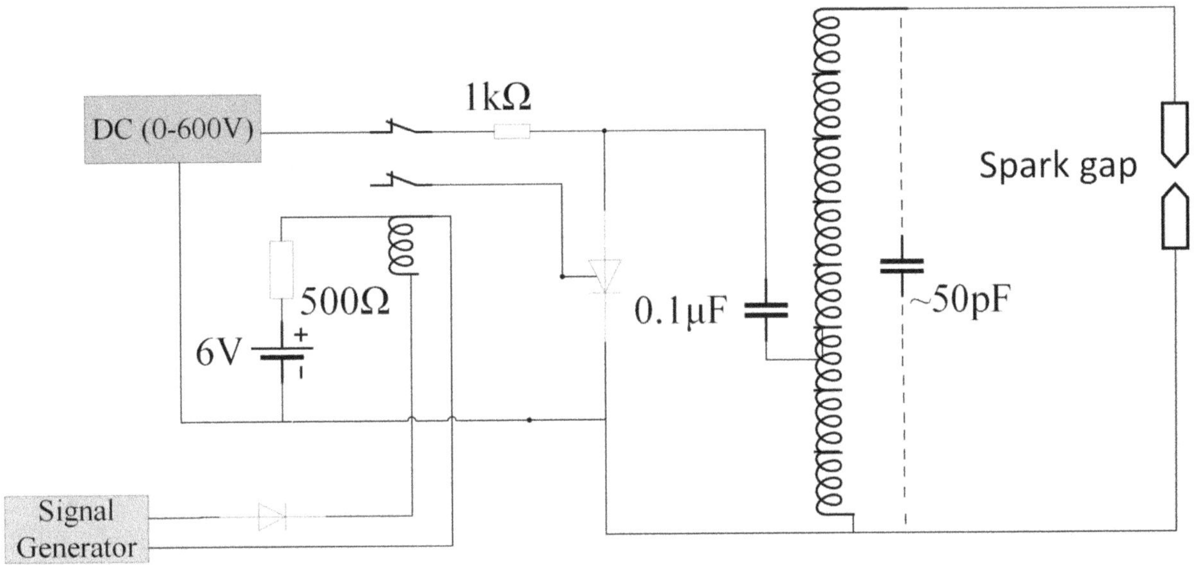

As shown in Table 5 (Department of the Air Force: United States of America, 1987), the only adjustable parameter of this test method is the spark gap spacing, therefore the only adjustable ESD parameter is the breakdown voltage. This limitation regarding the test waveform parameters of the pulse rise time and fall time/width was the main reason this specific test method became obsolete and is not referenced by any recent Military Standard version or newer space applications Standards.

Table 5. Gap-spacing and Breakdown levels according to MIL-STD-1541A (Department of the Air Force: United States of America, 1987)

Gap (mm)	V_b(kV)	Approximate Energy Dissipated (μJ)
1	1.5	56.5
2.5	3.5	305
5	6	900
7.5	9	2000

The aforementioned ESD test methods have been replaced by a magnetic-coupling ESD unit-level test method, introduced in the ISO 14302 Standard (International Standardization Organization [ISO], 2002) and adapted by the European Cooperation for Space Standardization (ECSS) in ECSS-E-ST-

20-07C (European Cooperation for Space Standardization [ECSS], 2012). According to the guidelines of ECSS-E-ST-20-07C this test should be performed on equipment designed to be installed under the following conditions:

- **Outside the main structure of the vehicle**
- **Connected to units located outside the vehicle, such as sensors, actuators etc.**
- **Comprising of a high voltage power source**

A critical point of ESD testing in space applications is the constraint regarding the testing on final versions of the equipment, ready to be used as in-flight equipment. Due to its potentially destructive effects and the possibility of latent failures, in combination with the limited availability of products to be tested, ESD testing is performed only in a representative manner, in equipment configurations that may differ from the actual flight configuration. Despite the efforts of simulating the conditions (interconnecting cables, auxiliary equipment etc.) as closely as possible, the magnetic coupling nature of the ISO 14302 test method makes the performance evaluation, under laboratory conditions, of the equipment immunity less credible.

ISO 14302

The testing configuration proposed by ISO 14302 is illustrated in Figure 7 (International Standardization Organization [ISO], 2002). The ESD current is coupled to the system harnesses in common mode running through a parallel wire. The examined cable harness may be either the harness of the interconnecting cables of two EUTs or the harness connecting the EUT to its ground support equipment.

The set-up of Figure 7 consists of the following components:

Figure 7. ESD test set-up according to ISO 14302 (ISO, 2002)

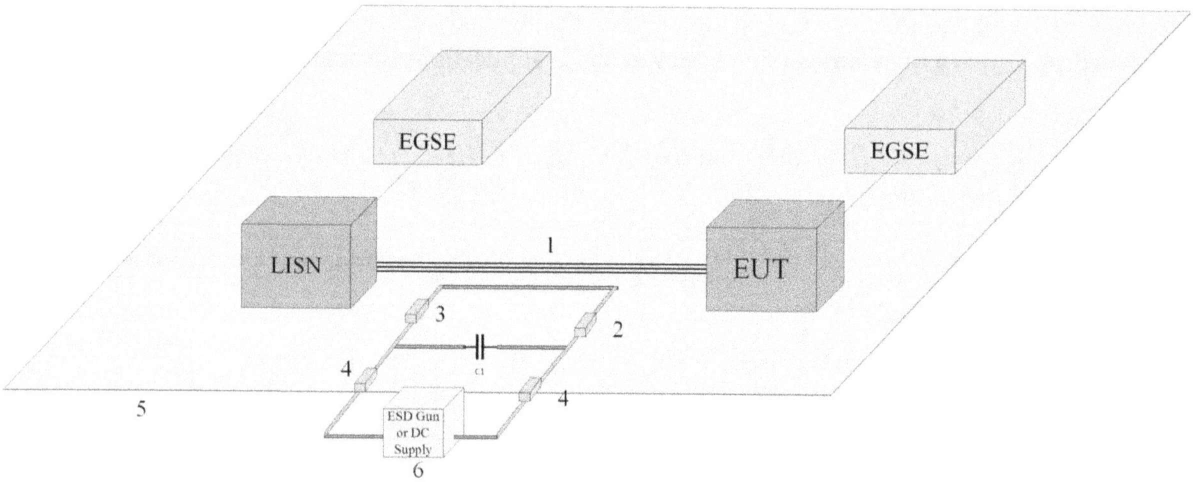

1. The length of the floating discharge circuit, which is firmly connected to the cable harness to be tested, is 20 cm.
2. This resistor, known as the damping resistor, is used in order to minimize the damping and thus the oscillations of the discharge circuit. In order to achieve critical damping, the value of the damping resistance should be selected according to the circuit capacitance and self-inductance values. The typical value of this resistor is 47 Ω.
3. A spark gap with a typical value of 6kV. Air spark gaps are not allowed by ISO 14302 forbids since the variation of the atmospheric conditions, and mainly of the pressure, could affect its discharge characteristics. Moreover, the selected spark gap should exhibit high *dI/dt* capabilities in order to adequately respond to the fast ESD pulse (Pelissou, Daout, & Wolf, 2015)
4. Choking resistors with a minimum value of 10Ω are used in order to block the high-frequency component of the discharge current from flowing in undesired directions.
5. Ground plane
6. Two configuration alternatives are provided for the high-voltage source. 1) A high-voltage DC supply in combination with a choke resistor with a resistance value of more than 10MΩ or 2) An ESD generator, as described in the IEC 61000-4-2 Standard (see Section "Personnel borne discharges"), in air discharge mode, where the two discharge tips are connected to the choke resistors of Figure 9 accordingly.

The typical value for the high-voltage capacitor is 100pF, while the resulting transient current has a peak value of 30A and duration of 30ns (interval between the two crossings at the 50% of the peak value). Before the adaptation of this test method by ECSS, many parameters of the testing procedure remained unclear. The separation distance between the coupling wire and the harness under test, a crucial parameter when a tight electromagnetic coupling is desired, the number and the repetition rate of the discharges, as well as the discharge circuit loop size, affecting the transient pulse duration, are some examples of undefined parameters. Combining these undefined parameters with the limited testing availability, described above, the repeatability and reproducibility levels of ESD testing according to ISO 14302 was significantly restricted until its adaptation by ECSS-E-ST-20-07C, described below.

ECSS-E-ST-20-07C Rev. 1

As mentioned above, the ESD test procedure of ISO 14302 was adapted and improved, largely, by ECSS, increasing its reproducibility, leaving nevertheless room for improvement, as it will be discussed later in this section. The test setup suggested by this EMC Standard is illustrated in Figure 8 (ECSS, 2012).

The EUT is mounted on a conductive ground plane via the spacecraft mount. The main and most important point of difference between the two configurations of Figures 7 and 8 is the requirement in ECSS-E-ST-20-07C for an insulating support. The "floating discharge circle" in ISO 14302 and the inconsistencies resulting from this ambiguous term regarding its stray capacitance are replaced in ECSS-E-ST-20-07C by the use of a 5cm non-conductive support with appropriate high-voltage insulating characteristics. Another specification of the testing setup as described in the ECSS standard is the distance of the tested wire harness -located closer to the front boundary- from the edge of the base plate, set at 10 cm. In contrast to ISO 14302, where the ESD discharges should be applied in the air discharge mode, in ECSS-E-ST-20-07C the discharges are applied in the contact discharge mode.

Figure 8. ESD Test setup according to ECSS-E-ST-20-07c Rev.1 (ECSS, 2012)

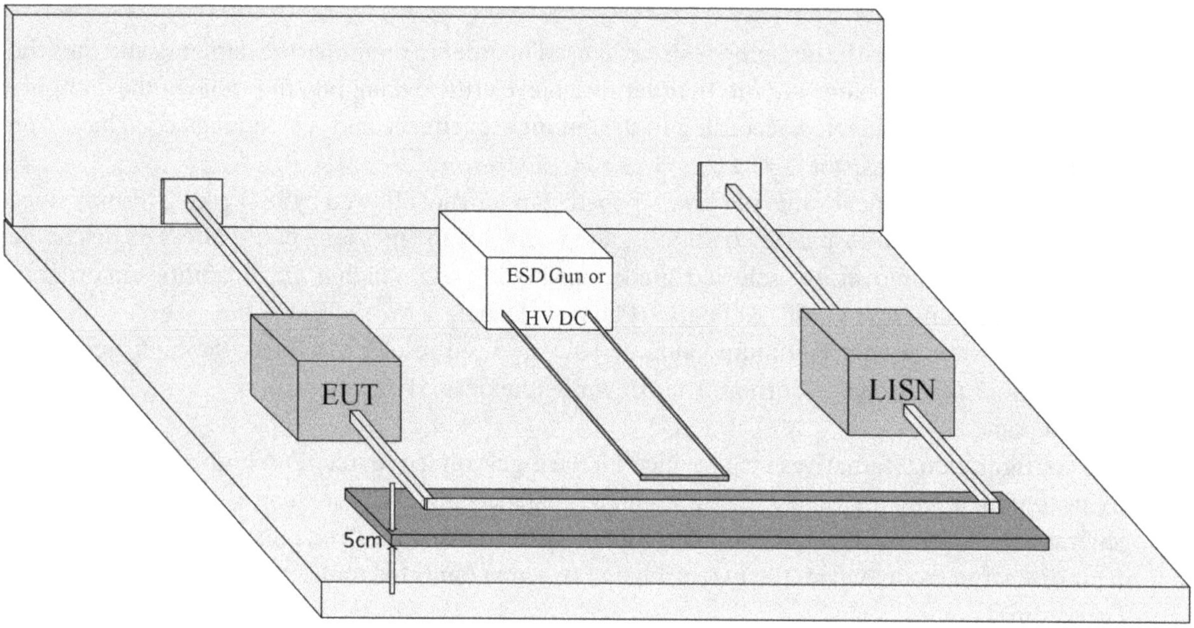

With regard to the discharge circuit, the differences are limited to different values for some components, such as the high-voltage capacitor, which in ECSS-E-ST-20-07C has a typical value of 50 pF with an inductance of less than 20 nH and the inductor resistances, which have a typical value of 10 kΩ. The spark gap may be an air gap, but its reduced stability and greater rise times must be taken into consideration. In contrast to ISO 14302, ECSS-E-ST-20-07C describes the calibration process of the test configuration as well as the properties of the required monitoring devices. The calibration setup is shown in Figure 9, while Figure 10 shows the ESD test configuration.

According to Figure 10 the following components are required: a high-voltage probe grounded on the side of the damping resistor with a 10kV range, a bandwidth of 1 MHz and a minimum input impedance, as well as a current monitor near the damping resistor on the capacitor side with 100A peak capability and bandwidth exceeding 100MHz. The desired waveform must have a minimum rise time, a peak value of 30 A and a 30 ns "time to half value" duration. The rise time of the pulse can be reduced by adjusting the damping resistor, minimizing the discharge loop size, and placing the choke resistors as close to the capacitor as possible.

Unfortunately, ECSS-E-ST-20-07C does not specify any tolerance values for the resulting pulse parameters, so the appropriateness of the pulse must be "ad hoc" assessed by the test engineers and the EUT manufacturer according to the previous hazard analysis and the estimated values if the real ESD Events. Reports from various EMC laboratories show a variation of the ESD duration ranging from 28 ns to 52 ns and of the ESD rise time ranging from 5.7 ns to 14 ns (Pelissou et al., 2015). Although these deviations may seem trivial and unimportant, it should be noted that the energy that is fed into the system by an impulse of almost twice the duration "at medium height" has almost the double value. In addition, the different rise time values lead to different interference frequency spectra. A pulse with 5.7ns rise time has a significant spectrum content up to 62 MHz, while a pulse with 14 ns rise time up

Figure 9. ESD calibration setup according to ECSS-E-ST-20-07C

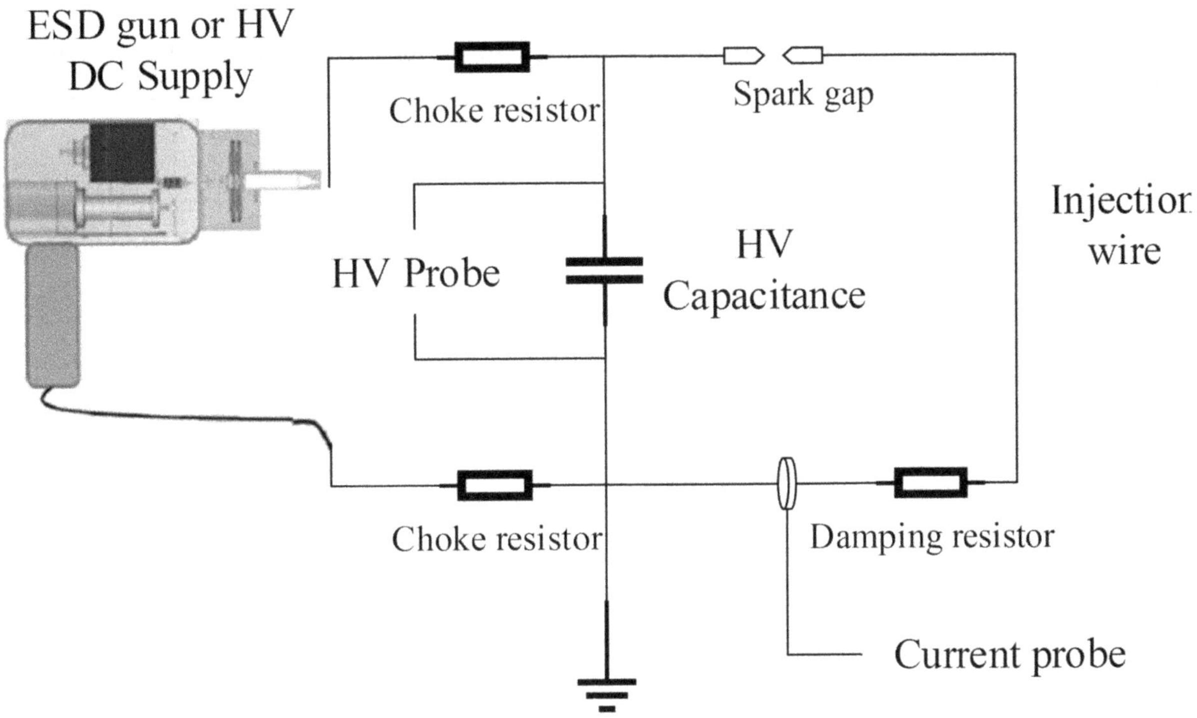

Figure 10. ESD test setup according to ECSS-E-ST-20-07C

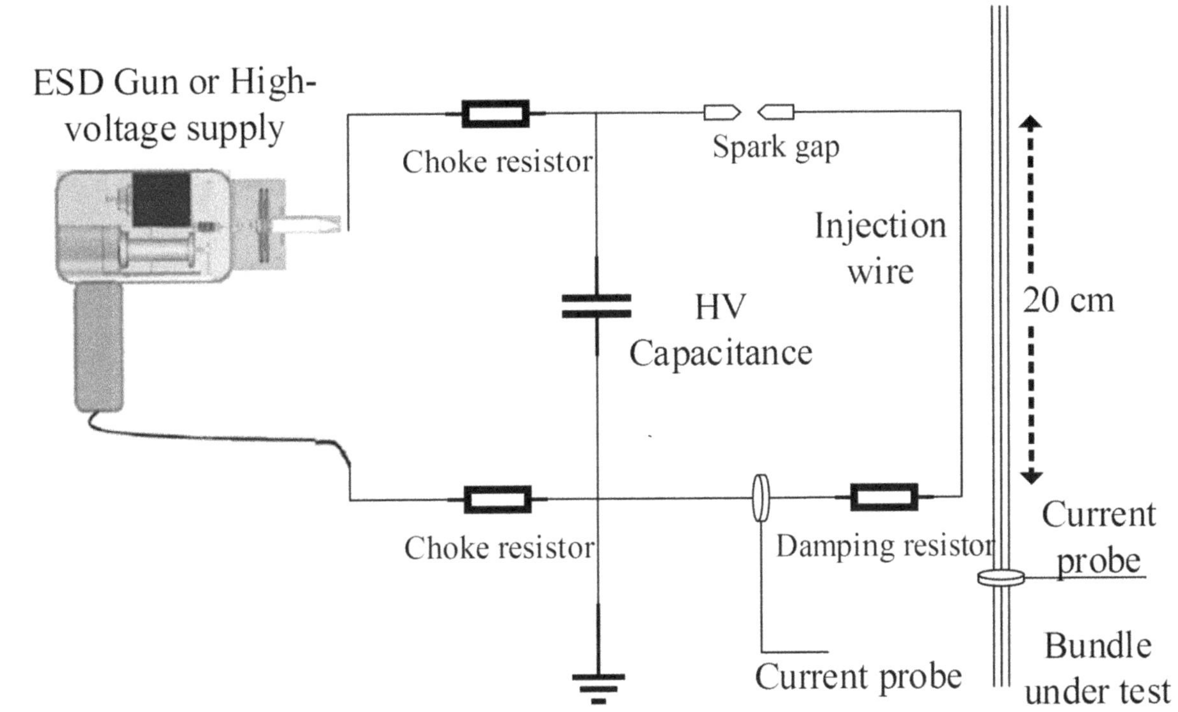

to 25 MHz. This deviation can potentially be critical when testing components and circuits that operate in these frequency bands.

In the test configuration of Figure 10, the high-voltage probe is removed and a second current probe is inserted in order to measure and monitor the coupled pulse on the bundle under test. This probe is located 5 cm from the EUT connector. The length of the cables to be tested running in parallel to the cable harness of the discharge circuit should not be greater than 20 cm. ECSS-E-ST-20-07C requires the application of at least 15 positive and 15 negative pulses at a rate of 1 Hz. A provision for tests at levels lower than the worst-case level estimated in the hazard analysis is not discussed in this standard.

Points for Future Improvements of the Reproducibility of ESD Testing for Space Applications

The ESD testing procedures suggested by ISO 14302 and ECSS-E-ST-20-07C are based on a magnetic coupling mechanism and therefore present increased levels of sensitivity and uncertainty regarding the adequate simulation of real ESD events and the reproducibility of the test method achieved by different EMC laboratories. Some points that require further investigation are the following:

- **Discharge circuit component placement:** The arbitrary placement of circuit parameters can lead to different total impedance values of the discharge circuit and thus to different magnetic coupling factors. Exact definition of the position of the components can be a difficult task especially given the limitations in the size of the loop dictated by the need to achieve an appropriate transient pulse duration.
- **Coupling wire test position and separation distance:** ECSS-E-ST-20-07C suggests a maximum distance of 1cm between the injection wire and the bundle under test so that a strong electromagnetic coupling is ensured. This range of values allows for different implementations of the injection wire placement and therefore for different resulting coupling factors.
- **Spark gap technology:** Probably the most important parameter that can affect the reproducibility of this ESD testing method is the selection of the proper spark gap technology. With the exception of air sparks, which are a poor option due to their reliance on atmospheric conditions, the use of hermetically sealed ceramic metal surge spark gaps is the most attractive solution. With break-

Figure 11. ESD waveforms at generator side (Pelissou et al., 2016)

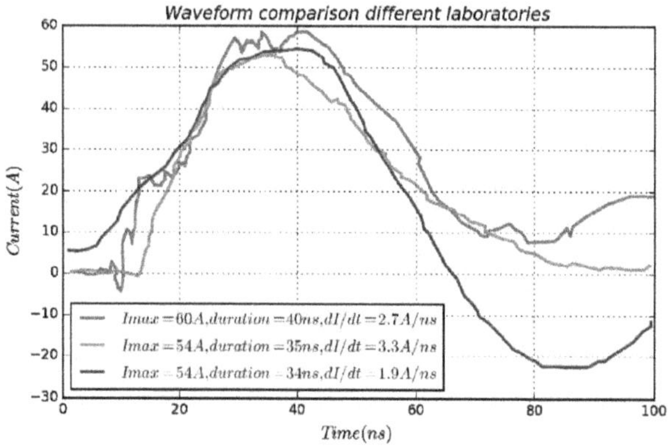

down current rise times between 1 ns and 5 ns and a current capacity of up to hundreds of kA, they offer a good alternative type of spark gap.

The aforementioned points have been extensively investigated by Pelissou, Daout, Romero & Wolf, (2016) in their review of the ECSS ECSS-E-ST-20-07C ESD test set-up published in the 2016 ESA Workspace on Aerospace EMC. The limited reproducibility of the ECSS ECSS-E-ST-20-07C ESD test set-up can be easily observed by the ESD waveform comparison between different EMC laboratories presented in Figure 11 (Pelissou et al., 2016).

Their review focuses on most of the parameters described above and include: the instability of the spark gap, the coupling wire geometry, the position of the coupling wire along the line under test and its distance from the line. The reproducibility results for each influence factor are presented below.

- **Spark gap instability**

The hermetically sealed, made of ceramic-metal, overvoltage spark gaps currently present the best alternative to the formerly used air sparks. Nevertheless, their large variance regarding the breakdown current rise times result in poor, regarding their reproducibility, ESD pulses as presented by the overlapping of 1000 discharges in the work of Pelissou et al. Results show a 28% variance for the measured rise times, 32% variance for the measured current rising slope and 18% variance for the current peak values. As it is expected, these variances affect the resulting induced current to the line under test significantly. Pelissou et al. report up to 100% variation on the induced current level (Pelissou et al., 2016).

- **Coupling wire geometry**

The coupling wire geometry is not strictly defined in ECSS-E-ST-20-07C allowing different implementations of the test set-up. Pelissou et al (2016). investigated four different wire geometries and their respective S_{21} transfer functions.

Figure 12 (Pelissou et al., 2016) presents the results for three different wire geometries highlighting the large variation of the resulting induced current waveforms, which further contributes to the limited reproducibility of the Standard test method.

Figure 12. Induced currents for three different wire geometries (Pelissou et al., 2016)

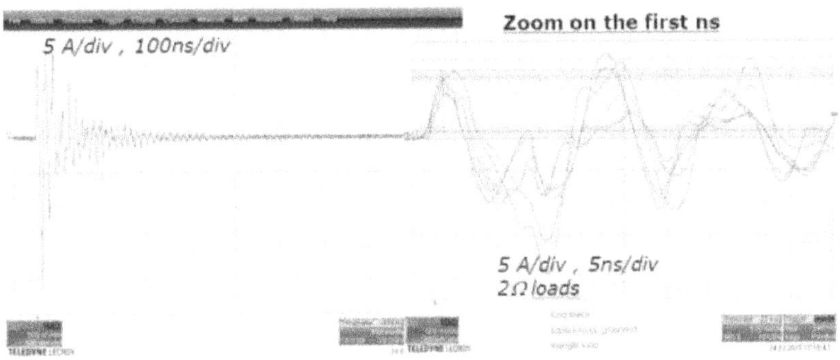

- **Position of the coupling wire along the line under test**

The position of the coupling wire along a 2m line under test has also been investigated by Pelissou et al. (2016) in their review of the ECSS-E-ST-20-07C test method. For different values of d_{near} and d_{far}, which are the distances to the Z_{near} and Z_{far} termination impedances respectively, the resulting induced current amplitude on the line under test displayed up to 50% variation according to the position of the coupling wire (Pelissou et al., 2016).

- **Distance between the coupling wire and the line under test**

The last influence factor investigated by Pelissou et al. (2016) was the distance between the coupling wire and the line under test. Measurements were taken for different length values of the feed wires while keeping the horizontal position of the coupling wire fixed. Results showed a wide variation of the induced current.

All the aforementioned influence factors insert a significant degree of uncertainty to the resulting induced current and limit the reproducibility of the ECSS-E-ST-20-07c test method. This observation led researchers to suggest new alternative ESD test methodologies, either by adjusting the test set-up introduced by the ECSS-E-ST-20-07c Standard or by suggesting new approaches to the ESD testing methodologies. These non-Standardized methods are presented in the next section.

Non-Standardized ESD Testing Methodologies for Spacecraft and Space Equipment

Based on their results for the limited reproducibility of the ECSS-E-ST-20-07c ESD test methodology Daout, Romero, Pelissou and Wolf (2016) went forward and suggested an alternative, modified, test set-up for the ESD testing of space equipment based on ECSS-E-ST-20-07c setup.

- **Alternative ECSS-E-ST-20-07c ESD test methodology**

Daout et al.(2016) made an effort to modify the ECSS test schematic in order to improve its reproducibility while keeping as much of the test topology intact as possible. Their modifications were focused mainly on a)replacing the spark gap, which as shown above shows significant instability regarding the current rise times, with pressurized relays, which according to their research show better reproducibility and a larger lifetime, being however more expensive, b)replacing the coupling wire methodology with BCI injection via suitable probes, which are commercially available since they are widely used for automotive EMC testing and bulk current injection procedures in generally, mitigating the uncertainty factors of coupling wire geometry and position and c) inserting a compensation circuit before the BCI probe so as to simulate the initial transfer function of the coupling wire methodology.The modified generator circuit as well as the alternative ECSS setup proposed by Daout et al.(2016) are presented in following Figure 13 (Daout et al., 2016).

According to Daout et al.(2016) conclusions, this alternative test set-up offers improved stability and reproducibility of the induced current waveform characteristics and utilizes components, which are commercially available and can be periodically calibrated.

Figure 13. Schematic of the proposed generator (Daout et al., 2016)

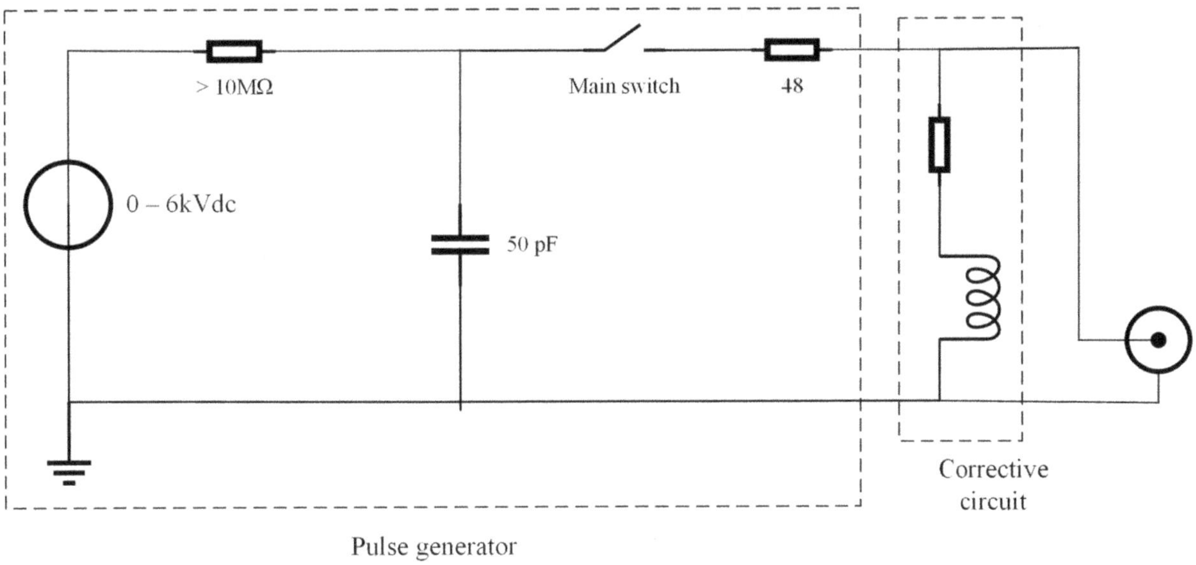

- **MIL-STD-1541A complementary testing methodology**

Iacovone, Scione and Scorzafava (2016) presented during the 2016 International Symposium on Electromagnetic Compatibility that was held in Wroclaw, Poland in September 5-9 2016, a satellite level ESD test approach based on non-Standardized test methodologies in order to better represent real in orbit ESD events and define a worst-case scenario for the injected current during ESD testing. This initiative was designed and implemented by the Thales Alenia Space Italia (TASO) team for the COSMO-SkyMed program. The scope of this effort was to complement the ESD testing procedures defined in the MIL-STD-1541A Standard, described in an earlier section of this chapter, with non-Standardized test setups and methodologies.

As mentioned above, the MIL-STD-1541A Standard defines two different ESD tests, performed separately: a) a conducted test via single point current discharges and b) a radiated test for the performance evaluation of the individual electronic sub-systems of the satellite. Iacovone et al. based on the findings of associated research suggest that the test procedure should also include a) a scenario of multiple electrostatic discharges covering the cases of large floating dielectrics exposed to Plasma and b) the simultaneous conducted and radiated ESD testing via a capacitive direct injection method.

The scenario of multiple electrostatic discharges was implemented by driving the required current pulses to a grid of wires so as to reproduce the multiple discharge points via a multi-probe discharge circuit. The suggested multi-probe conducted ESD test setup is presented in Figure 14 (Iacovone et al., 2016). The Capacitive Direct Injection method included a conducted ESD pulser, a capacitive coupling plane (with the central pin physically connected to the satellite) as well as a high voltage generator and a synchronizer. Scope of this method was to simultaneously produce the tangential magnetic field by the current injected and the electric field by the capacitive plane. The proposed Capacitive Direct Injection test set-up is presented in Figure 15 (Iacovone et al., 2016). Iacovone et al.(2016) have concluded that the non-Standardized ESD setups, that were briefly mentioned above and are extensively described in their paper, show to be more representative of the real in-orbit ESD events regarding the current injection representation and the current propagation over the satellite structure.

Power Bus Switching Transients

Figure 14. Multi-probe conduced ESD test setup (Iacovone et al., 2016)

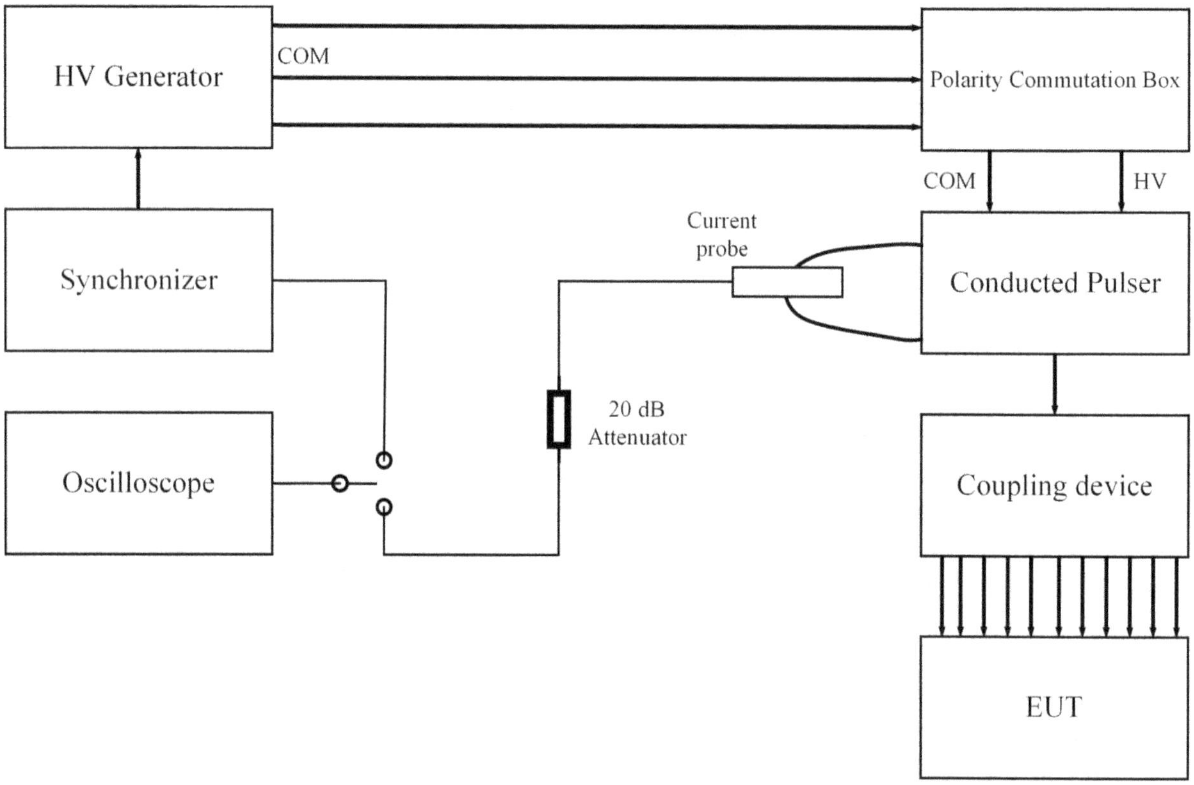

As it has been briefly described in the "Background" Section of this chapter, the performance evaluation of an equipment or system regarding power bus load-induced switching transients includes the three following procedures:

- The measurement of the load-induced switching transients
- The measurement of the peak inrush current
- The immunity testing against power-line switching transients

Load-Induced Switching Transients

Switching loads on and off on a power bus can cause a transient voltage variation due to the wiring impedance of the power bus described above. The method of measuring this voltage is shown in ISO 14302 and in the NASA Handbook for switching transients (NASA, 1999). This LISN-based method is preferable because the parameter of interest is the response of the bus wiring to rapid changes of cur-

Figure 15. Capacitive Direct Injection test setup (Iacovone et al., 2016)

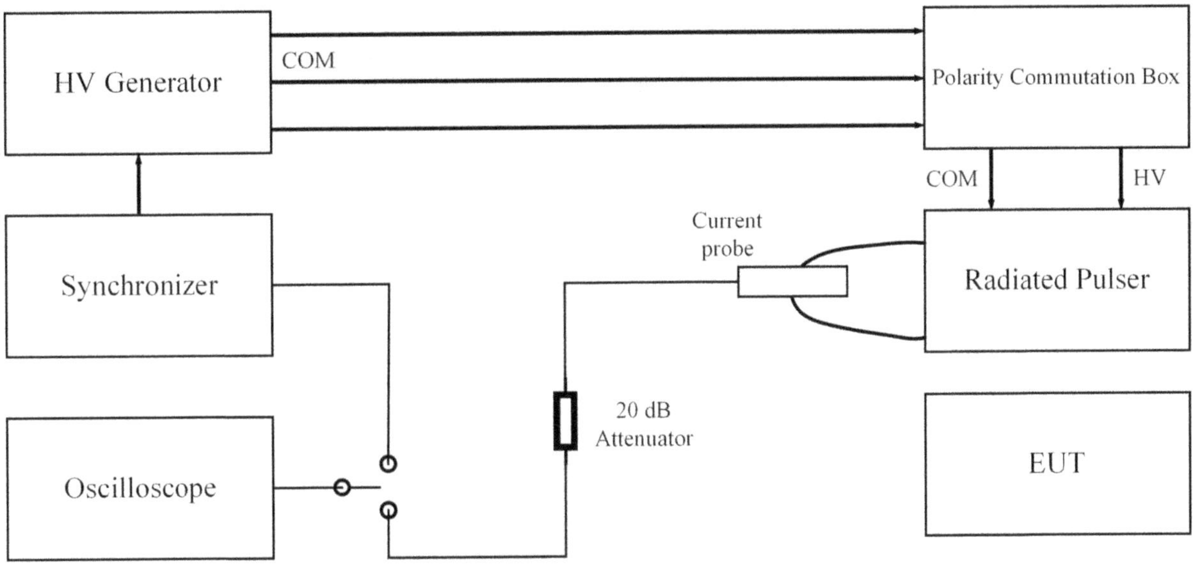

rent. In both cases, the selection of the correct LISN type -time domain or frequency domain- depends on the measured transient duration. Frequency domain LISNs, which are primarily used for performing conducted emissions measurements, do not adequately control impedance below 150 kHz. Therefore, the use of a time domain LISN is recommended as shown in Figure 16 (ISO, 2002). Additionally, Figure 17 (ISO, 2002) illustrates the suggested measurement setup.

The measuring apparatus can either be an oscilloscope with a bandwidth of at least 10 MHz or a suitable EMI meter. So that emission measurements can be carried out in the worst case, the selection of the switch should be done with special consideration. The maximum inrush current *dI/dt* is reached when the EUT is switched near the peak of the AC waveform.

Figure 16. Time domain LISN for the measurement of Load-induced switching transients (ISO, 2002)

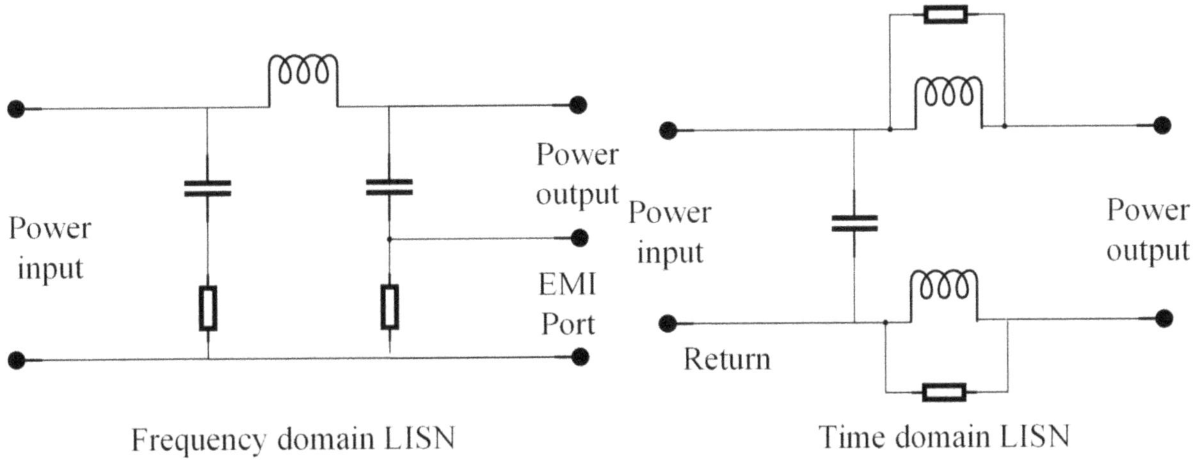

Figure 17. Measurement setup according to ISO 14302 (ISO, 2002)

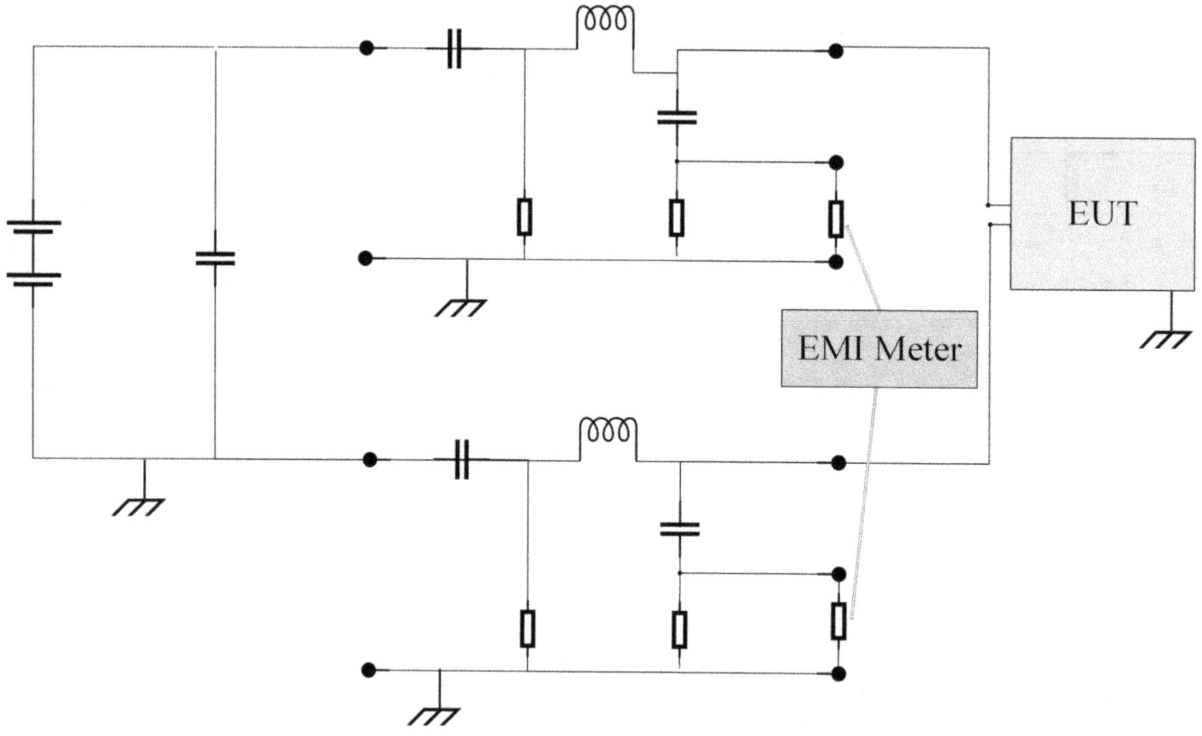

Measurement of the Peak Inrush Current

In the case of current-limited buses the peak inrush current of the EUT, which can result in voltage sags, shall remain within tolerance limits. These limits are derived from the following equation and are presented as a sample limit line in Figure 18 (ISO, 2002).

$$\Delta V_{trans}\left(V_{nom}-V_{min}\right)\leq\sqrt{\frac{I_{ss}}{I_{bus}}}$$

Where: ΔV_{trans}: maximum allowable voltage sag
V_{nom}: nominal bus voltage
V_{min}: minimum bus voltage
I_{ss}: EUT steady-state current draw
I_{bus}: power bus maximum steady-state current draw

The measurement setup suggested by ECSS in ECSS-E-ST-20-07C is presented in Figure 19 (ECSS, 2012).

Inrush current is measured directly with a current probe, while the on/off operations of the EUT are conducted either with a fast, bounce-free power switch or an actual power regulator if the on / off command is not implemented in the EUT. This method differs from the ISO 14302 test method, in which the measured parameter is the voltage drop across a series resistor.ECSS-E-ST-20-07C suggests the

Figure 18. Transient inrush current sample limit (ISO, 2002)

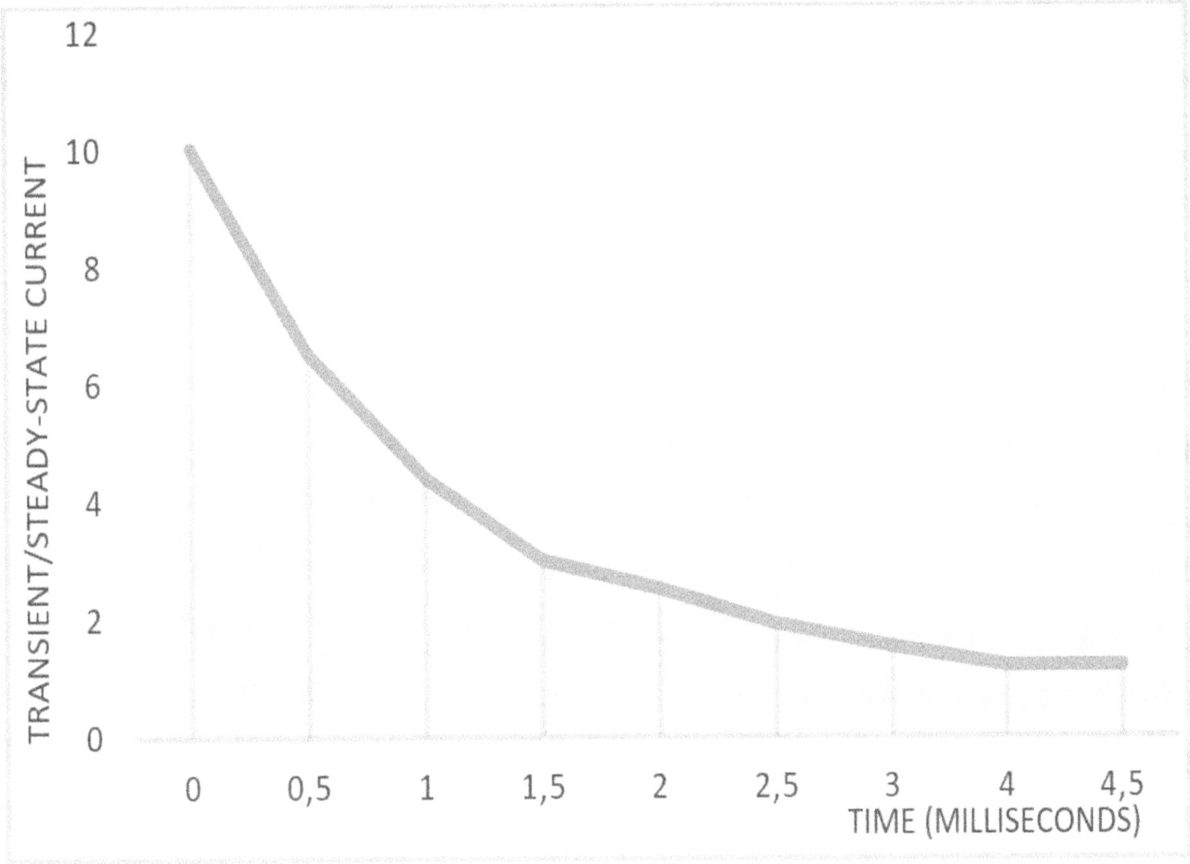

Figure 19. ECSS-E-ST-20-07C inrush current measurement setup (ECSS, 2012)

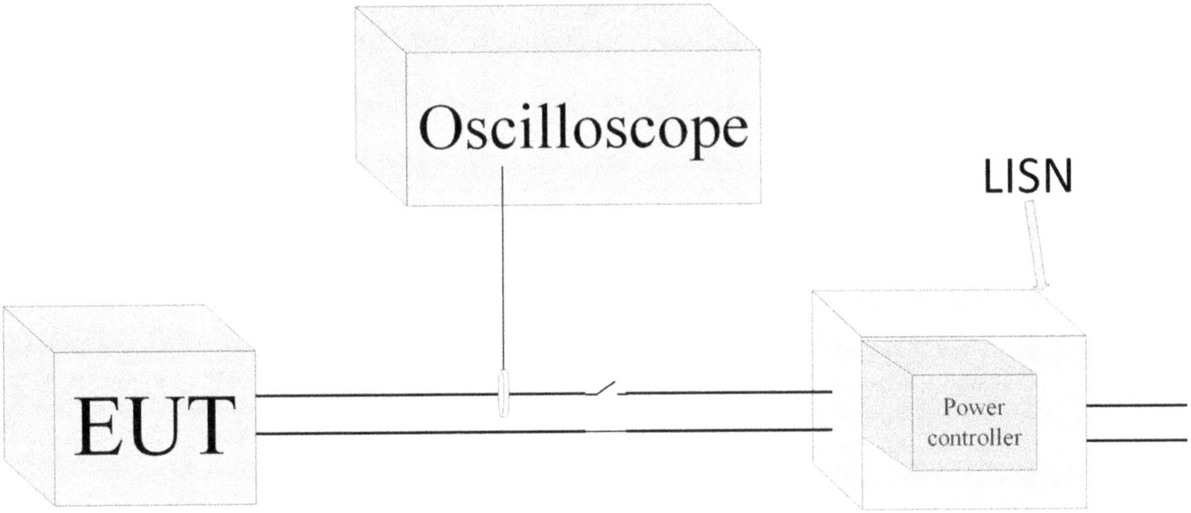

following limit values so that the inrush current of a device is limited and this way the resulting voltage drop on the power line is minimized:

- The rate of change of the current for any nominal change in operation should be less than 5 10^4 A/s.
- While switching on the EUT, this rate of change should be limited to 2×10^4 A s. These values may differ for pulsed radar or plasma engines.
- Parameters such as the duration of the inrush current (ms), the total transferred charge (mC) and the inrush current slope (A/μs) can also be limited according to the specifications of the spacecraft power system.

Immunity to Power-Line Switching Transients

The susceptibility of the EUT to positive or negative spikes superimposed on its input conductors is evaluated by the series or parallel injection of spikes with a predefined waveform to the EUT and its operational monitoring during the test. The suggested setups across all the aforementioned Standards involve the use of a spike generator in series or parallel to the EUT inputs, a LISN as the power source, an oscilloscope for the recording of the applied spikes and a monitoring system for the performance evaluation of the EUT. A typical setup is illustrated in Figure 20 and Figure 21 (ECSS, 2012).

Figure 20. Typical setups for immunity testing to power-line switching transients: Series configuration (ECSS, 2012)

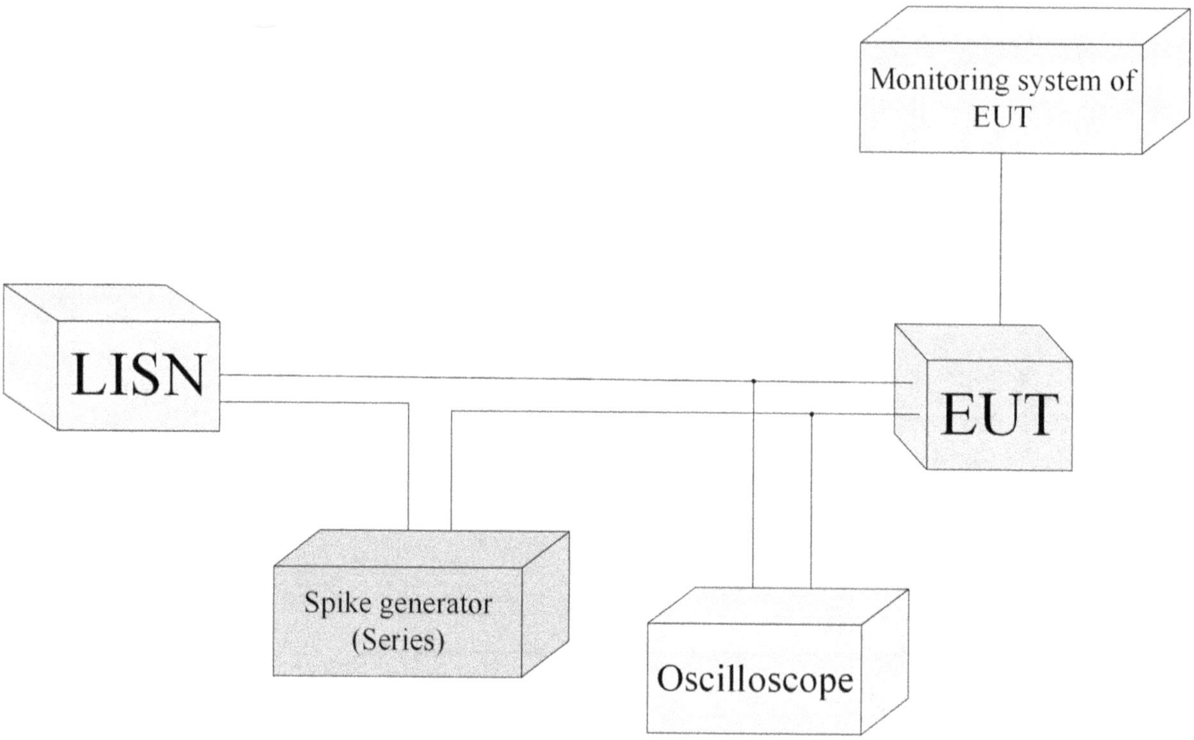

Figure 21. Typical setups for immunity testing to power-line switching transients: Parallel configuration (ECSS, 2012)

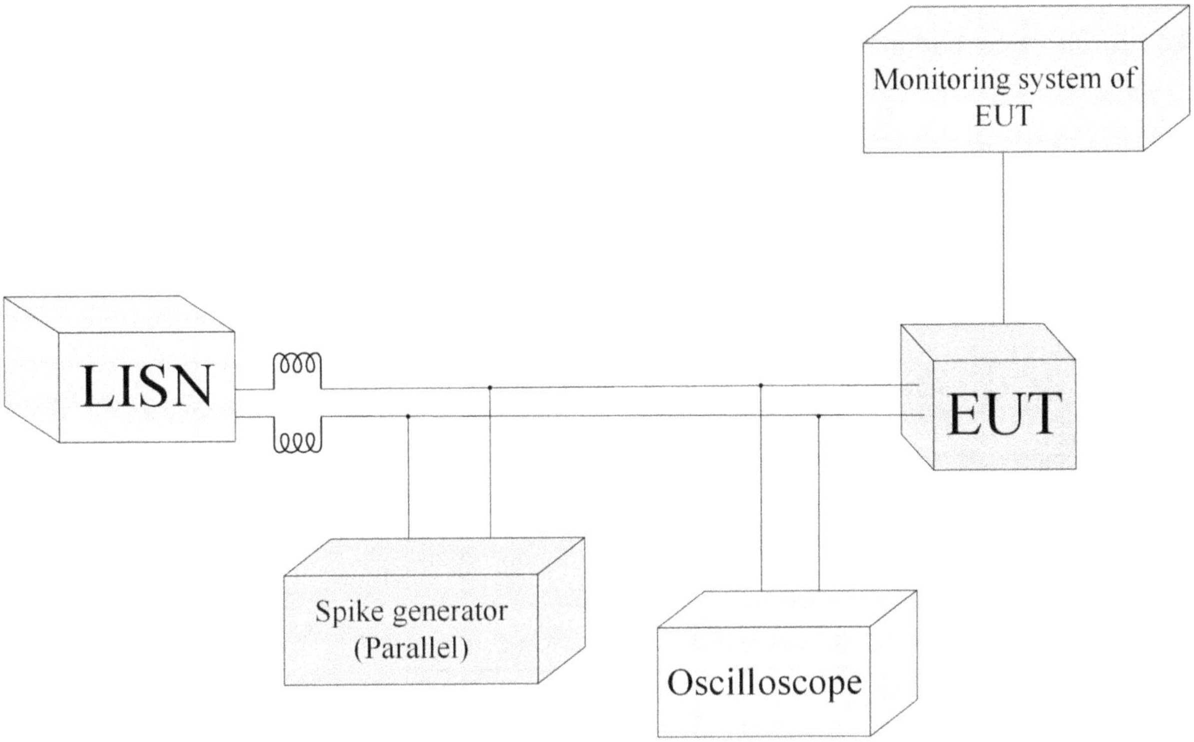

The parameters of the applied transient waveform shall be defined in accordance with the risk assessment that predicts the estimated properties of potential transient phenomena prior to the testing procedure. Parameters such as the expected bus-switched currents, the power bus LISN model, the type of loads connected to the bus and the hold-up capacitance must be considered when selecting the appropriate transient generator and the bandwidth of the oscilloscope for recording the pulses.

An important point when testing immunity against power line switching transients are the different parameters of the switch-on and switch-off transients. As shown in Figure 22 (ISO, 2002), the pulse amplitude and duration for transient pulses with a rise time of less than 1 μs vary considerably. It is therefore strongly recommended to select different values for pulse width and amplitude. ECSS-E-ST-20-07C suggests the use of pulses with a width value of 10μs and 0.15μs with a repetition rate of 10 pulses / second.

Furthermore, it should be noted that the aforementioned test method addresses the control of power-bus switching transients and does not cover the cases of externally induced transients, like field-to-wire coupled transients. The test methods for these types of phenomena are discussed and simulated using the bulk injection method in the CS115 and CS116 sections of MIL-STD-461G.

Having presented the various current or withdrawn/superseded test standards, the authors consider it necessary to comment on the preferred test method or combination of test methods according to the current equipment-level International Standardization in relation to transient electromagnetic phenomena in space applications. Based on the criteria of the best achieved accuracy of the phenomenon simulation, the repeatability and the reproducibility of the test method, the most reliable approach is the applica-

Figure 22. Switching transient envelopes on a 28Vdc bus (ISO, 2002)

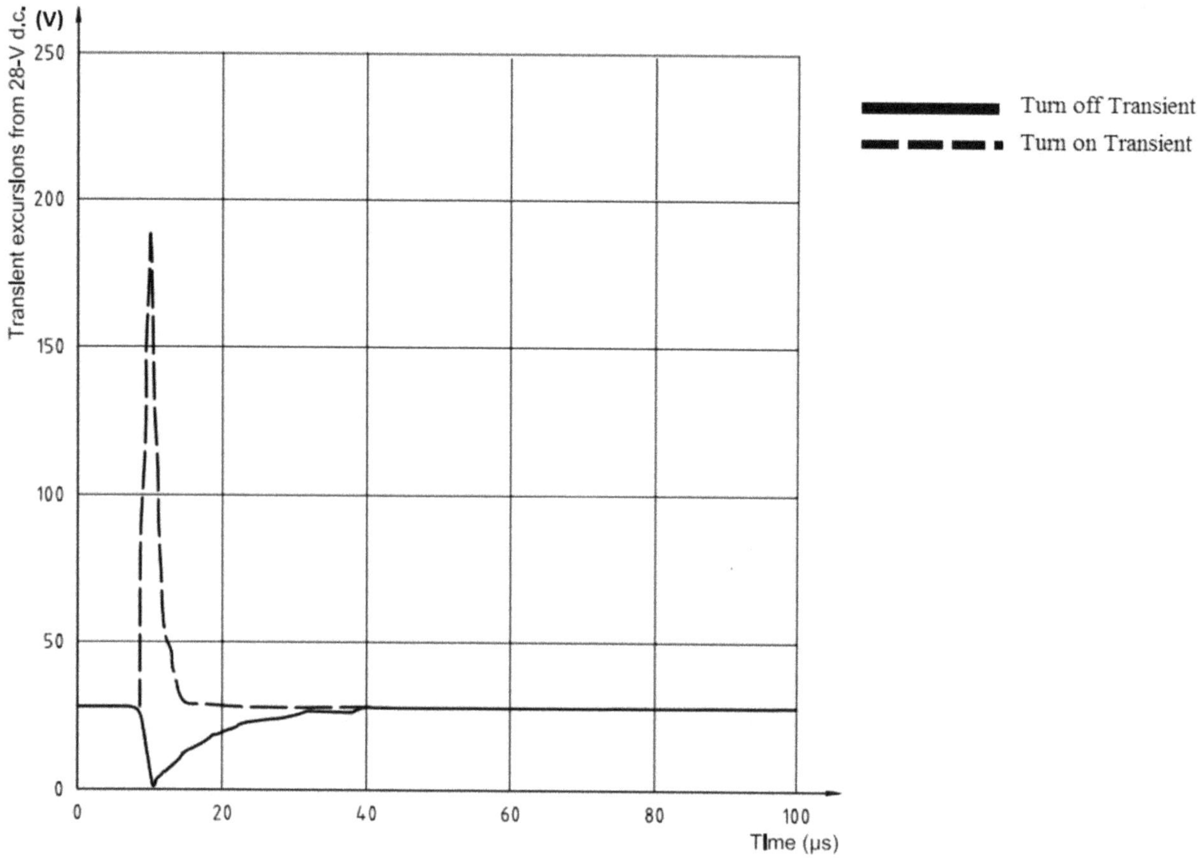

tion of the MIL-STD-461G Section CS118 test procedure for the human-induced ESD immunity test as well as the ECSS-E-ST -20-07C Test Method for Non-Human ESD Immunity Testing and Power Line Switching Transient Immunity Testing. Both standards, with their relatively recent revisions, provide the most accurate and analytical specifications for the immunity test procedures required that contribute to the desired repeatability and reproducibility. With regard to the required emission measurement methods, the authors propose the use of ISO 14302 for measuring the load-induced switching transients and the ECSS-E-ST-20-07C measurement method for the inrush current.

SOLUTIONS AND RECOMMENDATIONS

Failure Evaluation

Product manufacturers, EMC designer and engineers have always been concerned with the identification of the causes for an ESD immunity testing failure. Such failures can be characterized as either "Hard failures", that means some electric or thermal breakdown of the semiconductor materials has occurred mainly due to the lack, incorrect selection or aging of the ESD protection circuitry, or "Soft failures" which are defined as non-destructive change-of-state occurrences disrupting the normal operation of the equipment under test. These failures are attributed mainly to the high amplitude transient electromagnetic fields, resulting from the electrostatic discharge, which lead to the inductive or capacitive generation of transient voltages or currents on the traces and the wiring of the system, which potentially affect the voltage/current readings of the subsystems component, such as analog-to-digital converts, sensors etc. falsely changing the state of the system. This failure evaluation process is critical for the proper selection of the applicable protection approach and the appropriate set of mitigation techniques as discussed in the following section.

Mitigation Techniques

As in every stage of a space mission planning and design, the protection and mitigation analysis against the dangerous effects of surface and internal charging includes the application of a number of practical guidelines associated with the selection of protective circuit components and equipment, the selection of the most suitable materials and the correct arrangement of the system's cabling. These guidelines are categorized in the following axes:

1. **Orbit selection:** As discussed earlier in this chapter, the level of charging and as a result the threat level of the transient phenomena, depends on the electromagnetic characteristics of the selected orbit environment. Orbits with intense particle background should be avoided and mitigation measures should be adjusted for the orbit selected.
2. **Shielding:** The goal of the shielding procedure is the creation of an electrically and physically continuous enclosure, including the spacecraft main structure, the equipment casing and the cable shields to act as a Faraday cage and shield the sensitive electronic subsystems of the spacecraft. The desired attenuation for radiated electromagnetic fields induced by charging and transient phenomena is 40dB provided by the selected mitigation techniques, such as aluminum/magnesium foils either in the form of films, face sheets, blankets or honeycomb structures. Special attention should be paid for the identification of possible slits, apertures and any other opening that affects the continuity of the shielding enclosure, as well as the spacecraft cable penetration points, which should be adequately shielded along with exposed parts of the cabling (Garrett & Whittlesey, 2011). At equipment level, the first point of focus for the shielding process is the selection of the type of material of the equipment's enclosure. This material can be either metallic or plastic. Both of these types have advantages and disadvantages regarding their application. Metallic enclosures, which are mainly selected for effectively shielding the sensitive circuitry against radiated disturbances and reducing the emitted disturbances from the equipment electronic components can often present a challenge regarding the ESD immunity. An uncontrolled grounding path for the electrostatic discharge current in the proximity of sensitive circuitry nodes can jeopardize the stability of the system under test. Therefore, in cases of metallic enclosures, the provision of a "safe" grounding path for the ESD current is essential. On the other hand, plastic enclosures offer better ESD protection at the expense of limited radiated immunity and emission suppression capabilities. The same approach should be adopted for the screws and fixtures of the enclosure. The use of insulating plastic caps for the metallic screws has shown positive results during ESD immunity testing (Duvvury & Gossner, 2015).
3. **Bonding:** Important part of the aforementioned shielding process is the bonding of the various components and surfaces. The rationale of the bonding procedure is to connect all conductive struc-

tural elements on the surface as well as in the interior of the spacecraft, enclosures of electronic equipment and mechanical parts of the system to each other and to a common reference ground. The maximum bonding impedance to chassis value for internal conductive parts is $10^{12}\Omega$ and either directly or through a bleed-off resistor a low impedance path to the ground shall be established for the dissipation of electrostatic discharge currents. In cases where the bonding of surface materials is not structural but is used only for ESD mitigation purposes, the adequate surface-to-structure bonding impedance is $10^{9}\Omega$. Ground connections of signal or power ports should also be bonded to the metallic structure as directly as possible with special attention to not form loops.

Similar to the shielding process, proper material selection is of significant importance during the bonding analysis. Dielectrics with high insulating characteristics contribute negatively to the protection from internal charging, whereas metals, which are affected by surface charging can be used with an adequate charge-leakage path. For the exterior of the spacecraft, all connections, where possible, should be conductive and for the cases of sensitive areas, e.g. near antennas or receivers, where dielectrics are required, connections should be partially conductive and bonded to the structure of the spacecraft. For dielectrics installed in the interior of the spacecraft conductivity values in the range of 10^{4}-$10^{11}\Omega$cm are required.

Indicative materials suitable for this purpose are:

a. Combinations of deposited metals (aluminum, gold, silver) on dielectric substrates (Inconel® on Kapton®, Teflon®, Mylar® and fused silica) in the form of sheets, strips, or tiles
b. Thin, conductive front-surface coatings: ITO (250nm) is the most common material
c. Conductive paints (white Zinc orthotitanate is the most conductive paint). They are required also for conductive substrates.
d. Carbon-filled Teflon®, or carbon-filled polyester on Kapton®
e. Conductive adhesives, necessary for the bonding of a conductor to the ground reference
f. Exposed conductive facesheet materials (graphite/epoxy or metal).
g. Etched metal grids not simply stretched over dielectric surfaces but effectively bonded.

Anodizing procedures, fiberglass materials, uncoated Mylar®, Teflon® (even metalized Teflon® is extremely vulnerable to ESD threats) and Kapton®, silica cloth, quartz and glass surfaces should be avoided.

4. **PCB material selection:** Another approach towards ESD immunity of the on-board electronics of a spacecraft is the appropriate selection of the PCB dielectrics. As described earlier in this chapter, the PCB dielectrics are responsible for charge accumulation, which under extreme space conditions (e.g. geomagnetic storms) can result in electric field values large enough ($2*10^{7}$ V/m) to initiate internal electrostatic discharges. The main parameter of the chosen dielectric material associated with charge accumulation is the material conductivity. The general applicable rule is that with the increase of the material conductivity the charge accumulation and the associated Maxwell relaxation time decrease. The Maxwell relaxation time is also of significant importance for space equipment materials since large potential differences can appear under space conditions in a very small time window. An example of this sudden potential changes is when the spacecraft leaves the shadow.

The most promising material technology that could potentially eliminate the risks related to ESD events caused by charge accumulation is the nanoconducting material technology. Saenko, Tyutnev, Abrameshin and Belik (2017) have proposed a new approach to the accumulated charge ESD events in space utilizing the enhanced conductivity of these specific materials. With enhanced conductivity values possibly reaching the $5*10^{-9}\Omega^{-1}m^{-1}$ Saenko et al.(2017) research, based on computer simulations and mock-up testing, show promising results towards totally eliminating the possibility of electrostatic discharges for on-board electronics. It remains to be seen and investigated if nanoconducting materials can effectively be used for the design and development of space equipment electronics and to what extend these materials will affect their operating capabilities.

5. **Circuit filtering:** Low pass filters and CMOS circuits are some of the methods of electrical filtering in order to protect circuits from charge-generated ESD.
6. **Isolation of transformer primary windings from secondary windings.**
7. **Optimum cable harness layout:** Cable harnesses should be routed away from openings whereas cables coming from the exterior should be filtered and separated from internal cables to different bundles.

The aforementioned techniques, taking also into consideration the quantitative guidelines presented in the next section, can provide significant protection against transient phenomena and their effects of space systems, either at equipment level or at spacecraft level.

Quantitative Surface ESD Guidelines

Some qualitative design guidelines based on empirical equations for the achievement of the desired surface conductivity are the following (Garrett & Whittlesey, 2011; NASA, 1984):

1. The grounding of conductive components to the structure must be achieved through a grounding resistance R [Ω] with value:

$$R < 10^9 / A$$

Where: A = exposed surface area of the conductor [cm^2]

2. The combination of a partially conductive material with a grounded conductive surface (e.g. conductive paint coating) must comply with the following requirement regarding the resistivity of the coating and its thickness:

$$rt < 2 \bullet 10^9$$

Where: r = material resistivity [Ω•cm]
t = material thickness [cm]

3. 3. In the case of a partially conductive material combined with a grounded dielectric surface (e.g. conductive paint coating) the above requirement is modified as follows:

$rh^2 / t < 4 \bullet 10^9$

Where: r = material resistivity [Ω•cm]
t = material thickness [cm]
h = maximum distance of the conductive surface to a ground point [cm]

Depending on the specific geometry and application different limitations are formed based on the above rules. Some primary guidelines are: According to ECSS-E-ST-20-06C (ECSS, 2008) the grounding of isolated conductors must have resistance less than $10^6\Omega$. In the same context, conductive substrates require coating materials with bulk resistivities less than $10^{11}\Omega \bullet cm$ whereas dielectric materials allow coverage with surface materials of resistivity less than $10^9\Omega \bullet cm^2$.

Quantitative Internal ESD Guidelines

Similarly, some quantitative measures for the protection of the interior of the spacecraft from ESD due to charge penetration can also be applied (Garrett & Whittlesey, 2011; NASA, 1984):

- Conductive components that are not part of circuit boards, such as unused cables and conductor tracks with a surface area of more than 3 cm^2 or a length of more than 25 cm, should be earthed (the respective critical area for conductive areas on circuit boards is 0,3 cm^2). In ECSS-E-ST-20-06C (ECSS, 2008) stricter requirements are proposed allowing for a maximum area of ungrounded metal surfaces of 1 cm^2.
- Regarding the shielding measures to prevent electron fluxes from entering into the spacecraft, GEO environments are covered by 110mils equivalent aluminum shielding. Shielding based on the computed maximum flux of the orbit is based on the following guideline: calculated flux in the examined orbit less than 0.1pA/cm^2 requires no additional shielding, for fluxes between 0.1pA/cm^2 and 0.3pA/cm^2 the maximum remaining flux after shielding is 0.1pA/cm^2 for Class 1 ESD-sensitive circuitry (according to MIL-STD-88G3) (Department of Defense: United States of America, 1996), for fluxes between 0.3pA/cm^2 and 1pA/cm^2 shielding to a maximum level of 0.3pA/cm^2 is necessary for Class 2 ESD-sensitive circuitry (according to MIL-STD-88G3) (Department of Defense: United States of America, 1996), whereas for fluxes exceeding 1pA/cm^2 internal ESD phenomena cannot be mitigated through shielding.
- In addition to testing the immunity of components against ESD on an equipment level, MIL-STD-88G3 (Department of Defense: United States of America, 1996) proposes a categorization of the embedded microcircuits according to their susceptibility to ESD in order to determine the necessary packaging and handling specifications and define the abovementioned critical levels of maximum permissible flux. Various combinations of circuit pins are subjected to a double-exponential current waveform with rise time below 10ns and delay time 150±20ns derived for charging voltage levels 500V, 1kV, 2kV and 4kV. Equipment with immunity against voltages exceeding 4kV is

characterized as Class 3, withstand voltages between 2kV and 4kV are attributed to Class 2 and equipment with threshold up to 2kV belongs to Class 1.

- Filter circuits (usually RC circuits or diodes) should be dimensionalised according to the following specifications: withstand of a 5kV, 20ns width, pulse and of a 250μJ stress resulting from the charging of a 20pF capacitance with 100nC (ECSS, 2008). Immunity against ESD is preferable to exceed 20kV.
- All mitigation measures adopted should ensure limitation of the electric field stress developed within dielectrics below $4 \cdot 10^6$ V/m.
- Circuit boards should be coated with partially conductive dielectrics providing a leakage resistance of about 10^9-$10^{13}\Omega$.
- The reduction of the available volume of dielectrics allowing accumulation of ESD charge can be achieved by filling unused regions of circuit boards between dielectrics with grounded metal, although the method has not been validated.
- According to NASA practice (1993a) ESD threats near sensitive circuits and radio frequency (RF) receivers should be eliminated, whereas for other regions the energy of an ESD pulse depending on charge accumulation, geometry and material conductivity should be minimized to 3mJ.

An extensive and very detailed guideline regarding the mitigation of in-space charging effects is the NASA technical handbook 4002, which was revalidated on 19.10.2019. NASA-HDBK-4002A w/ CHANGE1 contains valuable information regarding the theoretical background of the charging and discharging mechanisms under space conditions and offers a variety of mitigation techniques, some of which were described earlier in this chapter, emphasizing especially on the requirements for grounding and proper material selection, not only for the conductors but mainly for the dielectrics. In this handbook the space charging/ESD risks associated with the solar arrays are also put in the foreground. Solar arrays, with their possibly high operating voltages can result into spacecraft charging issues that can cause not only surface damage but also affect the correct operation of measuring equipment inside the spacecraft etc. The specialized ESD mitigation techniques related to the solar arrays will not be discussed in this chapter since a lot of research is currently conducted regarding these issues and the results will be presented in the near future. However, NASA-HDBK-4002A contains a variety of mitigation methodologies for solar array related ESD protection and can be a very useful tool for space system designers.

The final step, supplementing the aforementioned guidelines, is the selection of the proper protection devices, presented in the next section, in order to eliminate, minimize or divert away from sensitive circuit the ESD energy.

Protection Devices

"Hard ESD Failures" are mainly avoided by placing protection devices at the entry points of sensitive circuits, reducing the resulting voltages to the overvoltage margins specified by the manufacturer of the components. These devices serve as the first stage of ESD protection, while on-chip ESD protection circuits serve as the last barrier for the reduction of the injected ESD energy. These first stage devices are divided in two categories: a) current limiting devices and b) voltage limiting devices.

Current limiting devices, having very low resistance for low amplitude currents and higher resistance for high amplitude currents, are places in series with the sensitive component, as presented in Figure 23. This type of protection devices cannot however provide adequate ESD protection alone, mainly due to their long reaction times, which are not compatible with the very fast rise times of ESD pulses and also due to the high energy requirements for their reaction, which are significantly larger than the ESD energy injected. These devices can however contribute positively to the overall transient protection of the system when used in combination with other types of protection devices, such as the Transient Voltage Suppressors, discussed below and illustrated in Figure 24.

Voltage limiting devices, commonly referred to as Transient Voltage Suppressors (TVS) should be placed in parallel to the circuit to be protected, most commonly connecting each line to ground, presenting a high impedance path at lower voltages (where protection is not required) and a low impedance path at higher voltages where the circuits need to be protected.

Figure 23. Current limiting devices used as protection against ESD

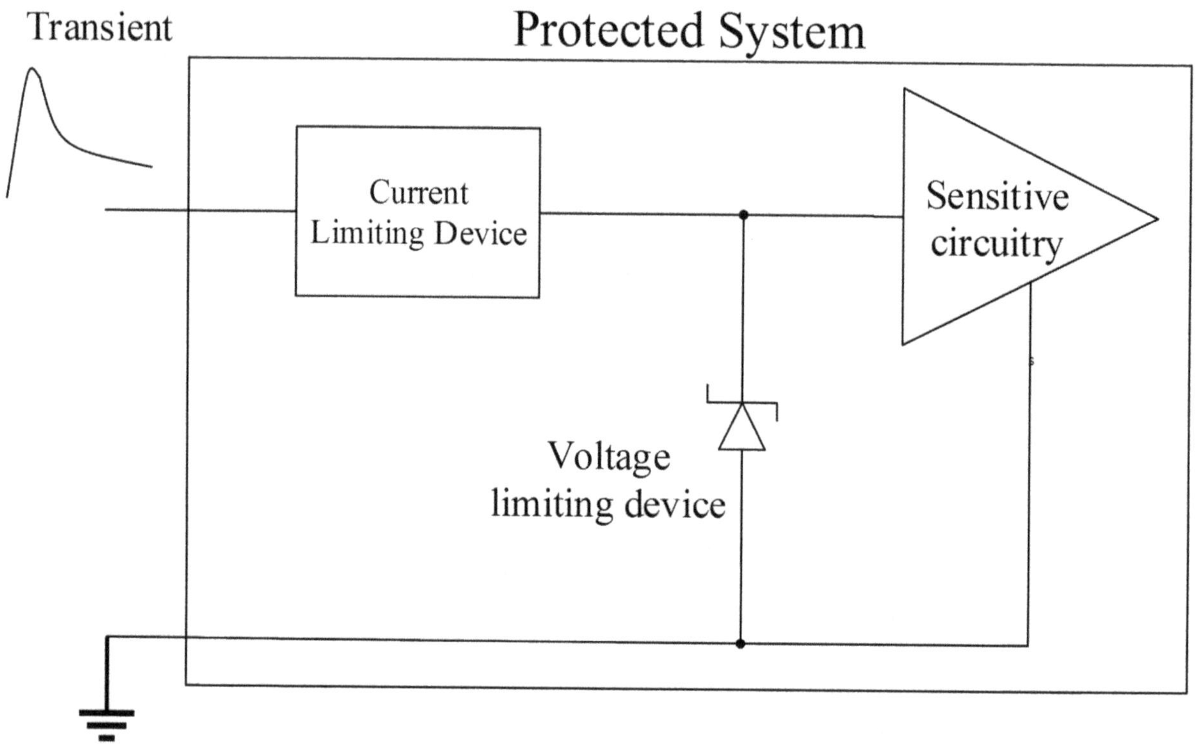

Voltage limiting devices, or Transient Voltage Suppressors (TVSs), are placed in parallel with the circuit to be protected, normally between a line and ground, and have a high impedance at low voltages and low impedance at higher voltages. At the low impedance mode, the injected ESD current is diverted away from the sensitive circuit inputs, ideally to ground. The most commonly used TVSs are listed and described briefly below (Duvvury & Gossner, 2015):

- **Silicon Based TVS devices**

Figure 24. Voltage limiting devices used as protection against ESD

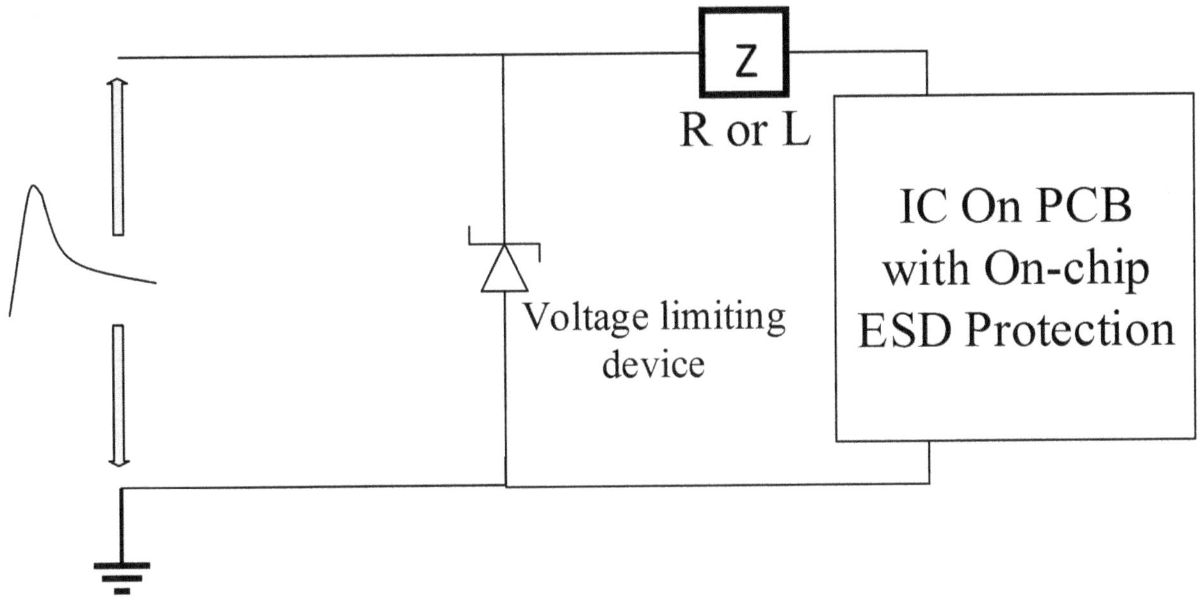

Diodes and more specifically Zener diodes are representative examples of these TVS devices. Despite being a cheap and easy to use measure, the Zener diode presents a big capacitance load and a high clamping voltage. The majority of the commercially available Zener diodes present a capacitance in the order of several pF, which is adequate for low or medium speed data lines, but is meanwhile too high for high or ultra-high speed lines, as they require a maximum capacitance value of < 1 pF (Jiang, 2012). A Zener diode might also be an inappropriate solution for low-voltage logic circuitry due to its relatively high clamping voltage (>5V). This is a serious drawback as the application of such circuits in modern equipment is constantly expanding.

- **Metal Oxide Varistors (MOVs) and Multilayer Varistors (MLVs)**

Varistors are another commonly used type of clamping devices, mainly fabricated by Zinc Oxide (ZnO) in combination with other oxides such as Bismuth, Cobalt and Manganese. Their major use is as a board level protection against most types of transient phenomena, due to their ability to withstand high currents. Their similarity to Zener diodes regarding their high capacitance makes them thereby a cost effective solution for power supply lines that can tolerate higher capacitance values and a good alternative for high wattage TVS diodes. Multi-line array configurations, the so called MLVs, are also commercially available providing additional high frequency noise filtering (Colby, 2011). Varistors are not recommended as the most suitable alternative for ESD protection due to their slow response time (>35ns) and their high ESD clamping voltage. Their nonlinear resistance characteristic poses an additional difficulty as their voltage drop increases dramatically with the increase of the current.

- **PCB Spark Gaps**

A very simple but also very cost-efficient solution for the protection against ESD events of a small PCB area is the installation of PCB level spark gaps, which are actually the bidirectional arrangement of two conducting electrodes, of different shapes, separated by air or a controlled gas (Gas Discharge Tubes, GDTs). For the suppression of the high breakdown current resulting from their high breakdown voltage (above 1.5kV) spark gaps require a series connection of a high resistance (of the kΩ or even MΩ order) thus increasing the required PCB area. A major disadvantage of spark gaps is the dependency on the environmental conditions. As it is known, breakdown voltage values increase with pressure. Other factors that may downgrade the insulating performance of spark gaps are factors related to the surface condition of the electrodes such as dust, dirt and other undesired traces on the electrodes.

All the aforementioned TVSs have been sufficient ESD protection measures over the last decades, but are starting to become obsolete in modern applications, mainly due to the emerge of high-speed data transmission and low-voltage IC logic. Each of them still finds applicability today but they are mostly used in conjunction with the two on-chip types of TVSs described below, the polymeric ESD Suppressors and the TVS Diode Arrays:

- **Polymeric ESD Suppressors**

Polymeric ESD Suppressors which are ideal for cases where ESD events are the main source of transient threat such as space applications and high-speed data, signal and control lines, present ultra-low capacitance values, in the range of 0.05-0.1pF, fast response times, in the range of 0.5-1ns, which are suitable for dealing with ESD pulses, and a very low leakage current reducing power consumption. Their only disadvantage in comparison to the TVS Diode arrays described below, is their relative high clamping voltage values (~ 30V).

- **TVS Diode Arrays**

A TVS Diode Array is an integrated multi-channel ESD protection device, with very small size, providing very low, but sufficient, peak values, as well as the lowest clamping voltage values available due to their low dynamic resistance. When installed together with a central TVS Diode, to which the ESD current will be directed, the absorbing capacity of these arrays is increased to 30kV ESD transients allowing multi-channel protection and a low capacitance load.

FUTURE RESEARCH DIRECTIONS

The transient electromagnetic phenomena and their effects on the spacecraft and the spacecraft electronic equipment have always presented challenges for product designers and EMC test engineers. The simulation of these phenomena in the space environment, contributing to the identification and better understanding of potential hazards, will play a significant role in the future research and experimentation efforts. International Standardization can also help considerably towards the effort of increasing the reproducibility of the proposed testing methodologies by specifying the testing requirements and parameters more strictly. The ESD test methodologies are an evident example of such cases, where reproducibility uncertainty affects the integrity of the testing procedure negatively. Finally, the collection of more real flight data, regarding power bus transient occurrences and ESD events would greatly benefit the overall effort of ensuring the transient immunity of the space system.

CONCLUSION

The performance evaluation of every space system to transient events which can either be induced externally, e.g. space ESD events or internally (e.g. power bus transients) has always been a concern during every space mission design. The most reliable approach towards this effort, as extensively described in this chapter, includes the identification of transient event sources and conditions, the implementation of the proper mitigation strategy (materials, protection devices, design guidelines) and the evaluation this strategy through reproducible, preferably Standardized, testing procedures. These approach aspects have been discussed in detail in this chapter, highlighting potential limitations and establishing the foundations for future research and experimentation. Summarizing, it should be noted that the mitigation procedure is a global process, ranging from investigation of a bonding connection at the furthest corner of the spacecraft's structure to the selection of the on-chip protection circuitry for an integrated circuit, leading therefor to the requirement of a solid, systematic testing and designing plan.

REFERENCES

Colby, J. (2011). *Application Note: ESD and Surge Circuit Protection*. Littelfuse, Inc.

Daout, B., Romero, C., Pelissou, P., & Wolf, K. J. (2016). New ESD test set-up for spacecraft equipment. *2016 ESA Workshop on Aerospace EMC (Aerospace EMC)*, 1-5. 10.1109/AeroEMC.2016.7504568

Department of Defense: United States of America. (1996). *MIL-STD-88G3, Test Method Standard: Microcircuits*. Author.

Department of Defense: United States of America. (2015). *MIL-STD-461G: Requirements for the Control of Electromagnetic Interference Characteristics of Subsystems and Equipment*. Author.

Department of the Air Force: United States of America. (1987). *MIL-STD-1541A (USAF), Electromagnetic compatibility, requirements for space systems*. Author.

Dunn, B. D. (2016). *Materials and processes - for Spacecraft and High Reliability Applications*. Springer.

Duvvury, C., & Gossner, H. (2015). *System level ESD CO-Design*. Wiley, IEEE Press. doi:10.1002/9781118861899

European Cooperation for Space Standardization (ECSS). (2008). *ECSS-E-ST-20-06C: Space Engineering Spacecraft charging*. Author.

European Cooperation for Space Standardization (ECSS). (2012). *ECSS-E-ST-20-07C Rev. 1: Space engineering, Electromagnetic compatibility*. Author.

Garrett, H. B., & Whittlesey, A. C. (2011). Guide to mitigating spacecraft charging effects. Jet Propulsion Laboratory, California Institute of Technology, JPL Space Science and Technology Series.

Iacovone, V., Scione, E., & Scorzafava, E. (2016). An ESD test approach for spacecraft applications. *2016 International Symposium on Electromagnetic Compatibility - EMC EUROPE*, 268-273.

International Electrotechnical Commission (IEC). (2009). IEC 61000-4-2:2009 - Electromagnetic compatibility (EMC). Testing and measurement techniques. Electrostatic discharge immunity test. Author.

International Standardization Organization (ISO). (2002). *ISO 14302:2002 Space systems -- Electromagnetic compatibility requirements.* ISO.

Jiang, R. H-C. (2012). *System Level ESD/EMI Issues and Protection Design.* Academic Press.

Lai, S. T. (2011). *Spacecraft and human electrostatic discharge: a comparison of the two phenomena, Fundamentals of Spacecraft Charging: Spacecraft Interactions with Space Plasmas*. Princeton University Press. doi:10.2307/j.ctvcm4j2n

National Aeronautics and Space Administration (NASA). (1984). *NASA Technical Paper 2361: Design Guidelines for Assessing and Controlling Spacecraft Charging Effects.* NASA.

National Aeronautics and Space Administration (NASA). (1993a). *NASA Preferred Reliability Practices, Practice No. PD-AP-1301: Surface Charging/ESD Analysis.* NASA.

National Aeronautics and Space Administration (NASA). (1993b). *NASA Preferred Reliability Practices, Practice No. PT-TE-1414: Electrostatic Discharge (ESD).* Test Practices.

National Aeronautics and Space Administration (NASA). (1999). *NASA Handbook on electrical transients Specification, Measurement, and Control of Electrical Switching Transients, NASA/CR-1999-209574.* NASA.

National Aeronautics and Space Administration (NASA). (2007). *Low Earth Orbit Spacecraft Charging Design Handbook, NASA-HDBK-4006.* NASA.

National Aeronautics and Space Administration (NASA). (2017) *Mitigating In-Space Charging Effects —A Guideline, NASA-HDBK-4002A w/CHANGE1.* NASA.

Pelissou, P., Daout, B., Romero, C., & Wolf, K.-F. J. (2016). Critical review of the ECSS-E-ST-20-07C ESD test set-up for testing spacecraft equipment. *2016 ESA Workshop on Aerospace EMC (Aerospace EMC)*, 1-4. 10.1109/AeroEMC.2016.7504567

Pelissou, P., Daout, B., & Wolf, K.-F. J. (2015). Improved ESD test method for testing spacecraft equipment. *Asia Pacific Symposium on Electromagnetic Compatibility (APEMC 2015).* 10.1109/APEMC.2015.7175316

Pisacane, V. L. (2016). *The Space Environment and its Effects on Space Systems.* AIAA Education Series. doi:10.2514/4.103537

Saenko, V., Tyutnev, A., Abrameshin, A., & Belik, G. (2017, August). Computer Simulations and Experimental Verification of the Nanoconductivity Concept for the Spacecraft Electronics. *IEEE Transactions on Plasma Science*, *45*(8), 1843–1846. doi:10.1109/TPS.2017.2688180

KEY TERMS AND DEFINITIONS

Electromagnetic Immunity (EMI): The ability of a system/apparatus to maintain its proper functionality under exposure to various sources of electromagnetic disturbance.

Electrostatic Discharge (ESD): A transient current pulse occurring during breakdown of a dielectric due to charge accumulation and different potential levels between conductive and/or non-conductive surfaces.

Geosynchronous Earth Orbits: An orbit around the Earth at 35786km altitude with a 24-hour period.

Power Line Switching Transients: Positive or negative spikes superimposed on the input terminals of equipment due to on/off switching of loads on a power bus.

Shielding: A method for the mitigation of electromagnetic disturbances which is based on the concept of creating an electrically continuous enclosure around the under protection structure, mainly by the use of conductive materials.

Spacecraft Charging Effects: The interaction of a spacecraft structure with the ambient space plasma translated either as charge deposition on the exterior surfaces of the spacecraft or as charge penetration into the interior.

Transient Protection Devices: Devices installed near the entry points of sensitive circuitry in order to limit overvoltages or overcurrents.

Compilation of References

Abramowitz, M., & Stegun, I. A. (1972). *Handbook of Mathematical Functions*. Dover.

Acuña, M. H. (2002). Space-based magnetometers. Review of Scientific Instruments. doi:10.1063/1.1510570

Ahmadi, A., & Mosallaei, H. (2010). Plasmonic nanoloop array antenna. *Optics Letters*, *35*(21), 3706–3708. doi:10.1364/OL.35.003706 PMID:21042398

Akhtar, M. J., Feher, L. E., & Thumm, M. (2006). A waveguide-based two-step approach for measuring complex permittivity tensor of uniaxial composite materials. *IEEE Transactions on Microwave Theory and Techniques*, *54*(5), 2011–2022. doi:10.1109/TMTT.2006.873623

Alberto, N., Domingues, M. F., Marques, C., André, P., & Antunes, P. (2018). Optical fiber magnetic field sensors based on magnetic fluid: A review. Sensors (Switzerland). doi:10.339018124325

Anastassiu, H. T. (2006). Fast, simple and accurate computation of the currents on an arbitrarily large circular loop antenna. *IEEE Transactions on Antennas and Propagation*, *54*(3), 860–866. doi:10.1109/TAP.2006.869929

Antonucci, F., Armano, M., Audley, H., Auger, G., Benedetti, M., Binetruy, P., Boatella, C., Bogenstahl, J., Bortoluzzi, D., Bosetti, P., Caleno, M., Cavalleri, A., Cesa, M., Chmeissani, M., Ciani, G., Conchillo, A., Congedo, G., Cristofolini, I., & Cruise, M., … Zweifel, P. (2011). LISA Pathfinder: Mission and status. *Classical and Quantum Gravity*. Advance online publication. doi:10.1088/0264-9381/28/9/094001

Ashton-Patton, M.M., Hall, M.M., & Shelby, J.E. (2006). Formation of Low Density Polyethylene/Hollow Glass Microspheres Composites. *Journal of Non-Crystalline Solids*, *352*(6-7), 615-619. doi: .2005.11.058 doi:10.1016/j.jnoncrysol

Badini, L., Grassi, F., Pignari, S. A., Spadacini, G., Bisognin, P., Pelissou, P., & Marra, S. (2016). Conducted-susceptibility testing as an alternative approach to unit-level radiated-susceptibility verifications. *2016 ESA Workshop on Aerospace EMC (Aerospace EMC), Valencia, 2016*, 1-5, 10.1109/AeroEMC.2016.7504582

Baklezos, A. T., Nikolopoulos, C. D., Capsalis, C. N., & Tsatalas, S. (2017, March). Effect of LVDS link speed and pattern length on spectrum measurements of a Spacewire harness. In *Antenna Technology: Small Antennas, Innovative Structures, and Applications (iWAT), 2017 International Workshop on* (pp. 38-41). IEEE. 10.1109/IWAT.2017.7915291

Baklezos, A. T., Nikolopoulos, C. D., Sigalas, I. R., & Capsalis, C. N. (2020, May 1). Measurement and modeling of spacewire radiation for electromagnetic compatibility assessment. *I2MTC 2020 - International Instrumentation and Measurement Technology Conference, Proceedings*. 10.1109/I2MTC43012.2020.9128502

Baklezos, A., Nikolopoulos, C., Vardiambasis, I., Kapetanakis, T., & Capsalis, C. (in press). Spacecraft Hull Effect on Radiated Emissions and Optimal Onboard Payload Allocation. *2020 International Symposium on Electromagnetic Compatibility (EMC EUROPE).*

Baklezos, A. T., Nikolopoulos, C. D., & Capsalis, C. N. (2017). An equivalent dipole method with novel measurement positioning for modeling electric emissions in space missions. *Electromagnetics*, *37*(7), 1–15. doi:10.1080/02726343.2017.1376902

Baklezos, A. T., Nikolopoulos, C. D., Katsouris, A. G., Koutantos, G. I., & Capsalis, C. N. (2016). Electromagnetic Emission Modeling in Case of Shielded Cabling with Respect to the Ground Dielectric Properties. *IEEE Transactions on Electromagnetic Compatibility*, *58*(6), 1694–1700. Advance online publication. doi:10.1109/TEMC.2016.2588583

Baklezos, A., Nikolopoulos, C. D., Spantideas, S., Chatzineofytou, E., Nicoletto, M., Marziali, I., & Capsalis, C. N. (2019). *Steady State Emissions Modeling of Low Frequency Magnetic and Electric Fields Generated by GOCE CDMU. In 2019 ESA Workshon on Aerospace EMC*. Aerospace EMC.

Balanis, C. A. (1982). *Antenna Theory, Analysis and Design*. John Wiley & Sons.

Bartington Inc. (2020a). *Operation Manual for HC1 Helmholtz Coils*. Retrieved September, 2020, from https://www.bartington.com/wp-content/uploads/pdfs/operation_manuals/HC1_OM3226.pdf

Bartington Inc. (2020b). *Operation Manual for HC2 Helmholtz Coils*. Retrieved September, 2020, from https://www.bartington.com/wp-content/uploads/pdfs/operation_manuals/HC2_OM3342.pdf

Bartington Inc. (2020c). *Operation Manual for TLMS-O Open and TLMS-C Capped Magnetic Shields*. Retrieved September, 2020, from https://www.bartington.com/wp-content/uploads/pdfs/operation_manuals/Magnetic_Shields_OM1938.pdf

Bayle, O., Lorenzoni, L., Blancquaert, T., Langlois, S., Walloschek, T., & Portigliotti, S. (2016). *Exomars entry descent and landing demonstrator mission and design overview*. Nasa Solar System.

Beale, M., Hagan, M., & Demuth, H. (2016). *Neural Network Toolbox: User's Guide (version 9)*. The MathWorks Inc.

Bedinger, K. L., Leach, R. D., & Alexander, M. B. (1996). *Spacecraft System Failures and Anomalies Attributed to the Natural Space Environment. NASA Reference Publication 1390, Marshall Space Flight Center (MSFC), National Aeronautics and Space Administration*. NASA.

Belk, C. A., Robinson, J. H., Alexander, M. B., Cooke, W. J., & Pavelitz, S. D. (1997). Meteoroids and Orbital Debris: Effects on Spacecraft. NASA Reference Publication 1408, Marshall Space Flight Center (MSFC), National Aeronautics and Space Administration (NASA).

Bel, T., Arslan, C., & Baydogan, N. (2019). Radiation Shielding Properties of Poly(Methyl Methacylate)/Colemanite Composite for the Use in Mixed Irradiation Fields of Neutrons and Gamma Rays. *Materials Chemistry and Physics*, *221*, 58–67. doi:10.1016/j.matchemphys.2018.09.014

Benkhoff, J., van Casteren, J., Hayakawa, H., Fujimoto, M., Laakso, H., Novara, M., Ferri, P., Middleton, H. R., & Ziethe, R. (2010). BepiColombo-Comprehensive exploration of Mercury: Mission overview and science goals. *Planetary and Space Science*, *58*(1-2), 2–20. Advance online publication. doi:10.1016/j.pss.2009.09.020

Bloch, F. (1946). Nuclear induction. *Physical Review*, *70*(7-8), 460–474. Advance online publication. doi:10.1103/PhysRev.70.460

Boschetti, D., Gervasio, G., & Marziali, I. (2012). Montecarlo Approach for Magnetic Cleanliness Evaluation on Spacecraft. *2012 ESA Workshop on Aerospace EMC*.

Bothmer, V., & Daglis, I. A. (2007). *Space Weather: Physics and Effects*. Springer. doi:10.1007/978-3-540-34578-7

Brownlee, J. (2011). Clever Algorithms - Nature-Inspired Programming Recipes. Search.

Brown, N. F. (2019). Space Scientific Instrument Taxonomy (SSIT). *IEEE Aerospace Conference Proceedings*. 10.1109/AERO.2019.8741740

Brown, P., Whiteside, B. J., Beek, T. J., Fox, P., Horbury, T. S., Oddy, T. M., Archer, M. O., Eastwood, J. P., Sanz-Hernández, D., Sample, J. G., Cupido, E., O'Brien, H., & Carr, C. M. (2014). Space magnetometer based on an anisotropic magnetoresistive hybrid sensor. *The Review of Scientific Instruments*, *85*(12), 125117. Advance online publication. doi:10.1063/1.4904702 PMID:25554336

Çakır, S., Şen, O., Çınar, M., Ayaydın, A., Çetintaş, M., & Üstüner, F. (2014). Effects of sensor positions on military radiated susceptibility tests. *29th Conference on Precision Electromagnetic Measurements (CPEM 2014)*, 584-585, 10.1109/CPEM.2014.6898520

Calders, S., Messios, N., Botek, E., De Donder, E., Kruglanski, M., Evans, H., & Rodgers, D. (2018, May). *Modeling the Space Environment and Its Effects on Spacecraft and Astronauts using SPENVIS*. Paper presented at 15th International Conference on Space Operations, Marseille, France. 10.2514/6.2018-2598

Capsalis, C. N., Nikolopoulos, C. D., Spantideas, S. T., Baklezos, A. T., Chatzineofytou, E. G., Koutantos, G. I., Boschetti, D., Marziali, I., Nicoletto, M., Tsatalas, S., Mehlem, K., & Junge, A. (2019). EMC Assessment for Pre-Verification of THOR Mission Electromagnetic Cleanliness Approach. *Proceedings of 2019 ESA Workshop on Aerospace EMC, Aerospace EMC 2019*. 10.23919/AeroEMC.2019.8788958

Cataldo, F., & Prata, M. (2019). Neutron Radiation Shielding with PUR Composites loaded with B_4C or Graphite. *Fullerenes and Carbon Nanostructures*, *27*(7), 531–537. doi:10.1080/1536383X.2019.1604512

Chancellor, J. C., Blue, R. S., Cengel, K. A., Auñón-Chancellor, S. M., Rubins, K. H., Katzgraber, H. G., & Kennedy, A. R. (2018). Limitations in Predicting the Space Radiation Health Risk for Exploration Astronauts. *NPJ Microgravity*, *4*(1), 8. doi:10.103841526-018-0043-2 PMID:29644336

Chatzineofytou, E. G., Spantideas, S. T., Nikolopoulos, C. D., Baklezos, A. T., Marziali, I., Nicoletto, M., Boschetti, D., & Capsalis, C. N. (n.d.). *Decoupling of ground plane effect on low frequency magnetic and electric field measurements & modeling*. Academic Press.

Choudhury, B., Thomas, S., & Jha, R. M. (2015). Implementation of soft computing optimization techniques in antenna engineering. *IEEE Antennas & Propagation Magazine*, *57*(6), 122–131. doi:10.1109/MAP.2015.2439612

Chowdhury, D. R., Singh, R., Reiten, M., Zhou, J., Taylor, A. J., & O'Hara, J. F. (2011). Tailored resonator coupling for modifying the terahertz metamaterial response. *Optics Express*, *19*(11), 10679–10685. doi:10.1364/OE.19.010679 PMID:21643323

Christodoulou, C., & Georgiopoulos, M. (2001). *Applications of Neural Networks in Electromagnetics*. Artech House.

Colby, J. (2011). *Application Note: ESD and Surge Circuit Protection*. Littelfuse, Inc.

Conway, J. T. (2005). New exact solution procedure for the near fields of the general thin circular loop antenna. *IEEE Transactions on Antennas and Propagation*, *53*(1), 509–517. doi:10.1109/TAP.2004.838804

Cook, B., & Walker, P. (2007). Reducing Electromagnetic Emissions From SpaceWire. In *DAta Systems in Aerospace*. DASIA.

Daniel, C. M. (1988). *The Theory of Magnetism I: Statics and Dynamics*. Springer.

Daout, B., Romero, C., Pelissou, P., & Wolf, K. J. (2016). New ESD test set-up for spacecraft equipment. *2016 ESA Workshop on Aerospace EMC (Aerospace EMC)*, 1-5. 10.1109/AeroEMC.2016.7504568

Dawar, D., & Ludwig, S. A. (2015). Differential evolution with dither and annealed scale factor. *IEEE SSCI 2014 - 2014 IEEE Symposium Series on Computational Intelligence - SDE 2014: 2014 IEEE Symposium on Differential Evolution, Proceedings*. 10.1109/SDE.2014.7031528

Department of Defense: United States of America. (1996). *MIL-STD-88G3, Test Method Standard: Microcircuits*. Author.

Department of Defense: United States of America. (2015). *MIL-STD-461G: Requirements for the Control of Electromagnetic Interference Characteristics of Subsystems and Equipment*. Author.

Department of the Air Force: United States of America. (1987). *MIL-STD-1541A (USAF), Electromagnetic compatibility, requirements for space systems*. Author.

Devabhaktuni, V., Yagoub, M. C. E., Fang, Y., Xu, J., & Zhang, Q. (2001). Neural networks for microwave modeling: Model development issues and nonlinear modeling techniques. *International Journal of RF and Microwave Computer-Aided Engineering*, *11*(1), 4–21. doi:10.1002/1099-047X(200101)11:1<4::AID-MMCE2>3.0.CO;2-I

Dever, J., Banks, B., de Groh, K., & Miller, S. (2012). Degradation of Spacecraft Materials. In M. Kutz (Ed.), *Handbook of Environmental Degradation of Materials* (2nd ed., pp. 717–770). Elsevier. doi:10.1016/B978-1-4377-3455-3.00024-9

Dias, R. A., Moreira, E. E., Alves, F. S., Mesquita, D., Gaspar, J., & Rocha, L. A. (2020). Design Optimization of a Lorentz Force, Amplitude Modulated, MEMS Space Magnetometer. *Proceedings of the IEEE International Conference on Micro Electro Mechanical Systems (MEMS)*. 10.1109/MEMS46641.2020.9056126

Díaz-Michelena, M. (2009). Small magnetic sensors for space applications. Sensors. doi:10.339090402271

Dragoi, E. N., Curteanu, S., & Vlad, D. (2012). Differential evolution applications in electromagnetics. *EPE 2012 - Proceedings of the 2012 International Conference and Exposition on Electrical and Power Engineering*. 10.1109/ICEPE.2012.6463801

Drinkwater, M. R., Floberghagen, R., Haagmans, R., Muzi, D., & Popescu, A. (2003). GOCE: ESA's first earth explorer core mission. *Space Science Reviews*, *108*(1/2), 419–432. Advance online publication. doi:10.1023/A:1026104216284

Drung, D., Abmann, C., & Beyer, J. (2007). Highly Sensitive and Easy-to-Use SQUID Sensors. *IEEE Transactions on Applied Superconductivity, Volume*, *17*(2), 699–704. doi:10.1109/TASC.2007.897403

Duffy, A. P., Martin, A. J., Orlandi, A., Antonini, G., Benson, T. M., & Woolfson, M. S. (2006). Feature selective validation (FSV) for validation of computational electromagnetics (CEM). part I-the FSV method. *IEEE Transactions on Electromagnetic Compatibility*, *48*(3), 449–459. doi:10.1109/TEMC.2006.879358

Dunn, B. D. (2016). *Materials and processes - for Spacecraft and High Reliability Applications*. Springer.

Duvvury, C., & Gossner, H. (2015). *System level ESD CO-Design*. Wiley, IEEE Press. doi:10.1002/9781118861899

ECSS-E50-12A vs ECSS-E-ST-50-12C, what's the difference? | STAR-Dundee. (2017). *Star-dundee.com*. Retrieved 15 June 2017, from https://www.star-dundee.com/knowledge-base/ecss-e50-12a-vs-ecss-e-st-50-12c-whats-difference

ECSS-E-ST-20-07C rev-1 Space engineering Electromagnetic compatibility, ECSS-E-ST-20-07C rev-1 (2012).

Elshafiey, I., Udpa, L., & Udpa, S. S. (1995). Solution of inverse problems in electromagnetics using Hopfield neural networks. *IEEE Transactions on Magnetics*, *31*(1), 852–861. doi:10.1109/20.364586

Enouf, K., & Hermant, S. (2016, October). Compact, impedance-matched SpaceWire connector development—"MicroMach SpaceWire": SpaceWire components, long paper. In SpaceWire Conference (SpaceWire), 2016 International (pp. 1-7). IEEE.

Escoubet, C. P., & Schmidt, R. (2000). Cluster II: Plasma measurements in three dimensions. *Advances in Space Research, 25*(7-8), 1305–1314. Advance online publication. doi:10.1016/S0273-1177(99)00639-0

European Cooperation for Space Standardization (ECSS) Secretariat. (2008). *ECSS- E-ST-50-12C, Space Engineering – SpaceWire- Links, nodes, routers and networks, ESA-ESTEC Noordwijk, The Netherlands, Requirements & Standards Division European Cooperation for Space Standardization (ECSS) Secretariat (2003), ESCC Detail Specification No. 3902/003, CABLE, "SPACEWIRE", ROUND, QUAD USING SYMMETRIC CABLES, FLEXIBLE, -200 TO +180 °C, ESA-ESTEC.* Requirements & Standards Division.

European Cooperation for Space Standardization (ECSS) Secretariat. (2012). *ECSS-E-ST-20-07C Rev. 1, Space Engineering –Electromagnetic compatibility, ESA-ESTEC Noordwijk.* Requirements & Standards Division.

European Cooperation for Space Standardization (ECSS) Secretariat. (2014). *ESCC Detail Specification No. 3902/003, CABLE, "SPACEWIRE", ROUND, QUAD USING SYMMETRIC CABLES, FLEXIBLE, -200 TO +180 °C, ESA-ESTEC.* Requirements & Standards Division.

European Cooperation for Space Standardization (ECSS) Secretariat. (2019). *ECSS- E-ST-50-12C Rev.1, Space Engineering – SpaceWire- Links, nodes, routers and networks, ESA-ESTEC Noordwijk.* Requirements & Standards Division.

European Cooperation for Space Standardization (ECSS). (2008). *ECSS-E-ST-20-06C: Space Engineering Spacecraft charging.* Author.

European Cooperation for Space Standardization (ECSS). (2012). *ECSS-E-HB-20-07A: Space engineering, Electromagnetic compatibility handbook.* Author.

European Cooperation for Space Standardization (ECSS). (2012). *ECSS-E-ST-20-07C Rev. 1: Space engineering, Electromagnetic compatibility.* Author.

Fakharzadeh, M., & Mohajer, M. (2014). An Integrated Wide-Band Circularly Polarized Antenna for Millimeter-Wave Applications. *IEEE Transactions on Antennas and Propagation, 62*(2), 925–929. doi:10.1109/TAP.2013.2290038

Fayazbakhsh, K., & Abedian, A. (2012). Materials Selection for Electronic Enclosures in Space Environment considering Electromagnetic Interference Effect. *Advances in Space Research, 49*(3), 586–593. doi:10.1016/j.asr.2011.10.025

Fikioris, G., Papakanellos, P. J., & Anastassiu, H. T. (2008). On the Use of Nonsingular Kernels in Certain Integral equations for Thin-Wire Circular-Loop Antennas. *IEEE Transactions on Antennas and Propagation, 56*(1), 151–157. doi:10.1109/TAP.2007.913076

Fikioris, G., Papakanellos, P. J., & Anastassiu, H. T. (2010). Corrections to "On the Use of Nonsingular Kernels in Certain Integral equations for Thin-Wire Circular-Loop Antennas". *IEEE Transactions on Antennas and Propagation, 58*(10), 3436. doi:10.1109/TAP.2010.2055816

Foner, S. (1959). Versatile and Sensitive Vibrating-Sample Magnetometer. *The Review of Scientific Instruments, 30*(7), 548–557. doi:10.1063/1.1716679

Fraden, J. (2010). *Handbook of Modern Sensors Physics, Designs, and Applications.* Springer-Verlag. doi:10.1007/978-1-4419-6466-3

Fratantuono, L. (2015). *A Reading of Lucretius' De Rerum Natura.* Lexington Books.

Freitas, P. P., Ferreira, R., Cardoso, F., & Cardoso, F. (2007). Magnetoresistive sensors. *Journal of Physics Condensed Matter, 19*(16), 165221. doi:10.1088/0953-8984/19/16/165221

Gandolfo, A., Azaro, R., & Festa, D. (2017). Improving the accuracy of radiated emission measurements for frequency below 30 MHz by using a fiber optic isolated rod antenna. *2017 IEEE International Symposium on Electromagnetic Compatibility & Signal/Power Integrity (EMCSI)*, 63-68, 10.1109/ISEMC.2017.8077842

Garrett, H. B., & Whittlesey, A. C. (2011). Guide to mitigating spacecraft charging effects. Jet Propulsion Laboratory, California Institute of Technology, JPL Space Science and Technology Series.

Gehani, A., & Pujara, D. A. (2015). Predicting the return loss performance of a hexa-band PIFA using ANFIS. *Microwave and Optical Technology Letters*, *57*(9), 2072–2075. doi:10.1002/mop.29277

Geng, C., Tang, X., Gong, C., Guan, F., Johns, J., Shu, D., & Chen, D. (2015). A Monte-Carlo-based Radiation Safety Assessment for Astronauts in an Environment with Confined Magnetic Field Shielding. *Journal of Radiological Protection*, *35*(4), 777–788. doi:10.1088/0952-4746/35/4/777 PMID:26484984

Gordon, D., Lundsten, R., & Chiarodo, R. (1965). Factors affecting the sensitivity of gamma-level ring-core magnetometers. *IEEE Transactions on Magnetics, Volume*, *1*(4), 330–337. doi:10.1109/TMAG.1965.1062987

Gosal, G., Almajali, E., McNamara, D., & Yagoub, M. (2016). Transmit array antenna design using forward and inverse neural network modeling. *IEEE Antennas and Wireless Propagation Letters*, *15*, 1483–1486. doi:10.1109/LAWP.2015.2514065

Gradshteyn, I. S., & Ryzhik, I. M. (1980). *Table of Integrals, Series, and Products*. Academic Press.

Griffin, D. K., Masseglia, O., Hall, M., Trougnou, L., Hewitson, M., Howe, C., Poyntz-Wright, O., Leopoldi, M., Ding, L., Turner, S., & Harmon, S. (2012). Design and calibration of a compact low-noise magnetic gradiometer. *Proceedings of the 2012 ESA Workshop on Aerospace EMC 2012*.

Grosz, A., Haji-Sheikh, J. M., & Mukhopadhyay, C. S. (2017). *High Sensitivity Magnetometers*. Springer. doi:10.1007/978-3-319-34070-8

Guetersloh, S., Zeitlin, C., Heilbronn, L., Miller, J., Komiyama, T., Fukumura, A., Iwata, Y., Murakami, T., & Bhattacharya, M. (2006). Polyethylene as a Radiation Shielding Standard in Stimulated Cosmic-Ray Environments. *Nuclear Instruments & Methods in Physics Research. Section B, Beam Interactions with Materials and Atoms*, *252*(2), 319–332. doi:10.1016/j.nimb.2006.08.019

Hadjigeorgiou, N., Hristoforou, E., & Sotiriadis, P. P. (2017). Closed-loop current-feedback, signal-chopped, low noise AMR sensor with high linearity. *IEEE Modern Circuits and Systems Technologies, 6*.

Hadjigeorgiou, N., Tsalikidou, A.C., Hristoforou, E., & Sotiriadis, P. P. (2017). Highly Linear and Low Noise AMR Sensor Using Closed Loop and Signal-Chopped Architecture. *World Academy of Science, Engineering and Technology, 11*(8).

Hadjigeorgiou, N. (2014). Experimental Study Noise of HMR2003 Sensor with or without the Use of Amplifier AMP04. *Key Engineering Materials*, *605*, 629–632. doi:10.4028/www.scientific.net/KEM.605.629

Hadjigeorgiou, N. G., & Sotiriadis, P. P. (2020). Parasitic Capacitances, Inductive Coupling, and High-Frequency Behavior of AMR Sensors. *IEEE Sensors Journal*, *20*(5), 2339–2347. Advance online publication. doi:10.1109/JSEN.2019.2953351

Hadjigeorgiou, N., Konstantopoulos, C., & Masxas, D. (2015). Fourier Analysis for Orthogonal and Parallel Fluxgate. *Key Engineering Materials*, *644*, 270–273. doi:10.4028/www.scientific.net/KEM.644.270

Hallen, E. (1938). Theoretical investigations into transmitting and receiving qualities of antennae. *Nova Acta Upsaliensis series IV*, *11*, 1–43.

Hamed, S. M. A. (2013). Exact field expressions for circular loop antennas using spherical functions expansion. *IEEE Transactions on Antennas and Propagation*, *61*(6), 2956–2963. doi:10.1109/TAP.2013.2250241

Han, F., Harada, S., & Sasada, I. (2012). Fluxgate and search coil hybrid: A low-noise wide-band magnetometer. *IEEE Transactions on Magnetics*, *48*(11), 3700–3703. Advance online publication. doi:10.1109/TMAG.2012.2196762

Harrison, C., Weaver, S., Bertelsen, C., Burgett, E., Hertel, N., & Grulke, E. (2008). Polyethylene/Boron Nitride Composites for Space Radiation Shielding. *Journal of Applied Polymer Science*, *109*(4), 2529–2538. doi:10.1002/app.27949

Harrison, E. P., Turney, G. L., & Rowe, H. (1935). Electrical Properties of Wires of High Permeability. *Nature*, *135*(3423), 961. doi:10.1038/135961a0

Hartman, E., Keeler, J., & Kowalski, J. (1990). Layered neural networks with gaussian hidden units as universal approximations. *Neural Computation*, *2*(2), 210–215. doi:10.1162/neco.1990.2.2.210

Hastings, D., & Garrett, H. (1996). *Spacecraft-Environment Interactions*. Cambridge University Press. doi:10.1017/CBO9780511525032

Haykin, S. (1999). *Neural Networks: A Comprehensive Foundation*. Pearson Education.

Hernandez, M. A., Lorenzo, J. M., Escolar, G. J., Sanz, L. D., Trillol, A. A., & Cancela, A. M. (2018). EMC Challenges for ESA Scientific Missions. *2018 International Symposium on Electromagnetic Compatibility (EMC EUROPE), Amsterdam, 2018*, 34-39. 10.1109/EMCEurope.2018.8485137

Herr, J. L., & McCollum, M. B. (1994). Spacecraft Environments Interactions: Protecting Against the Effects of Spacecraft Charging. NASA Reference Publication 1354, Marshall Space Flight Center (MSFC), National Aeronautics and Space Administration (NASA).

Honewell Inc. (2011). *Hall effect sensing and application*. Retrieved September 2017, from www.honeywell.com/sensing

Hornik, K., Stinchcombe, M., & White, H. (1989). Multilayer Feedforward Networks are Universal Approximators. *Neural Networks*, *2*(5), 359–366. doi:10.1016/0893-6080(89)90020-8

Hospodarsky, G. B. (2016). Spaced-based search coil magnetometers. In *Journal of Geophysical Research*. Space Physics. doi:10.1002/2016JA022565

Hristoforou, E., Chiriac, H., & Neagu, M. (1997). A New Magnetic Field Sensor Based on Magnetostrictive Delay Lines. *IEEE Transactions on Instrumentation and Measurement, Vol*, *46*(2), 632–635. doi:10.1109/19.571942

Hristoforou, E., & Reilly, R. E. (1991). Nonuniformity in amorphous ribbon delay lines after stress and current annealing. *Journal of Physics*, *69*, 5008.

Iacovone, V., Scione, E., & Scorzafava, E. (2016). An ESD test approach for spacecraft applications. *2016 International Symposium on Electromagnetic Compatibility - EMC EUROPE*, 268-273.

International Electrotechnical Commission (IEC). (2009). IEC 61000-4-2:2009 - Electromagnetic compatibility (EMC). Testing and measurement techniques. Electrostatic discharge immunity test. Author.

International Standardization Organization (ISO). (2002). *ISO 14302:2002 Space systems -- Electromagnetic compatibility requirements*. ISO.

Ishiyama, A., & Hirooka, H. (1991). Magnetic Shielding for MRI Superconducting Magnets. *IEEE Transactions on Magnetics, Volume*, *27*(2), 1692–1695. doi:10.1109/20.133514

Jackson, J. D. (2003). Electrodynamics, Classical. Digital Encyclopedia of Applied Physics. doi:10.1002/3527600434.eap109

James, B. F., Norton, O. W., & Alexander, M. B. (1994). The Natural Space Environment: Effects on Spacecraft. NASA Reference Publication 1350. Marshall Space Flight Center (MSFC), National Aeronautics and Space Administration (NASA).

Jang, J.-S. R. (1993). ANFIS: Adaptive-Network-Based Fuzzy Inference Systems. *IEEE Transactions on Systems, Man, and Cybernetics*, *23*(3), 665–685. doi:10.1109/21.256541

Jang, T. S., Rhee, J., Hyun, B.-S., Seo, H.-S., Kim, T. Y., & Seo, J. K. (2016). A Study of Grid-Stiffened Multi-Functional Composite Structures with Radiation Spot Shielding. *Aerospace Science and Technology*, *54*, 330–339. doi:10.1016/j.ast.2016.05.002

Jiang, R. H-C. (2012). *System Level ESD/EMI Issues and Protection Design.* Academic Press.

Jiles, D. (2015). *Introduction to Magnetism and Magnetic Materials*. CRC Press. doi:10.1201/b18948

Jiles, D. C., & Atherton, D. L. (1984). Theory of ferromagnetic hysteresis. *Journal of Applied Physics*, *55*(6), 2115–2120. doi:10.1063/1.333582

Jordan, C. E., & Balmain, G. K. (1968). *Electromagnetic Waves and Radiation Systems*. Prentice-Hall, Inc.

Josephson, B. D. (1962). Possible new effects in superconductive tunnelling. *Elsevier Physics Letters, Volume*, *1*(7), 251–254.

Junge, A., & Marliani, F. (2011). Prediction of DC Magnetic Fields for Magnetic Cleanliness on Spacecraft. *2011 IEEE International Symposium on Electromagnetic CompatibilityWorkshop on Aerospace EMC*. 10.1109/ISEMC.2011.6038424

Kabir, H., Wang, Y., Yu, M., & Zhang, Q. J. (2008). Neural network inverse modeling and applications to microwave filter design. *IEEE Transactions on Microwave Theory and Techniques*, *56*(4), 2811–2813. doi:10.1109/TMTT.2008.919078

Kamsali, N., Chakravarty, S. C., & Basuvaraj, P. K. (2019). Investigation of HZE Particle Fluxes as a Space Radiation Hazard for Future Mars Missions. *Heliyon*, *5*(12), e02972. doi:10.1016/j.heliyon.2019.e02972 PMID:31867459

Kapetanakis, T. N., Vardiambasis, I. O., Ioannidou, M. P., & Maras, A. (2018b). Neural network modeling for the solution of the inverse loop antenna radiation problem. *IEEE Transactions on Antennas and Propagation*, *66*(11), 6283–6290. doi:10.1109/TAP.2018.2869136

Kapetanakis, T. N., Vardiambasis, I. O., Liodakis, G., & Maras, A. (2012a). Neural network solution of the circular loop antenna radiation problem. In *Proceedings of the 20th Telecommunications Forum* (pp. 1193-1196). 10.1109/TELFOR.2012.6419428

Kapetanakis, T. N., Vardiambasis, I. O., Liodakis, G., & Maras, A. (2012b). Solving the inverse loop antenna radiation problem using a hybrid neuro-fuzzy system. In *Proceedings of the 20th Telecommunications Forum* (pp. 1189-1192). 10.1109/TELFOR.2012.6419427

Kapetanakis, T. N., Vardiambasis, I. O., Lourakis, E. I., & Maras, A. (2018a). Applying neuro-fuzzy soft computing techniques to the circular loop antenna radiation problem. *IEEE Antennas and Wireless Propagation Letters*, *17*(9), 1673–1676. doi:10.1109/LAWP.2018.2862939

Karlsson, I. (2015, May). Common mode effects in high speed serial links. In *2015 IEEE 19th Workshop on Signal and Power Integrity (SPI)* (pp. 1-4). IEEE. 10.1109/SaPIW.2015.7237394

Kartashov, D., & Shurshakov, V. (2018). Analysis of Space Radiation Exposure Level at Different Shielding Configurations by Ray-Tracing Dose Estimation Method. *Acta Astronautica*, *144*, 320–330. doi:10.1016/j.actaastro.2018.01.013

Kayabasi, A., Toktas, A., Akdagli, A., Bicer, M. B., & Ustun, D. (2014). Applications of ANN and ANFIS to predict the resonant frequency of L-shaped compact microstrip antennas. *Applied Computational Electromagnetics Society Journal, 29*, 460–469.

King, R. W. P. (1969). The loop antenna for transmission and reception. In R. E. Collin & F. J. Zucker (Eds.), *Antenna Theory* (pp. 458–481). McGraw-Hill.

Kleiman, J. I., Iskanderova, Z. A., Pérez, F. J., & Tennyson, R. C. (1995). Protective Coatings for LEO Environments in Spacecraft Applications. *Surface and Coatings Technology, 76-77*(Part 2), 827–834. doi:10.1016/0257-8972(95)02497-2

Korth, H., Strohbehn, K., Tejada, F., Andreou, A. G., Kitching, J., Knappe, S., Lehtonen, S. J., London, S. M., & Kafel, M. (2016). Miniature atomic scalar magnetometer for space based on the rubidium isotope 87Rb. *Journal of Geophysical Research. Space Physics, 121*(8), 7870–7880. Advance online publication. doi:10.1002/2016JA022389 PMID:27774373

Kountantos, G., Nikolopoulos, C., Baklezos, A., & Capsalis, C. N. (2019). Proper Equipment Ordinance for Achieving EM Cleanliness in Space Missions: The case of ELF Electric Sources. *IEEE Transactions on Electromagnetic Compatibility, 62*(5), 1686–1692. doi:10.1109/TEMC.2019.2937379

Koutantos, G. I., Nikolopoulos, C. D., Baklezos, A. T., & Capsalis, C. N. (2017). On the Modeling of ELF Electric Fields for Space Mission Equipment. *IEEE Transactions on Electromagnetic Compatibility, 59*(5), 1–8. doi:10.1109/TEMC.2017.2685199

Lai, S. T. (2011). *Spacecraft and human electrostatic discharge: a comparison of the two phenomena, Fundamentals of Spacecraft Charging: Spacecraft Interactions with Space Plasmas*. Princeton University Press. doi:10.2307/j.ctvcm4j2n

Lapeña, E., Herranz, J. L., Gómez-Carpintero, F., & Rodríguez, M. (2017). The LEO PCDU EVO - A Modular and Flexible Concept for Low to Medium Power LEO & Scientific Missions. *E3S Web of Conferences*. doi:10.1051/e3sconf/20171618009

Lassakeur, A., & Underwood, G. (2019). Magnetic Cleanliness Program on Cubesats For Improved Attitude Stability. *2019 9th International Conference on Recent Advances in Space Technologies (RAST).*

Le Contel, O., Leroy, P., Roux, A., Coillot, C., Alison, D., Bouabdellah, A., Mirioni, L., Meslier, L., Galic, A., Vassal, M. C., Torbert, R. B., Needell, J., Rau, D., Dors, I., Ergun, R. E., Westfall, J., Summers, D., Wallace, J., Magnes, W., … de la Porte, B. (2016). The Search-Coil Magnetometer for MMS. Space Science Reviews. doi:10.100711214-014-0096-9

Leininger, M., Thurecht, F., Pfeiffer, E., & Ruddle, A. (2012, May). Advanced grounding methods in the presence of carbon fibre reinforced plastic structures. In *Aerospace EMC, 2012 Proceedings ESA Workshop on* (pp. 1-6). IEEE.

Lellouch, G., Mishra, A. K., & Inggs, M. (2016). Design of OFDM radar pulses using genetic algorithm based techniques. *IEEE Transactions on Aerospace and Electronic Systems, 52*(4), 1953–1966. doi:10.1109/TAES.2016.140671

Lenz, J. E. (1990). A Review of Magnetic Sensors. *Proceedings of the IEEE.* 10.1109/5.56910

Lenz, J., & Edelstein, A. S. (2006). Magnetic sensors and their applications. *IEEE Sensors Journal, 6*(3), 631–649. Advance online publication. doi:10.1109/JSEN.2006.874493

Letfullin, R. R., George, T. F., & Ramazanov, A. K. (2019). Multifunctional Cosmic-Ray Shielding of Spacecraft with Elements of Systems Engineering Design. *Journal of Spacecraft and Rockets, 56*(5), 1312–1321. doi:10.2514/1.A34440

Li, L. W., Leong, M. S., Kooi, P. S., & Yeo, T. S. (1997). Exact solutions of electromagnetic fields in both near and far zones radiated by thin circular-loop antennas: A general representation. *IEEE Transactions on Antennas and Propagation, 45*(12), 1741–1748. doi:10.1109/8.650191

Lin, Q., Yang, B., Li, J., Meng, X., & Shen, J. (2000). Synthesis, Characterization and Property Studies of Pb^{2+}-containing Optical Resins. *Polymer*, *41*(23), 8305–8309. doi:10.1016/S0032-3861(00)00212-3

Liu, C., & Kang, X. (2019). Facile Fabrication of Conductive Silver Films on Carbon Fiber Fabrics via Two Components Spray Deposition Technique for Electromagnetic Interference Shielding. *Applied Surface Science*, *487*, 1245–1252. doi:10.1016/j.apsusc.2019.04.186

Liu, L., He, L., Yang, C., Zhang, W., Jin, R.-G., & Zhang, L.-Q. (2004). In situ Reaction and Radiation Protection Properties of $Gd(AA)_3$/NR Composites. *Macromolecular Rapid Communications*, *25*(12), 1197–1202. doi:10.1002/marc.200400077

Lowrie, W. (2007). *Fundamentals of Geophysics*. Cambridge University Press. doi:10.1017/CBO9780511807107

Low, T. S., & Chao, B. (1992). The use of finite elements and neural networks for the solution of inverse electromagnetic problems. *IEEE Transactions on Magnetics*, *28*(5), 2811–2813. doi:10.1109/20.179635

Lu, B. Q., Nagar, J., Yue, T., Pantoja, M. F., & Werner, D. H. (2017). Closed-form expressions for the radiation properties of nanoloops in the Terahertz, infrared and optical regimes. *IEEE Transactions on Antennas and Propagation*, *65*(1), 121–133. doi:10.1109/TAP.2016.2624150

Mager, J. A. (1970). Magnetic Shields. *IEEE Transactions on Magnetics, Volume*, *6*(1), 67–75. doi:10.1109/TMAG.1970.1066714

Mahdi, A. E., Panina, L., & Mapps, D. (2003). Some new horizons in magnetic sensing: High-Tc SQUIDs, GMR and GMI materials. *Sensors and Actuators. A, Physical*, *105*(3), 271–285. Advance online publication. doi:10.1016/S0924-4247(03)00106-7

McCloskey, J. (2016). EMC testing on the Integrated Science Instrument Module (ISIM): A summary of the EMC test campaign for the science payload of the James Webb Space Telescope (JWST). *2016 IEEE International Symposium on Electromagnetic Compatibility (EMC)*, 138-143. 10.1109/ISEMC.2016.7571632

McKinley, A. (2019). *The Analytical Foundations of Loop Antennas and Nano-scale Rings*. Springer. doi:10.1007/978-981-13-5893-7

McKinley, A. F., White, T. P., & Catchpole, K. R. (2013a). Theory of the circular closed loop antenna in the terahertz, infrared, and optical regions. *Journal of Applied Physics*, *114*(4), 044317-1, 044317–10. doi:10.1063/1.4816619

McKinley, A. F., White, T. P., & Catchpole, K. R. (2013b). Designing nanoloop antenna arrays for light-trapping in solar cells. *Proceedings of Photovolt. Specialists Conference (PVSC)*, 1894-1896.

Memarzadeh, B., & Mosallaei, H. (2011). Array of planar plasmonic scatterers functioning as light concentrator. *Optics Letters*, *36*(13), 2569–2571. doi:10.1364/OL.36.002569 PMID:21725482

Merrill, R. T., & McElhinny, M. W. (1983). The Earth's magnetic field: Its history, origin and planetary perspective. San Francisco: Academic Press.

Metglas Ind. (2011). *Magnetic Alloy*. Retrieved October, 2017, from www.metglas.com

Michelena, M. D., Rivero, M. A., Frutos, J. d., Ordóñez-Cencerrado, A., & Mesa, J. (2019). Adaption of Magnetic Cleanliness Facilities and Procedures to Overcome the New Challenges of the Scientific Missions. *2019 ESA Workshop on Aerospace EMC (Aerospace EMC)*. 10.23919/AeroEMC.2019.8788959

Miles, D. M., Mann, I. R., Ciurzynski, M., Barona, D., Narod, B. B., Bennest, J. R., Pakhotin, I. P., Kale, A., Bruner, B., Nokes, C. D. A., Cupido, C., Haluza-DeLay, T., Elliott, D. G., & Milling, D. K. (2016). A miniature, low-power scientific fluxgate magnetometer: A stepping-stone to cube-satellite constellation missions. *Journal of Geophysical Research. Space Physics*, *121*(12), 11,839–11,860. Advance online publication. doi:10.1002/2016JA023147

Mishra, R. K. (2001). An overview of neural network methods in computational electromagnetics. *International Journal of RF and Microwave Computer-Aided Engineering*, *12*(1), 98–108. doi:10.1002/mmce.10009

Mishra, S., Yadav, R. N., & Singh, R. P. (2015). Directivity estimations for short dipole antenna arrays using radial basis function neural networks. *IEEE Antennas and Wireless Propagation Letters*, *14*, 1219–1222. doi:10.1109/LAWP.2015.2399453

Morse, P. M., & Feshbach, H. (1953). *Methods of Theoretical Physics, Part II*. McGraw-Hill.

Müller, D., Marsden, R. G., St. Cyr, O. C., & Gilbert, H. R. (2013). Solar Orbiter: Exploring the Sun-Heliosphere Connection. *Solar Physics*. Advance online publication. doi:10.100711207-012-0085-7

Nagar, J., Lu, B. Q., Pantoja, M. F., & Werner, D. H. (2017). Analytical expressions for the mutual coupling of loop antennas valid from the RF to optical regimes. *IEEE Transactions on Antennas and Propagation*, *65*(12), 6889–6903. doi:10.1109/TAP.2017.2754411

National Aeronautics and Space Administration (NASA). (1984). *NASA Technical Paper 2361: Design Guidelines for Assessing and Controlling Spacecraft Charging Effects*. NASA.

National Aeronautics and Space Administration (NASA). (1993a). *NASA Preferred Reliability Practices, Practice No. PD-AP-1301: Surface Charging/ESD Analysis*. NASA.

National Aeronautics and Space Administration (NASA). (1993b). *NASA Preferred Reliability Practices, Practice No. PT-TE-1414: Electrostatic Discharge (ESD)*. Test Practices.

National Aeronautics and Space Administration (NASA). (1999). *NASA Handbook on electrical transients Specification, Measurement, and Control of Electrical Switching Transients, NASA/CR-1999-209574*. NASA.

National Aeronautics and Space Administration (NASA). (2007). *Low Earth Orbit Spacecraft Charging Design Handbook, NASA-HDBK-4006*. NASA.

National Aeronautics and Space Administration (NASA). (2017) *Mitigating In-Space Charging Effects —A Guideline, NASA-HDBK-4002A w/CHANGE1*. NASA.

Nenarokomov, A. V., Alifanov, O. M., Krainova, I. V., Titov, D. M., & Morzhukhina, A. V. (2019). Estimation of Environmental Influence on Spacecraft Materials Radiative Properties by Inverse Problems Technique. *Acta Astronautica*, *160*, 323–330. doi:10.1016/j.actaastro.2019.04.014

Nicoletto, M., Boschetti, D., & Savi, P. (2014, September). High Speed Digital Lines routed on non-metallic Spacecraft structures. In *Electromagnetic Compatibility (EMC Europe), 2014 International Symposium on* (pp. 785-789). IEEE. 10.1109/EMCEurope.2014.6931011

Nikolopoulos, C. D., Baklezos, A. T., & Capsalis, C. N. (2019). Measuring Transient and Steady State Electric Field Emissions of Space Equipment for EMC and Cleanliness Purposes. *ICHVE 2018 - 2018 IEEE International Conference on High Voltage Engineering and Application*. 10.1109/ICHVE.2018.8642200

Nikolopoulos, C. D., Baklezos, A. T., Tsatalas, S., & Capsalis, C. N. (2019). Verification of Radiated Emissions Modeling for SpaceWire/ LVDS Links Routed on CFRP Ground. *IEEE Transactions on Aerospace and Electronic Systems*, 1–1. doi:10.1109/taes.2019.2914540

Nikolopoulos, C. D., Baklezos, A. T., Tsatalas, S., & Capsalis, C. N. (2020). Verification of Radiated Emissions Modeling for SpaceWire/LVDS Links Routed on CFRP Ground. *IEEE Transactions on Aerospace and Electronic Systems*, *56*(1), 393–402. doi:10.1109/TAES.2019.2914540

Nikolopoulos, C., Baklezos, A., & Capsalis, C. (2018). Measuring Transient and Stedy State Electric Field Emissions of Space Equipment for EMC and Cleanliness Purposes. *IEEE International Conference on High Voltage Engineering and Application (ICHVE)*, 1-4.

Nikolopoulos, C., Baklezos, A., & Capsalis, C. (2020). On Achieving Spacecraft Level Magnetic Cleanliness with Proper Equipment Ordinance of DC and ELF Magnetic Sources. *IEEE Transactions on Electromagnetic Compatibility*, 1–11. doi:10.1109/TEMC.2020.2992682

Novotny, D., Petrucha, V., Dressler, M., & Platil, A. (2020). AMR magnetometer with digital feedback for space applications. *I2MTC 2020 - International Instrumentation and Measurement Technology Conference, Proceedings*. 10.1109/I2MTC43012.2020.9129039

Obiekezie, C. S., Thomas, D. W. P., Nothofer, A., Greedy, S., Arnaut, L. R., & Sewell, P. (2014). Complex locations of equivalent dipoles for improved characterization of radiated emissions. *IEEE Transactions on Electromagnetic Compatibility*, *56*(5), 1087–1094. Advance online publication. doi:10.1109/TEMC.2014.2313406

Obiekezie, C., Thomas, D. W., Nothofer, A., Greedy, S., Arnaut, L. R., & Sewell, P. (2013). Extended scheme using equivalent dipoles for characterizing edge currents along a finite ground plane. *Applied Computational Electromagnetics Society Journal*.

Oliveri, G., Rocca, P., & Massa, A. (2012). Differential evolution as applied to electromagnetics: Advances, comparisons, and applications. *Proceedings of 6th European Conference on Antennas and Propagation, EuCAP 2012*. 10.1109/EuCAP.2012.6206056

Orlandi, A., Duffy, A. P., Archambeault, B., Antonini, G., Coleby, D. E., & Connor, S. (2006). Feature selective validation (FSV) for validation of computational electromagnetics (CEM). part II-assessment of FSV performance. *IEEE Transactions on Electromagnetic Compatibility*, *48*(3), 460–467. doi:10.1109/TEMC.2006.879360

Overfelt, P. L. (1996). Near fields of the constant current thin circular loop antenna of arbitrary radius. *IEEE Transactions on Antennas and Propagation*, *44*(2), 166–171. doi:10.1109/8.481643

Pantoja, M. F., Nagar, J., Lu, B., & Werner, D. H. (2017). Existence of superdirective radiation models in thin-wire nanoloops. *ACS Photonics*, *4*(3), 509–516. doi:10.1021/acsphotonics.6b00486

Parkes, S. (2012). SpaceWire User's Guide. STAR-Dundee Limited.

Park, J., & Sandberg, I. (1991). Universal approximation using radial-basis-function networks. *Neural Computation*, *3*(2), 246–257. doi:10.1162/neco.1991.3.2.246 PMID:31167308

Paul, C. R. (2006). Introduction to Electromagnetic Compatibility: Second Edition. doi:10.1002/0471758159

Pelissou, P., Daout, B., Romero, C., & Wolf, K.-F. J. (2016). Critical review of the ECSS-E-ST-20-07C ESD test setup for testing spacecraft equipment. *2016 ESA Workshop on Aerospace EMC (Aerospace EMC)*, 1-4. 10.1109/AeroEMC.2016.7504567

Pelissou, P., Daout, B., & Wolf, K.-F. J. (2015). Improved ESD test method for testing spacecraft equipment. *Asia Pacific Symposium on Electromagnetic Compatibility (APEMC 2015)*. 10.1109/APEMC.2015.7175316

Pendry, J. B., Holden, A. J., Robbins, D. J., & Stewart, W. J. (1999). Megnetism from conductors and enhanced nonlinear phenomena. *IEEE Transactions on Microwave Theory and Techniques*, *47*(11), 2075–2084. doi:10.1109/22.798002

Pisacane, V. L. (2016). *The Space Environment and its Effects on Space Systems*. AIAA Education Series. doi:10.2514/4.103537

Piscane, V. L. (2008). The Space Environment and Its Effects on Space Systems. Reston, VA: American Institute of Aeronautics and Astronautics (AIAA). doi:10.2514/4.862533

Pocklington, H. C. (1897). Electrical oscillations in wire. *Cambridge Philosophical Society Proceedings*, *9*, 324.

Polirpo, A., & Cucca, M. (2012). New Facility For S/C Cleanliness Program. *2012 ESA Workshop on Aerospace EMC*.

Powell, M. (1987). Radial basis functions for multivariable interpolation: a review. In J. Mason & M. Cox (Eds.), *Algorithms for Approximation* (pp. 143–167). Clarendon Press.

Prudenziati, M. (1999). *Handbook of Sensors and Actuators. Thick Film Sensors*. Elsevier.

Pujara, D., Modi, A., Pisharody, N., & Mehta, J. (2014). Predicting the performance of pyramidal and corrugated horn antennas using ANFIS. *IEEE Antennas and Wireless Propagation Letters*, *13*, 293–296. doi:10.1109/LAWP.2014.2305518

Rask, J., Vercoutere, W., Navarro, B.J. & Krause, A. (2008). *Space Faring: The Radiation Challenge, An Interdisciplinary Guide on Radiation and Human Space Flight*. NASA.

Rawal, S. (2018). Materials and Structures Technology Insertion into Spacecraft Systems: Successes and Challenges. *Acta Astronautica*, *146*, 151–160. doi:10.1016/j.actaastro.2018.02.046

Ripka, P. (1993). Race-track fluxgate sensors. *Sensors and Actuators A. Physical, Volume*, *37-38*, 417–421.

Ripka, P. (2001). *Magnetic Sensors and Magnetometers*. Artech House Publishers.

Ripka, P., & Tipek, A. (2007). *Modern Sensors Handbook*. ISTE. doi:10.1002/9780470612231

Roberge, V., Tarbouchi, M., & Labonté, G. (2018). Fast genetic algorithm path planner for fixed-wing military UAV using GPU. *IEEE Transactions on Aerospace and Electronic Systems*, *54*(5), 2105–2117. doi:10.1109/TAES.2018.2807558

Saenko, V., Tyutnev, A., Abrameshin, A., & Belik, G. (2017, August). Computer Simulations and Experimental Verification of the Nanoconductivity Concept for the Spacecraft Electronics. *IEEE Transactions on Plasma Science*, *45*(8), 1843–1846. doi:10.1109/TPS.2017.2688180

Samwel, S. W., El-Aziz, E. A., Garrett, H. B., Hady, A. A., Ibrahim, M., & Amin, M. Y. (2019). Space Radiation Impact on Smallsats during Maximum and Minimum Solar Activity. *Advances in Space Research*, *64*(1), 239–251. doi:10.1016/j.asr.2019.03.025

Scheeres, D. J., Marzari, F., Tomasella, L., & Vanzani, V. (1998). ROSETTA mission: Satellite orbits around a cometary nucleus. *Planetary and Space Science*, *46*(6-7), 649–671. Advance online publication. doi:10.1016/S0032-0633(97)00200-6

Schnecker, M. (2016). Verifying MIL-STD-461G CS101 using frequency domain measurements on an oscilloscope. *2016 IEEE International Symposium on Electromagnetic Compatibility (EMC)*, 847-851. 10.1109/ISEMC.2016.7571760

Secretariat, E. C. S. S. (2019). *ECSS-E-ST-50-12C Rev. 1, Space Engineering – SpaceWire- Links, nodes, routers and networks*. ESA-ESTEC Requirements & Standards Division.

Sen, O., Cakir, S., Celep, M., Cinar, M., Hamid, R., & Cetintas, M. (2016). Influence of dielectric support on military radiated emission tests above 30 MHz. *2016 Asia-Pacific International Symposium on Electromagnetic Compatibility (APEMC)*, 709-711. 10.1109/APEMC.2016.7522843

Shelby, R. A., Smith, D. R., Nemat-Nasser, S. C., & Schultz, S. (2001). Microwave transmission through a two-dimensional, isotropic, left-handed metamaterial. *Applied Physics Letters*, *78*(4), 489–491. doi:10.1063/1.1343489

Shuvalov, V. A., Gorev, N. B., Tokmak, N. A., & Kuchugurnyi, Y. P. (2020). Drag on a Spacecraft produced by the Interaction of Its Magnetic Field with the Earth's Ionosphere. Physical Modelling. *Acta Astronautica*, *166*, 41–51. doi:10.1016/j.actaastro.2019.10.018

Silverman, E.M. (1995a). *Space Environmental Effects on Spacecraft: LEO Materials Selection Guide*. Langley Research Center, NASA Contractor Report 4661 Part 1.

Silverman, E.M. (1995b). *Space Environmental Effects on Spacecraft: LEO Materials Selection Guide*. Langley Research Center, NASA Contractor Report 4661 Part 2.

Simovski, C., Morits, D., Voroshilov, P., Guahva, M., Belov, P., & Kivshar, Y. (2013). Enhanced efficiency of light-trapping nanoantenna arrays for thin-film solar cells. *Optics Express*, *21*(S4), A714–A725. doi:10.1364/OE.21.00A714 PMID:24104498

Smith, D. R., Padilla, W. J., Vier, D. C., Nemat-Nasser, S. C., & Schultz, S. (2000). Composite medium with simultaneous negative permeability and permittivity. *Physical Review Letters*, *84*(18), 4184–4187. doi:10.1103/PhysRevLett.84.4184 PMID:10990641

Smith, G. S. (2007). Loop antennas. In R. C. Johnson & H. Jasik (Eds.), *Antenna Engineering Handbook* (pp. 5-1–5-24). McGraw-Hill.

SpaceWire - The Standard. (2017). *Spacewire.esa.int*. Retrieved 15 June 2017, from http://spacewire.esa.int/content/Standard/Standard.php

Specht, D. F. (1991). A general regression neural network. *IEEE Transactions on Neural Networks*, *2*(6), 568–576. doi:10.1109/72.97934 PMID:18282872

Spillantini, P., Taccetti, F., Papini, P., & Rossi, L. (2000). Radiation Shielding of Spacecraft in Manned Interplanetary Flights. *Nuclear Instruments & Methods in Physics Research. Section A, Accelerators, Spectrometers, Detectors and Associated Equipment*, *443*(2-3), 254–263. doi:10.1016/S0168-9002(99)01091-8 PMID:11543201

Storer, J. E. (1956). Impedance of thin-wire loop antennas. *AIEE Transactions. Part I Communications and Electronics*, *9*, 606–619.

Storm, R., & Price, K. (1995). Differential Evolution - a simple and efficient adaptive scheme for global optimization over continuous spaces. *Journal of Global Optimization*, 23.

Storn, R., & Price, K. (1997). Differential Evolution - A Simple and Efficient Heuristic for Global Optimization over Continuous Spaces. *Journal of Global Optimization*, *11*(4), 341–359. Advance online publication. doi:10.1023/A:1008202821328

Telecommunications Industry Association. (2001). *ANSI/TIA/EIA-644, Electrical Characteristics of Low Voltage Differential Signaling (LVDS) Interface Circuits*. Standards and Technology Department.

Tishkevich, D. I., Grabchikov, S. S., Lastovskii, S. B., Trukhanov, S. V., Vasin, D. S., Zubar, T. I., Kozlovskiy, A. L., Zdorovets, M. V., Sivakov, V. A., Muradyan, T. R., & Trukhanov, A. V. (2019). Function Composites Materials for Shielding Applications: Correlation between Phase Separation and Attenuation Properties. *Journal of Alloys and Compounds*, *771*, 238–245. doi:10.1016/j.jallcom.2018.08.209

Treviso, F., Trinchero, R., & Canavero, F. G. (2019, May). Validation of a Physical-Based Model for a Spacewire Cable. In *2019 ESA Workshop on Aerospace EMC (Aerospace EMC)* (pp. 1-6). IEEE. 10.23919/AeroEMC.2019.8788945

Tripathi, R. K., Wilson, J. W., Cucinotta, F. A., Anderson, B. M., & Simonsen, L. C. (2003). Materials Trade Study for Lunar/Gateway Missions. *Advances in Space Research*, *31*(11), 2383–2388. doi:10.1016/S0273-1177(03)00551-9 PMID:14696588

Tripathi, R. K., Wilson, J. W., & Youngquist, R. C. (2008). Electrostatic Space Radiation Shielding. *Advances in Space Research, 42*(6), 1043–1049. doi:10.1016/j.asr.2007.09.015

Tumanski, S. (2007). Induction coil sensors - A review. *Measurement Science & Technology, 18*(3), R31–R46. Advance online publication. doi:10.1088/0957-0233/18/3/R01

Tumanski, S. (2011). *Handbook of Magnetic Measurements*. CRC Press.

U.S.A. Department of Defence. (2015). *MIL-STD-461G - Requirements for The Control of Electromagnetic Interference Characteristics of Subsystems and Equipment*. Author.

Vaivads, A., Retinò, A., Soucek, J., Khotyaintsev, Y. V., Valentini, F., Escoubet, C. P., Alexandrova, O., André, M., Bale, S. D., Balikhin, M., Burgess, D., Camporeale, E., Caprioli, D., Chen, C. H. K., Clacey, E., Cully, C. M., De Keyser, J., Eastwood, J. P., & Fazakerley, A. N. … Wimmer-Schweingruber, R. F. (2016). Turbulence Heating ObserveR - Satellite mission proposal. *Journal of Plasma Physics*. Advance online publication. doi:10.1017/S0022377816000775

Wanasinghe, D., & Aslani, F. (2019). A Review on Recent Advancement of Electromagnetic Interference Shielding Novel Metallic Materials and Processes. *Composites. Part B, Engineering, 176*, 107207. doi:10.1016/j.compositesb.2019.107207

Warden, D., & Bayazitoglu, Y. (2019). New Comparative Metric for Evaluating Spacecraft Radiation Shielding. *Journal of Spacecraft and Rockets, 56*(4), 1024–1038. doi:10.2514/1.A34360

Weikert, S., Mehlem, K., & Wiegand, A. (2012). Spacecraft Magnetic Cleanliness Prediction and Control. *2012 ESA Workshop on Aerospace EMC*.

Werner, D. H. (1996). An exact integration procedure for vector potentials of thin circular loop antennas. *IEEE Transactions on Antennas and Propagation, 44*(2), 157–165. doi:10.1109/8.481642

Werner, D. H. (2000). Near-field and far-field expansions for traveling-wave circular loop antennas. *Progress in Electromagnetic Research, 28*, 29–42. doi:10.2528/PIER99100103

William, G. (1991). *De Magnete*. Dover Publications.

Wilson, J.W., Clowdsley, M.S., Cucinotta, F.A., Tripathi, R.K., Nealy, J.E. & De Angelis, G. (2004). Deep Space Environments for Human Exploration. *Advances in Space Research, 34*(6), 1281-1287. doi: 101016/j.asr.2003.10.052

Wu, T. T. (1962). Theory of the thin circular loop antenna. *Journal of Mathematical Physics, 3*(6), 1301–1304. doi:10.1063/1.1703875

Zhang, Q. J., & Gupta, K. C. (2000). *Neural Networks for RF and Microwave Design*. Artech House.

Zhang, Q., Gupta, K. C., & Devabhaktuni, V. K. (2003). Artificial neural networks for RF and microwave design: From theory to practice. *IEEE Transactions on Microwave Theory and Techniques, 51*(4), 339–1350.

Zhou, Z.-H., Liang, Y., Huang, H.-D., Li, L., Yang, B., Li, M. Z., Yan, D.-X., Lei, J., & Li, Z.-M. (2019). Structuring Dense Three-Dimensional Sheet-like Skeleton Networks in Biomass-derived Carbon Aerogels for Efficient Electromagnetic Interference Shielding. *Carbon, 152*, 316–324. doi:10.1016/j.carbon.2019.06.027

About the Contributors

Christos D. Nikolopoulos was born in Athens, Greece, in 1981. He receives the B.Sc. & M.Sc. degree in Physics and Informatics from the National & Kapodistrian University of Athens (UOA), and PhD degree from the National Technical University of Athens (NTUA). He is an Assistant Professor at the School of Engineering of the Hellenic Mediterranean University, School of Electronic Engineering and also working at the Wireless and Long Distance Communication Lab., School of Electrical and Computer Engineering, National Technical University of Athens as an Associate Researcher. His main research interests are in the field of Electromagnetic Compatibility, Inverse EM Scattering, Smart Antennas Design and Mobile Communications. He participates as a member in various workgroups in Hellenic Ministries. He is a licensed Physicist-Radioelectrologist Grade A' since 2014. Past positions include Hellenic Telecommunications Organization and various research projects in collaboration with National and Kapodistrian University of Athens, National Technical University of Athens, European Research Agencies, European Space Agency, and the Hellenic Ministry of Finance. He also acts as a consultant to EMC issues in EMTech SPACE P.C.

* * *

Anargyros T. Baklezos was born in Lamia, Greece, in 1981. He holds a B.Sc. degree in Physics and an M.Sc. diploma in Electronics & Radioelectrology from the National & Kapodistrian University of Athens (UOA). He holds a Ph.D. in Analysis & Modeling of SpaceWire System Link Electromagnetic Radiation for Space Applications from the School of Electrical and Computer Engineering of the National & Technical University of Athens (NTUA). He is currently an adjunct lecturer in the Dept. of Electronic Engineering of the Hellenic Mediterranean University (HMU) and a researcher at the School of Electrical &Computer Engineering School of National Technical University of Athens (NTUA). His main research interests are in the field of electromagnetic compatibility, inverse scattering problems, electromagnetic field modeling for space applications and EM cleanliness, and antenna design for radiometry applications. He has taken part in a number of research projects in collaboration with the European Space Agency, various Greek Ministries, and companies from the private sector. He is a published author in international peer-reviewed journals and scientific books.

Alexandros D. Bechrakis Triantafyllos was born in Athens, Greece in 1996. He received his integrated M.sC. degree from the department of Electrical and Computer Engineering of the National Technical University of Athens in July 2020. Currently, he is pursuing his first research steps in the field of Electromagnetic Compatibility by acting as a research assistant in the Wireless & Long Distance Communications Laboratory of the National Technical University of Athens.

Christos N. Capsalis was born in Nafplion, Greece, in 1956. He received the Diploma in electrical and mechanical engineering from the National Technical University of Athens (NTUA), in 1979, the B.S. degree in economics from the University of Athens, in 1983, and the Ph.D. degree in electrical engineering from NTUA, in 1985. He is currently a Professor in the Department of Electrical and Computer Engineering, NTUA. His current scientific activity concerns satellite and mobile communications, antenna theory and design, and electromagnetic compatibility.

Ioannis Gonos is Associate Professor at the School of Electrical and Computer Engineering of the National Technical University of Athens. At his university he serves since 2019 as the Director of the High Voltage and Electric Measurements Laboratory and of the High Voltage Service Providing Laboratory Unit. He is member of the Technical Committees ELOT TE/80 "Electromagnetic Compatibility" and ELOT TE/82 "Electrical installations of buildings" of the Hellenic Organization for Standardization (ELOT). Dr Gonos is member of IEEE, IET, CIGRE and IEEE DEIS. He is member of the reviewers committee of the International Symposium on High Voltage Engineering (ISH) and of the Steering Committee of the IEEE International Conference on High Voltage Engineering and Application (ICHVE). In 2018 he was the Chairman of the ICHVE 2018 which was held in Athens, September 10-13. He is reviewer in many scientific journals, Associate Editor in IET High Voltage and Guest Editor in Energies, MDPI AG. His research interests are high voltages, grounding, power systems protection, lightning and electromagnetic compatibility. His lifetime publication record includes more than 240 scientific publications, 70 of which are in international scientific journals.

Neoclis Hadjigeorgiou received his Diploma degree in Electrical and Computer Engineering from the National Technical University of Athens, Greece in (2011), his M.S. degree in Mechanical Engineering from National Technical University of Athens, Greece in (2013). He is currently pursuing a Ph.D degree in the subject "Integrated Architectures for Magnetic Sensors" in Electrical and Computer Engineering from the National Technical University of Athens, Greece. From 2013 to 2014, he was a Research Assistant in the laboratory of Metallurgy in Mining and Metallurgical Engineering in from the National Technical University of Athens, Greece. He has authored and coauthored some technical papers in IEEE conferences and some non-IEEE journals. He published a book chapter in the book Electromagnetic Compatibility for Systems Design (IGI-Global,2018). His research interests include circuit design, magnetic sensor design, parallel programming, and GPU programming.

Melina P. Ioannidou was born in Thessaloniki, Greece. She is currently an Associate Professor on "Wireless Communications, Antennae and Communication Theory" in the Department of Information and Electronic Engineering, International Hellenic University (I.H.U.). She received the M.Sc. and Ph.D. degrees from the Department of Electrical and Computer Engineering, Faculty of Engineering, Aristotle University of Thessaloniki, Greece, in 1989 and 1997, respectively. Her research interests are in propagation, scattering and diffraction of electromagnetic waves, wireless communications and antennas, biological effects of radiowaves, radar meteorology and remote sensing. She serves as director of the postgraduate programme "Master of Science (M.Sc.) in Applied Electronic Systems" offered by the I.H.U.. She is a member of the Technical Chamber of Greece.

Theodoros N. Kapetanakis was born in Chania, Crete, Greece, in 1984. He holds a BSc degree in Electronic Engineering and an MSc in Telecommunication & Automation Systems from the Department of Electronic Engineering of the Hellenic Mediterranean University (HMU), and a PhD from the Department of Informatics & Telecommunications of the University of Peloponnese. He is currently an adjunct lecturer at the Department of Electronic Engineering of HMU and a researcher at the Telecommunications & Electromagnetic Applications Laboratory of the same department. He has published several papers in international peer review journals and conferences. He is a reviewer in IEEE and Springer journals. His research interests include applications of artificial intelligence algorithms in radiation and scattering of electromagnetic waves, computational electromagnetics, bio-electromagnetics, design and implementation of wearable microwave devices.

Antonios Konstantaras was born in Portaria Halkidiki, Greece. He is currently an Associate Professor in "Software Engineering" at the School of Engineering of the Hellenic Mediterranean University, Head of the Department of Electronic Engineering, and Director of the Institutionalized Research Laboratory of Computer Technology, Informatics & Electronic Devices. He received his first-degree BEng (Hons) in Electronics Engineering from the University of Central Lancashire, UK, in 1999, his MSc in Mobile Robotics from the University of Portsmouth, UK, in 2000, and his PhD in Hybrid Intelligent Systems from the University of Central Lancashire, UK, in 2004. His main research interests are in the fields of data mining, machine learning and heterogeneous parallel processing with applications in physical sciences and educational robotics. His publications include 48 scientific articles (20 international journals, 27 international conferences and 1 book chapter). He has been a lead researcher in 2 research projects, a senior researcher in 11 other projects and has also received a postdoctoral fellowship from the Hellenic Scholarships Foundation. Dr. A. Konstantaras is an editor for the Earth Sciences and the Artificial Intelligence Evolution journals.

Alexandra P. Mavropoulou was born in Athens, Greece in 1997. She received her integrated MSc degree from the department of Electrical and Computer Engineering of the National Technical University of Athens in July 2020. Currently, she is pursuing her first research steps in the field of Electromagnetic Compatibility by acting as a research assistant in the Wireless & Long Distance Communications Laboratory of the National Technical University of Athens.

Eleni Nicolopoulou received the diploma and the PhD degree in Electrical and Computer Engineering from the School of Electrical and Computer Engineering of the National Technical University of Athens. Since 2010 she is a laboratory assistant at the High Voltage Laboratory of the National Technical University of Athens where she participates in research activities dealing with electromagnetic compatibility tests of electrical and electronic equipment. Her research interests concern high voltages, lightning protection and electromagnetic compatibility.

Panagiotis Papastamatis was born in 1986 in Athens where he completed his school education. He graduated from the Department of Electrical and Computer Engineering of the National Technical University of Athens in 2013 with the grade «Very Good». Currently he is working on his PhD in the field of Electromagnetic Compatibility. He speaks English (Certificate of Proficiency in English, 2001) and German (Abitur-German School of Athens graduate) fluently. Career: During his military service (2013-2014) he worked for the Greek Ministry of Defense in the field of Information Systems. Since

2015 he participates in tests and measurements, and he is also a laboratory assistant at the High Voltage Laboratory of the National Technical University of Athens. Author Activities: He is the author of 1 diploma thesis entitled «Electromagnetic Compatibility Tests on Telecommunication Lines». Other Activities: He participates in research activities of the NTUA, financially supported by public and private companies, regarding electrical tests and tests of Electromagnetic Compatibility of electrical and electronic equipment.

Ioannis Vardiambasis was born in Athens, Greece, in 1968. He is currently an Associate Professor on "Satellite & Microwave Communications" in the Department of Electronic Engineering of the Hellenic Mediterranean University (HMU). He has served in several administrative positions at HMU for many years, being the Dean of the Faculty of Applied Sciences (09/2018-05/2019), the Vice-Dean of the Faculty of Applied Sciences (09/2014-08/2018), the Head of the Department of Electronic Engineering (07/2012-08/2018), the Director of the Division of Telecommunications (12/2002-present), the Director of the Telecommunications & Electromagnetic Applications Laboratory (12/2002-present), and the Director of the Master's program in "Telecommunication & Automation Systems" (09/2018-present). He received the Diploma of Electrical Engineering and the PhD degree in Electrical & Computer Engineering from the National Technical University of Athens (NTUA), Greece, in 1991 and 1996, respectively. He has published more than 90 scientific papers in referred journals and international conference proceedings. He has participated as research engineer in more than 30 Greek and EU-funded research and technological development ICT projects, being scientific coordinator in more than 10 of them. His fields of interest include antennas, microwave technology/applications, bioelectromagnetics, applications of radiometry in biomedicine, analytical and computational methods/techniques, electromagnetic wave propagation, radiation and scattering from isotropic/anisotropic media structures, wireless communications, telecommunication and sensor networks, smart antennas and devices, electromagnetic compatibility, neural networks, artificial intelligence, machine learning, Internet of Things (IoT) applications and data analysis.

Rafael Vargas-Bernal received a Bachelor's degree in Communications and Electronics Engineering from the University of Guanajuato in 1995, and the degrees of Master of Science and Doctorate in Science with Specialty in Electronics from the National Institute of Astrophysics, Optics and Electronics (INAOE) in 1997 and 2000, respectively. Since January 2002, he has been a professor-researcher at the Higher Technological Institute of Irapuato (ITESI), Mexico, and, particularly since 2006, he has worked in the Department of Materials Engineering where he has established himself as a researcher. He has authored 15 articles in journals, 32 chapters in books, and about 100 conference articles. He is a member of the National System of Researchers (SNI-Mexico). He regularly serves as a reviewer of scientific articles in RSC Advances and Royal Society Open Science as well as Standards in Semiconductor Equipment and Materials International (SEMI). His research interests include two-dimensional materials, nanomaterials, space materials, composite materials, gas sensors, and biosensors.

Index

H

I

J

L

M

N

P

R

S

T

V

www.ingramcontent.com/pod-product-compliance
Lightning Source LLC
LaVergne TN
LVHW080526300926
849162LV00007B/1549